W9-AVJ-726

MATRIX METHODS

An Introduction

MATRIX

METHODS AN INTRODUCTION

Richard Bronson

FAIRLEIGH DICKINSON UNIVERSITY

ACADEMIC PRESS, INC.
(Harcourt Brace Jovanovich, Publishers)
Orlando San Diego San Francisco New York London
Toronto Montreal Sydney Tokyo São Paulo

COPYRIGHT © 1970, BY ACADEMIC PRESS, INC.
ALL RIGHTS RESERVED
NO PART OF THIS BOOK MAY BE REPRODUCED IN ANY FORM,
BY PHOTOSTAT, MICROFILM, RETRIEVAL SYSTEM, OR ANY
OTHER MEANS, WITHOUT WRITTEN PERMISSION FROM
THE PUBLISHERS.

ACADEMIC PRESS, INC.
Orlando, Florida 32887

United Kingdom Edition Published by
ACADEMIC PRESS, INC. (LONDON) LTD.
24/28 Oval Road, London NW1 7DX

LIBRARY OF CONGRESS CATALOG CARD NUMBER: *70 97490*

PRINTED IN THE UNITED STATES OF AMERICA

to Evy

CONTENTS

2. DETERMINANTS

3. THE INVERSE

4. SIMULTANEOUS LINEAR EQUATIONS

5. EIGENVALUES AND EIGENVECTORS

6. MATRIX CALCULUS

PREFACE

This book is intended for an introductory course in matrices similar to one I have been giving for some time to sophomore and junior engineering students at Fairleigh Dickinson University. Although the book is basically designed for students of science, engineering, and applied mathematics, I hope there is sufficient material to interest mathematics majors as well.

The presentation of this book has been kept as clear and straightforward as I could make it in the belief that this should be a text for the student rather than the teacher. Notes and cautions are used throughout to warn the reader of certain pitfalls and to draw his attention to particular subtleties that might easily be overlooked. Furthermore, an unusual number of completely worked out problems are interspersed; and, spurred by the conviction that illustrative examples presented in detail serve as an invaluable aid to understanding, I have tried to clarify each new concept by at least one or, where warranted, several examples.

As the title suggests, the emphasis of this book is on methodology rather than theory. Proofs of theorems are given in the main body of the text only if they are simple. Otherwise they are deferred to the appendices or omitted and referenced. At the end of each section there is a set of exercises which is basically routine in nature and serves primarily to enhance the reader's ability to use the methods just presented. On occasion, however, problems are assigned that will extend or complete topics previously introduced.

The only prerequisite for understanding all the material is calculus. Chapters 1–5 require merely a knowledge of high school algebra. For Chapter 7, a familiarity with differential equations would be helpful but is by no means necessary.

This book differs from other texts to a greater or lesser extent in three ways:

1. New concepts are developed in a very detailed and deliberate manner. I have tried not to omit a single step in the presentation of new ideas and techniques with the hope that the book can be used by readers not having available the assistance of an instructor.
2. Topics are developed more completely than is usually the case. This is particularly true of the chapters dealing with differential equations and Jordan canonical forms.
3. Certain topics are included which are generally not available in other introductory texts—for example, the transition matrix, techniques for obtaining the exponential of a matrix, and a number of special matrices.

This book is divided into two distinct parts, the first five chapters, which are elementary in nature, constituting the first part. The concluding four delve more deeply into the subject matter and develop material of fundamental importance to both engineers and scientists. Chapters 1–7 provide enough material for a complete one-semester course. Chapters 1–8 or Chapters 1–7 and Chapter 9 can also be used for a one-semester course if certain sections in the beginning chapters are either omitted or assigned as outside reading. I have tried both approaches with equal success.

RICHARD BRONSON

ACKNOWLEDGMENTS

I wish to express my grateful appreciation to the many people who helped make this book a reality. In particular, I acknowledge my debt to those students who used this book in its various prepublication forms and whose comments proved of immeasurable value.

Special thanks must go also to P. D. Ritger and N. Rose under whose tutelage much of the material used in this book was first introduced to me, to W. Ames, J. Benson, E. Harriet, and P. D. Ritger for reviewing the original manuscript and offering many invaluable suggestions and criticisms, to W. Kempton and J. Benson for proofreading the final galleys, and to M. Mutz, M. Page, T. Ranzan, V. Restivo, and S. Vico who typed various sections of the manuscript. I should like to acknowledge the assistance given to me by my publisher, Academic Press. My greatest debt, however, is to my wife Evelyn to whom this book is dedicated. In truth, she has every right to claim coauthorship for the countless hours she has spent in editing and proofreading the text from the original draft to the final galleys.

MATRIX METHODS

An Introduction

Chapter I MATRICES

1.1 MATRICES

Definition 1 A *matrix* is a rectangular array of elements arranged in horizontal rows and vertical columns. Thus,

$$\begin{bmatrix} 1 & 3 & 5 \\ 2 & 0 & -1 \end{bmatrix}, \tag{1}$$

$$\begin{bmatrix} 4 & 1 & 1 \\ 3 & 2 & 1 \\ 0 & 4 & 2 \end{bmatrix}, \tag{2}$$

and

$$\begin{bmatrix} \sqrt{2} \\ \pi \\ 19.5 \end{bmatrix} \tag{3}$$

are all examples of a matrix.

The matrix given in (1) has two rows and three columns; it is said to have *order* (or size) *2* × *3* (read two by three). By convention, the row index is always given first. The matrix in (2) has order *3* × *3*, while that in (3) has order *3* × *1*. The entries of a matrix are called *elements*.

In general, a matrix **A** (matrices will always be designated by uppercase boldface letters) of order $p \times n$ is given by

$$\mathbf{A} = \begin{bmatrix} a_{11} & a_{12} & a_{13} & \cdots & a_{1n} \\ a_{21} & a_{22} & a_{23} & \cdots & a_{2n} \\ a_{31} & a_{32} & a_{33} & \cdots & a_{3n} \\ \vdots & \vdots & \vdots & & \vdots \\ a_{p1} & a_{p2} & a_{p3} & \cdots & a_{pn} \end{bmatrix}, \tag{4}$$

which is often abbreviated to $[a_{ij}]_{p \times n}$ or just $[a_{ij}]$. In this notation, a_{ij} represents the general element of the matrix and appears in the ith row and the jth column. The subscript i, which represents the row, can have any value 1 through p, while the subscript j, which represents the column, runs 1 through n. Thus, if $i = 2$ and $j = 3$, a_{ij} becomes a_{23} and designates the element in the second row and third column. If $i = 1$ and $j = 5$, a_{ij} becomes a_{15} and signifies the element in the first row, fifth column. Note again that the row index is always given before the column index.

If the matrix has as many rows as columns, $p = n$, it is called a *square matrix*; in general it is written as

$$\begin{bmatrix} a_{11} & a_{12} & a_{13} & \cdots & a_{1n} \\ a_{21} & a_{22} & a_{23} & \cdots & a_{2n} \\ a_{31} & a_{32} & a_{33} & \cdots & a_{3n} \\ \vdots & \vdots & \vdots & & \vdots \\ a_{n1} & a_{n2} & a_{n3} & \cdots & a_{nn} \end{bmatrix}. \tag{5}$$

In this case, the elements $a_{11}, a_{22}, a_{33}, \ldots, a_{nn}$ lie on and form the *main* (or principal) *diagonal*.

It should be noted that the elements of a matrix need not be numbers; they can be, and quite often arise physically as, functions, operators or, as we shall see later, matrices themselves. Hence,

$$\begin{bmatrix} \int_0^1 (t^2 + 1)\,dt & t^2 & \sqrt{3t} & 2 \end{bmatrix}, \qquad \begin{bmatrix} \sin\theta & \cos\theta \\ -\cos\theta & \sin\theta \end{bmatrix},$$

and

$$\begin{bmatrix} x^2 & x \\ e^x & \dfrac{d}{dx}\ln x \\ 5 & x+2 \end{bmatrix}$$

are good examples of matrices. Finally, it must be noted that a matrix is an entity unto itself; it is not a number. If the reader is familiar with determinants, he will undoubtedly recognize the similarity in form between the two. *Warning*: the similarity ends there. Whereas a determinant can be evaluated to yield a number, a matrix cannot. A matrix is a rectangular array, period.

1.2 OPERATIONS

The simplest relationship between two matrices is equality. Intuitively one feels that two matrices should be equal if their corresponding elements are equal. This is the case, providing the matrices are of the same order.

Definition 1 Two matrices $\mathbf{A} = [a_{ij}]_{p \times n}$ and $\mathbf{B} = [b_{ij}]_{p \times n}$ are equal if they have the same order and if $a_{ij} = b_{ij}$ $(i = 1, 2, 3, \ldots, p; j = 1, 2, 3, \ldots, n)$. Thus, the equality

$$\begin{bmatrix} 5x + 2y \\ x - 3y \end{bmatrix} = \begin{bmatrix} 7 \\ 1 \end{bmatrix}$$

implies that $5x + 2y = 7$ and $x - 3y = 1$.

The intuitive definition for matrix addition is also the correct one.

Definition 2 If $\mathbf{A} = [a_{ij}]$ and $\mathbf{B} = [b_{ij}]$ are both of order $p \times n$, then $\mathbf{A} + \mathbf{B}$ is a $p \times n$ matrix $\mathbf{C} = [c_{ij}]$ where $c_{ij} = a_{ij} + b_{ij}$ $(i = 1, 2, 3, \ldots, p; j = 1, 2, 3, \ldots, n)$. Thus,

$$\begin{bmatrix} 5 & 1 \\ 7 & 3 \\ -2 & -1 \end{bmatrix} + \begin{bmatrix} -6 & 3 \\ 2 & -1 \\ 4 & 1 \end{bmatrix} = \begin{bmatrix} 5 + (-6) & 1 + 3 \\ 7 + 2 & 3 + (-1) \\ (-2) + 4 & (-1) + 1 \end{bmatrix} = \begin{bmatrix} -1 & 4 \\ 9 & 2 \\ 2 & 0 \end{bmatrix}$$

and

$$\begin{bmatrix} t^2 & 5 \\ 3t & 0 \end{bmatrix} + \begin{bmatrix} 1 & -6 \\ t & -t \end{bmatrix} = \begin{bmatrix} t^2 + 1 & -1 \\ 4t & -t \end{bmatrix};$$

but the matrices

$$\begin{bmatrix} 5 & 0 \\ -1 & 0 \\ 2 & 1 \end{bmatrix} \quad \text{and} \quad \begin{bmatrix} -6 & 2 \\ 1 & 1 \end{bmatrix}$$

cannot be added since they are not of the same order.

It is not difficult to show that the addition of matrices is both commutative and associative: that is, if **A**, **B**, **C** represent matrices of the same order, then

(A1) $\mathbf{A} + \mathbf{B} = \mathbf{B} + \mathbf{A}$,

(A2) $\mathbf{A} + (\mathbf{B} + \mathbf{C}) = (\mathbf{A} + \mathbf{B}) + \mathbf{C}$.

We define the zero matrix **0** to be the matrix consisting of only zero elements. We then have the additional property

(A3) $\mathbf{A} + \mathbf{0} = \mathbf{A}$.

Subtraction of matrices is defined in a manner analogous to addition; the orders of the matrices involved must be identical and the operation is performed elementwise.
Thus,

$$\begin{bmatrix} 5 & 1 \\ -3 & 2 \end{bmatrix} - \begin{bmatrix} 6 & -1 \\ 4 & -1 \end{bmatrix} = \begin{bmatrix} -1 & 2 \\ -7 & 3 \end{bmatrix}.$$

Another simple operation is that of multiplying a scalar times a matrix. Intuition guides one to perform the operation elementwise, and once again intuition is correct. Thus, for example,

$$7\begin{bmatrix} 1 & 2 \\ -3 & 4 \end{bmatrix} = \begin{bmatrix} 7 & 14 \\ -21 & 28 \end{bmatrix} \quad \text{and} \quad t\begin{bmatrix} 1 & 0 \\ 3 & 2 \end{bmatrix} = \begin{bmatrix} t & 0 \\ 3t & 2t \end{bmatrix}.$$

Definition 3 If $\mathbf{A} = [a_{ij}]$ is a $p \times n$ matrix and if λ is a scalar, then $\lambda\mathbf{A}$ is a $p \times n$ matrix $\mathbf{B} = [b_{ij}]$ where $b_{ij} = \lambda a_{ij}$ $(i = 1, 2, 3, \ldots, p; j = 1, 2, 3, \ldots, n)$.

EXAMPLE 1 Find $5\mathbf{A} - \frac{1}{2}\mathbf{B}$ if

$$\mathbf{A} = \begin{bmatrix} 4 & 1 \\ 0 & 3 \end{bmatrix} \quad \text{and} \quad \mathbf{B} = \begin{bmatrix} 6 & -20 \\ 18 & 8 \end{bmatrix}$$

Solution

$$\begin{aligned} 5\mathbf{A} - \tfrac{1}{2}\mathbf{B} &= 5\begin{bmatrix} 4 & 1 \\ 0 & 3 \end{bmatrix} - \tfrac{1}{2}\begin{bmatrix} 6 & -20 \\ 18 & 8 \end{bmatrix} \\ &= \begin{bmatrix} 20 & 5 \\ 0 & 15 \end{bmatrix} - \begin{bmatrix} 3 & -10 \\ 9 & 4 \end{bmatrix} = \begin{bmatrix} 17 & 15 \\ -9 & 11 \end{bmatrix}. \end{aligned}$$

It is not difficult to show that if λ_1 and λ_2 are scalars, and if **A** and **B** are matrices of identical order, then

(S1) $\lambda_1\mathbf{A} = \mathbf{A}\lambda_1$,

(S2) $\lambda_1(\mathbf{A} + \mathbf{B}) = \lambda_1\mathbf{A} + \lambda_1\mathbf{B}$,

(S3) $(\lambda_1 + \lambda_2)\mathbf{A} = \lambda_1\mathbf{A} + \lambda_2\mathbf{A}$,

(S4) $\lambda_1(\lambda_2\mathbf{A}) = (\lambda_1\lambda_2)\mathbf{A}$.

The reader is cautioned that there is *no* such operation as matrix division. We will, however, define a somewhat analogous operation, namely matrix inversion, in a later chapter.

PROBLEMS 1.2

1. Find $5\mathbf{A} - 3\mathbf{B}$ if

$$\mathbf{A} = \begin{bmatrix} 2 & 1 \\ 3 & 0 \end{bmatrix} \quad \text{and} \quad \mathbf{B} = \begin{bmatrix} -1 & 2 \\ 9 & -6 \end{bmatrix}.$$

2. Find **C** if $4\mathbf{A} + 3\mathbf{C} = \mathbf{B}$ where

$$\mathbf{A} = \begin{bmatrix} 2 & -1 & 0 \\ 3 & 2 & -4 \\ 5 & 1 & 9 \end{bmatrix}, \quad \mathbf{B} = \begin{bmatrix} 17 & -1 & 3 \\ -24 & -1 & -16 \\ -7 & 1 & 1 \end{bmatrix}.$$

3. Find $6\mathbf{A} - \theta\mathbf{B}$ if

$$\mathbf{A} = \begin{bmatrix} \theta^2 & 2\theta - 1 \\ 4 & 1/\theta \end{bmatrix} \quad \text{and} \quad \mathbf{B} = \begin{bmatrix} \theta^2 - 1 & 6 \\ 3/\theta & \theta^3 + 2\theta + 1 \end{bmatrix}.$$

4. Find $3\mathbf{A} + 2\mathbf{B}$ if

$$\mathbf{A} = \begin{bmatrix} 1 & 2 \\ 0 & -1 \\ 1 & 1 \end{bmatrix} \quad \text{and} \quad \mathbf{B} = \begin{bmatrix} 1 & -1 \\ 2 & 1 \\ 0 & 2 \\ -1 & 0 \end{bmatrix}.$$

1.3 MATRIX MULTIPLICATION—I

Matrix multiplication is the first operation we encounter where our intuition fails. First, two matrices are *not* multiplied together elementwise. Secondly, it is not always possible to multiply matrices of the same order while it is possible to multiply certain matrices of different orders. Thirdly, if **A** and **B** are two matrices for which multiplication is defined, it is generally not the case that $\mathbf{AB} = \mathbf{BA}$; that is, *matrix multiplication is not a commutative operation.* There are other properties of matrix multiplication, besides the three mentioned that defy our intuition, and we shall have occasion to illustrate them in the next section. In this section, we determine which matrices can be multiplied together and how this operation is to be performed.

RULE 1 The product of two matrices **AB** is defined if the number of columns of **A** equals the number of rows of **B**.

Thus, if **A** and **B** are given by

$$\mathbf{A} = \begin{bmatrix} 6 & 1 & 0 \\ -1 & 2 & 1 \end{bmatrix} \quad \text{and} \quad \mathbf{B} = \begin{bmatrix} -1 & 0 & 1 & 0 \\ 3 & 2 & -2 & 1 \\ 4 & 1 & 1 & 0 \end{bmatrix}, \tag{6}$$

then the product **AB** is defined since **A** has three columns and **B** has three rows. The product **BA**, however, is not defined since **B** has four columns while **A** has only two rows.

When the product is written **AB**, **A** is said to *premultiply* **B** while **B** is said to *postmultiply* **A**.

RULE 2 If the product **AB** is defined, then the resultant matrix will have the same number of rows as **A** and the same number of columns as **B**.

Thus, the product **AB**, where **A** and **B** are given in (6), will have two rows and four columns since **A** has two rows and **B** has four columns.

An easy method of remembering these two rules is the following: write the orders of the matrices on paper in the sequence in which the multiplication is to be carried out; that is, if **AB** is to be found where **A** has order 2×3 and **B** has order 3×4, write

$$(2 \times \underset{\frown}{3)(3} \times 4) \tag{7}$$

If the two adjacent numbers (indicated in (7) by the curved arrow) are both equal (in this case they are both three), the multiplication is defined. The order of the product matrix is obtained by canceling the adjacent numbers and using the two remaining numbers. Thus in (7), we cancel the adjacent 3's and are left with 2×4, which in this case, is the order of **AB**.

As a further example, consider the case where **A** is 4×3 matrix while **B** is a 3×5 matrix. The product **AB** is defined since, in the notation $(4 \times 3)(3 \times 5)$, the adjacent numbers denoted by the curved arrow are equal. The product will be a 4×5 matrix. The product **BA** however is not defined since in the notation $(3 \times 5)(4 \times 3)$ the adjacent numbers are not equal. In general, one may schematically state the method as

$$(k \times \underset{\frown}{n)(n} \times p) = (k \times p).$$

RULE 3 If the product **AB** = **C** is defined, where **C** is denoted by $[c_{ij}]$, then the element c_{ij} is obtained by multiplying the elements in ith row of **A** by the corresponding elements in the jth column of **B** and adding. Thus, if **A** has order $k \times n$, and **B** has order $n \times p$, and

$$\begin{bmatrix} a_{11} & a_{12} & \cdots & a_{1n} \\ a_{21} & a_{22} & \cdots & a_{2n} \\ \vdots & \vdots & & \vdots \\ a_{k1} & a_{k2} & \cdots & a_{kn} \end{bmatrix} \begin{bmatrix} b_{11} & b_{12} & \cdots & b_{1p} \\ b_{21} & b_{22} & \cdots & b_{2p} \\ \vdots & \vdots & & \vdots \\ b_{n1} & b_{n2} & \cdots & b_{np} \end{bmatrix} = \begin{bmatrix} c_{11} & c_{12} & \cdots & c_{1p} \\ c_{21} & c_{22} & \cdots & c_{2p} \\ \vdots & \vdots & & \vdots \\ c_{k1} & c_{k2} & \cdots & c_{kp} \end{bmatrix},$$

then c_{11} is obtained by multiplying the elements in the first row of **A** by the corresponding elements in the first column of **B** and adding; hence,

$$c_{11} = a_{11}b_{11} + a_{12}b_{21} + \cdots + a_{1n}b_{n1}.$$

The element c_{12} is found by multiplying the elements in the first row of **A** by the corresponding elements in the second column of **B** and adding; hence,

$$c_{12} = a_{11}b_{12} + a_{12}b_{22} + \cdots + a_{1n}b_{n2}.$$

The element c_{kp} is obtained by multiplying the elements in the kth row of **A** by the corresponding elements in the pth column of **B** and adding; hence,

$$c_{kp} = a_{k1}b_{1p} + a_{k2}b_{2p} + \cdots + a_{kn}b_{np}.$$

EXAMPLE 1 Find **AB** and **BA** if

$$\mathbf{A} = \begin{bmatrix} 1 & 2 & 3 \\ 4 & 5 & 6 \end{bmatrix} \quad \text{and} \quad \mathbf{B} = \begin{bmatrix} -7 & -8 \\ 9 & 10 \\ 0 & -11 \end{bmatrix}.$$

Solution

$$\mathbf{AB} = \begin{bmatrix} 1 & 2 & 3 \\ 4 & 5 & 6 \end{bmatrix} \begin{bmatrix} -7 & -8 \\ 9 & 10 \\ 0 & -11 \end{bmatrix}$$

$$= \begin{bmatrix} 1(-7) + 2(9) + 3(0) & 1(-8) + 2(10) + 3(-11) \\ 4(-7) + 5(9) + 6(0) & 4(-8) + 5(10) + 6(-11) \end{bmatrix}$$

$$= \begin{bmatrix} -7 + 18 + 0 & -8 + 20 - 33 \\ -28 + 45 + 0 & -32 + 50 - 66 \end{bmatrix} = \begin{bmatrix} 11 & -21 \\ 17 & -48 \end{bmatrix},$$

$$\mathbf{BA} = \begin{bmatrix} -7 & -8 \\ 9 & 10 \\ 0 & -11 \end{bmatrix} \begin{bmatrix} 1 & 2 & 3 \\ 4 & 5 & 6 \end{bmatrix}$$

$$= \begin{bmatrix} (-7)1 + (-8)4 & (-7)2 + (-8)5 & (-7)3 + (-8)6 \\ 9(1) + 10(4) & 9(2) + 10(5) & 9(3) + 10(6) \\ 0(1) + (-11)4 & 0(2) + (-11)5 & 0(3) + (-11)6 \end{bmatrix}$$

$$= \begin{bmatrix} -7 - 32 & -14 - 40 & -21 - 48 \\ 9 + 40 & 18 + 50 & 27 + 60 \\ 0 - 44 & 0 - 55 & 0 - 66 \end{bmatrix} = \begin{bmatrix} -39 & -54 & -69 \\ 49 & 68 & 87 \\ -44 & -55 & -66 \end{bmatrix}.$$

The preceding three rules can be incorporated into the following formal definition:

Definition 1 If $\mathbf{A} = [a_{ij}]$ is a $k \times n$ matrix and $\mathbf{B} = [b_{ij}]$ is an $n \times p$ matrix, then the product $\mathbf{AB}$ is defined to be a $k \times p$ matrix $\mathbf{C} = [c_{ij}]$ where $c_{ij} = \sum_{l=1}^{n} a_{il} b_{lj} = a_{i1}b_{1j} + a_{i2}b_{2j} + \cdots + a_{in}b_{nj}$ $(i = 1, 2, \ldots, k;$ $j = 1, 2, \ldots, p)$.

1.4 MATRIX MULTIPLICATION—II

EXAMPLE 1 Find **AB** if

$$\mathbf{A} = \begin{bmatrix} 2 & 1 \\ -1 & 0 \\ 3 & 1 \end{bmatrix} \quad \text{and} \quad \mathbf{B} = \begin{bmatrix} 3 & 1 & 5 & -1 \\ 4 & -2 & 1 & 0 \end{bmatrix}.$$

Solution

$$\mathbf{AB} = \begin{bmatrix} 2 & 1 \\ -1 & 0 \\ 3 & 1 \end{bmatrix} \begin{bmatrix} 3 & 1 & 5 & -1 \\ 4 & -2 & 1 & 0 \end{bmatrix}$$

$$= \begin{bmatrix} 2(3) + 1(4) & 2(1) + 1(-2) & 2(5) + 1(1) & 2(-1) + 1(0) \\ -1(3) + 0(4) & -1(1) + 0(-2) & -1(5) + 0(1) & -1(-1) + 0(0) \\ 3(3) + 1(4) & 3(1) + 1(-2) & 3(5) + 1(1) & 3(-1) + 1(0) \end{bmatrix}$$

$$= \begin{bmatrix} 10 & 0 & 11 & -2 \\ -3 & -1 & -5 & 1 \\ 13 & 1 & 16 & -3 \end{bmatrix}.$$

Note that in this example the product **BA** is not defined.

EXAMPLE 2 Find **AB** and **BA** if

$$\mathbf{A} = \begin{bmatrix} 2 & 1 \\ -1 & 3 \end{bmatrix} \quad \text{and} \quad \mathbf{B} = \begin{bmatrix} 4 & 0 \\ 1 & 2 \end{bmatrix}.$$

Solution

$$\mathbf{AB} = \begin{bmatrix} 2 & 1 \\ -1 & 3 \end{bmatrix} \begin{bmatrix} 4 & 0 \\ 1 & 2 \end{bmatrix} = \begin{bmatrix} 2(4) + 1(1) & 2(0) + 1(2) \\ -1(4) + 3(1) & -1(0) + 3(2) \end{bmatrix}$$

$$= \begin{bmatrix} 9 & 2 \\ -1 & 6 \end{bmatrix};$$

$$\mathbf{BA} = \begin{bmatrix} 4 & 0 \\ 1 & 2 \end{bmatrix} \begin{bmatrix} 2 & 1 \\ -1 & 3 \end{bmatrix} = \begin{bmatrix} 4(2) + 0(-1) & 4(1) + 0(3) \\ 1(2) + 2(-1) & 1(1) + 2(3) \end{bmatrix}$$

$$= \begin{bmatrix} 8 & 4 \\ 0 & 7 \end{bmatrix}.$$

This, therefore, is an example where both products **AB** and **BA** are defined but unequal.

EXAMPLE 3 Find **AB** and **BA** if

$$\mathbf{A}=\begin{bmatrix}3 & 1\\0 & 4\end{bmatrix} \quad \text{and} \quad \mathbf{B}=\begin{bmatrix}1 & 1\\0 & 2\end{bmatrix}.$$

Solution

$$\mathbf{AB}=\begin{bmatrix}3 & 1\\0 & 4\end{bmatrix}\begin{bmatrix}1 & 1\\0 & 2\end{bmatrix}=\begin{bmatrix}3 & 5\\0 & 8\end{bmatrix},$$

$$\mathbf{BA}=\begin{bmatrix}1 & 1\\0 & 2\end{bmatrix}\begin{bmatrix}3 & 1\\0 & 4\end{bmatrix}=\begin{bmatrix}3 & 5\\0 & 8\end{bmatrix}.$$

This, therefore, is an example where both products **AB** and **BA** are defined and equal.

In general, it can be shown that matrix multiplication has the following properties:

(M1) $\mathbf{A}(\mathbf{BC}) = (\mathbf{AB})\mathbf{C}$ (Associative Law)

(M2) $\mathbf{A}(\mathbf{B} + \mathbf{C}) = \mathbf{AB} + \mathbf{AC}$ (Left Distributive Law)

(M3) $(\mathbf{B} + \mathbf{C})\mathbf{A} = \mathbf{BA} + \mathbf{CA}$ (Right Distributive Law)

providing that the matrices **A**, **B**, **C** have the correct order so that the above multiplications and additions are defined. The one basic property that matrix multiplication does not possess is commutativity; that is, in general, **AB** does not equal **BA** (see Example 2). We hasten to add, however, that while matrices in general do not commute, it may very well be the case that, given two particular matrices, they do commute as can be seen from Example 3.

Commutativity is not the only property that matrix multiplication lacks. We know from our experiences with real numbers that if the product $xy = 0$, then either $x = 0$ or $y = 0$ or both are zero. Matrices do not possess this property as the following example shows:

EXAMPLE 4 Find **AB** if

$$\mathbf{A}=\begin{bmatrix}4 & 2\\2 & 1\end{bmatrix} \quad \text{and} \quad \mathbf{B}=\begin{bmatrix}3 & -4\\-6 & 8\end{bmatrix}.$$

Solution

$$\mathbf{AB}=\begin{bmatrix}4 & 2\\2 & 1\end{bmatrix}\begin{bmatrix}3 & -4\\-6 & 8\end{bmatrix}=\begin{bmatrix}4(3)+2(-6) & 4(-4)+2(8)\\2(3)+1(-6) & 2(-4)+1(8)\end{bmatrix}$$

$$=\begin{bmatrix}0 & 0\\0 & 0\end{bmatrix}.$$

Thus, even though neither **A** nor **B** is zero, their product is zero.

One final "unfortunate" property of matrix multiplication is that the equation $\mathbf{AB} = \mathbf{AC}$ does not imply $\mathbf{B} = \mathbf{C}$.

EXAMPLE 5 Find **AB** and **AC** if

$$\mathbf{A}=\begin{bmatrix}4 & 2\\ 2 & 1\end{bmatrix}, \qquad \mathbf{B}=\begin{bmatrix}1 & 1\\ 2 & 1\end{bmatrix}, \qquad \mathbf{C}=\begin{bmatrix}2 & 2\\ 0 & -1\end{bmatrix}.$$

Solution

$$\mathbf{AB}=\begin{bmatrix}4 & 2\\ 2 & 1\end{bmatrix}\begin{bmatrix}1 & 1\\ 2 & 1\end{bmatrix}=\begin{bmatrix}4(1)+2(2) & 4(1)+2(1)\\ 2(1)+1(2) & 2(1)+1(1)\end{bmatrix}=\begin{bmatrix}8 & 6\\ 4 & 3\end{bmatrix};$$

$$\mathbf{AC}=\begin{bmatrix}4 & 2\\ 2 & 1\end{bmatrix}\begin{bmatrix}2 & 2\\ 0 & -1\end{bmatrix}=\begin{bmatrix}4(2)+2(0) & 4(2)+2(-1)\\ 2(2)+1(0) & 2(2)+1(-1)\end{bmatrix}=\begin{bmatrix}8 & 6\\ 4 & 3\end{bmatrix}.$$

Thus, cancellation is not a valid operation in the matrix algebra.

The reader has no doubt wondered why this seemingly complicated procedure for matrix multiplication has been introduced when the more obvious methods of multiplying matrices termwise could be used. The answer lies in systems of simultaneous linear equations. Consider the set of simultaneous linear equations given by

$$\begin{aligned} 5x - 3y + 2z &= 14,\\ x + y - 4z &= -7,\\ 7x \qquad - 3z &= 1. \end{aligned} \tag{8}$$

This system can easily be solved by the method of substitution. Matrix algebra, however, will give us an entirely new method for obtaining the solution.

Consider the matrix equation

$$\mathbf{Ax}=\mathbf{b} \tag{9}$$

where

$$\mathbf{A}=\begin{bmatrix}5 & -3 & 2\\ 1 & 1 & -4\\ 7 & 0 & -3\end{bmatrix}, \qquad \mathbf{x}=\begin{bmatrix}x\\ y\\ z\end{bmatrix}, \qquad \text{and} \qquad \mathbf{b}=\begin{bmatrix}14\\ -7\\ 1\end{bmatrix}.$$

Here **A**, called the *coefficient matrix*, is simply the matrix whose elements are the coefficients of the unknowns x, y, z in (8). (Note that we have been very careful to put all the x coefficients in the first column, all the y coefficients in the second column, and all the z coefficients in the third column. The zero in the (3, 2) entry appears because the y coefficient in the third equation of system (8) is zero.) **x** and **b** are obtained in the obvious manner. One note of warning: there is a basic difference between the unknown matrix **x** in (9) and the unknown variable x. The reader should be especially careful not to confuse their respective identities.

Now using our definition of matrix multiplication, we have that

$$\mathbf{Ax} = \begin{bmatrix} 5 & -3 & 2 \\ 1 & 1 & -4 \\ 7 & 0 & -3 \end{bmatrix} \begin{bmatrix} x \\ y \\ z \end{bmatrix} = \begin{bmatrix} (5)(x) + (-3)(y) + (2)(z) \\ (1)(x) + (1)(y) + (-4)(z) \\ (7)(x) + (0)(y) + (-3)(z) \end{bmatrix}$$

$$= \begin{bmatrix} 5x - 3y + 2z \\ x + y - 4z \\ 7x \quad - 3z \end{bmatrix} = \begin{bmatrix} 14 \\ -7 \\ 1 \end{bmatrix}. \tag{10}$$

Using the definition of matrix equality, we see that (10) is precisely system (8). Thus (9) is an alternate way of representing the original system. It should come as no surprise, therefore, that by redefining the matrices **A**, **x**, **b**, appropriately, we can represent any arbitrary system of simultaneous linear equations by the matrix equation **Ax** = **b**.

EXAMPLE 6 Put the following system into matrix form:

$$\begin{aligned} x - y + z + w &= 5, \\ 2x + y - z \quad &= 4, \\ 3x + 2y \quad + 2w &= 0, \\ x - 2y + 3z + 4w &= -1. \end{aligned}$$

Solution Define

$$\mathbf{A} = \begin{bmatrix} 1 & -1 & 1 & 1 \\ 2 & 1 & -1 & 0 \\ 3 & 2 & 0 & 2 \\ 1 & -2 & 3 & 4 \end{bmatrix}, \qquad \mathbf{x} = \begin{bmatrix} x \\ y \\ z \\ w \end{bmatrix}, \qquad \mathbf{b} = \begin{bmatrix} 5 \\ 4 \\ 0 \\ -1 \end{bmatrix}.$$

The above system is then equivalent to the matrix system **Ax** = **b**.

Unfortunately, we are not yet in a position to solve systems that are in the matrix form **Ax** = **b**. One method of solution depends upon the operation of inversion, and we must postpone a discussion of it until the inverse has been defined. For the present, however, we hope that the reader will be content with the knowledge that matrix multiplication, as we have defined it, does serve some useful purpose.

PROBLEMS 1.4

1. Find **AB** and **BA** if

$$\mathbf{A} = \begin{bmatrix} 2 & 1 & -1 \\ 0 & 3 & 2 \end{bmatrix} \quad \text{and} \quad \mathbf{B} = \begin{bmatrix} 4 & 2 \\ 1 & -1 \\ 1 & 3 \end{bmatrix}.$$

2. Find **AB** and **BA** if

$$\mathbf{A}=\begin{bmatrix} 1 & 3 & -1 \\ -2 & 0 & 1 \\ 1 & 2 & 2 \end{bmatrix} \quad \text{and} \quad \mathbf{B}=\begin{bmatrix} 5 & 7 & 9 \\ 2 & 3 & -1 \\ -2 & 1 & 0 \end{bmatrix}.$$

3. Find **AB** if

$$\mathbf{A}=\begin{bmatrix} 2 & 1 \\ 1 & 0 \end{bmatrix} \quad \text{and} \quad \mathbf{B}=\begin{bmatrix} 3 & 1 & 2 \\ 0 & 1 & 1 \end{bmatrix}.$$

What conclusion can be made about **BA**?

4. Find **AB** if

$$\mathbf{A}=\begin{bmatrix} 2 & 6 \\ 3 & 9 \end{bmatrix} \quad \text{and} \quad \mathbf{B}=\begin{bmatrix} 3 & -6 \\ -1 & 2 \end{bmatrix}.$$

Note that $\mathbf{AB} = \mathbf{0}$ but neither **A** nor **B** equals the zero matrix.

5. Find **AB** and **CB** if

$$\mathbf{A}=\begin{bmatrix} 3 & 2 \\ 1 & 0 \end{bmatrix}, \quad \mathbf{B}=\begin{bmatrix} 2 & 4 \\ 1 & 2 \end{bmatrix}, \quad \mathbf{C}=\begin{bmatrix} 1 & 6 \\ 3 & -4 \end{bmatrix}.$$

Thus show that $\mathbf{AB} = \mathbf{CB}$ but $\mathbf{A} \neq \mathbf{C}$

6. Verify property (M1) for

$$\mathbf{A}=\begin{bmatrix} 2 & 1 \\ 1 & 3 \end{bmatrix}, \quad \mathbf{B}=\begin{bmatrix} 0 & 1 \\ -1 & 4 \end{bmatrix}, \quad \mathbf{C}=\begin{bmatrix} 5 & 1 \\ 2 & 1 \end{bmatrix}.$$

7. In the preceding problem, show that $\mathbf{A}(\mathbf{B} + \mathbf{C}) = \mathbf{AB} + \mathbf{AC}$.

8. If **A**, **B**, and **C** are of orders 2×3, 3×2, and 3×3 respectively, determine the order of:

(a) **C(BA)** (b) **B(AC)**

(c) **(AC)B** (d) **(AB)C**

9. Put the following system of equations into matrix form:

$$\begin{aligned} x + \quad z + y &= 2, \\ 3z + 2x + y &= 4, \\ y + x &= 0. \end{aligned}$$

10. Put the following system of equations into matrix form:

$$\begin{aligned} 5x + 3y + 2z + 4w &= 5, \\ x + \ y + \quad\ w &= 0, \\ 3x + 2y + 2z \quad &= -3, \\ x + \ y + 2z + 3w &= 4. \end{aligned}$$

1.5 SPECIAL MATRICES

There are certain types of matrices that occur so frequently that it becomes advisable to discuss them separately. One such type is the *transpose*. Given a matrix **A**, the transpose of **A**, denoted by **A**′ and read *A*-transpose, is obtained by changing all the rows of **A** into the columns of **A**′ while preserving the order; hence, the first row of **A** becomes the first column of **A**′, while the second row of **A** becomes the second column of **A**′, and the last row of **A** becomes the last column of **A**′. Thus if

$$\mathbf{A} = \begin{bmatrix} 1 & 2 & 3 \\ 4 & 5 & 6 \\ 7 & 8 & 9 \end{bmatrix}, \qquad \text{then } \mathbf{A}' = \begin{bmatrix} 1 & 4 & 7 \\ 2 & 5 & 8 \\ 3 & 6 & 9 \end{bmatrix}$$

and if

$$\mathbf{A} = \begin{bmatrix} 1 & 2 & 3 & 4 \\ 5 & 6 & 7 & 8 \end{bmatrix}, \qquad \text{then } \mathbf{A}' = \begin{bmatrix} 1 & 5 \\ 2 & 6 \\ 3 & 7 \\ 4 & 8 \end{bmatrix}.$$

Definition 1 If **A**, denoted by $[a_{ij}]$ is an $n \times p$ matrix, then the transpose of **A**, denoted by $\mathbf{A}' = [a'_{ij}]$ is a $p \times n$ matrix where $a'_{ij} = a_{ji}$.

It can be shown that the transpose possesses the following properties:

(1) $(\mathbf{A}')' = \mathbf{A}$,

(2) $(\lambda\mathbf{A})' = \lambda\mathbf{A}'$ where λ represents a scalar,

(3) $(\mathbf{A} + \mathbf{B})' = \mathbf{A}' + \mathbf{B}'$,

(4) $(\mathbf{A} + \mathbf{B} + \mathbf{C})' = \mathbf{A}' + \mathbf{B}' + \mathbf{C}'$,

(5) $(\mathbf{AB})' = \mathbf{B}'\mathbf{A}'$,

(6) $(\mathbf{ABC})' = \mathbf{C}'\mathbf{B}'\mathbf{A}'$

Transposes of sums and products of more than three matrices are defined in the obvious manner. We caution the reader to be alert to the ordering of properties (5) and (6). In particular, one should be aware that the transpose of a product is not the product of the transposes but rather the *commuted* product of the transposes.

EXAMPLE 1 Find $(\mathbf{AB})'$ and $\mathbf{B}'\mathbf{A}'$ if

$$\mathbf{A} = \begin{bmatrix} 3 & 0 \\ 4 & 1 \end{bmatrix} \qquad \text{and} \qquad \mathbf{B} = \begin{bmatrix} -1 & 2 & 1 \\ 3 & -1 & 0 \end{bmatrix}.$$

Solution

$$\mathbf{AB} = \begin{bmatrix} -3 & 6 & 3 \\ -1 & 7 & 4 \end{bmatrix}, \qquad (\mathbf{AB})' = \begin{bmatrix} -3 & -1 \\ 6 & 7 \\ 3 & 4 \end{bmatrix};$$

$$\mathbf{B}'\mathbf{A}' = \begin{bmatrix} -1 & 3 \\ 2 & -1 \\ 1 & 0 \end{bmatrix} \begin{bmatrix} 3 & 4 \\ 0 & 1 \end{bmatrix} = \begin{bmatrix} -3 & -1 \\ 6 & 7 \\ 3 & 4 \end{bmatrix}.$$

Note that $(\mathbf{AB})' = \mathbf{B}'\mathbf{A}'$ but $\mathbf{A}'\mathbf{B}'$ is not defined.

For the remainder of this section, we concern ourselves with square matrices; that is, matrices that have the same number of rows and columns. A *diagonal* matrix is a square matrix all of whose entries are zero except possibly for those on the main diagonal. (Recall that the main diagonal was defined to be the diagonal running from the upper left to the lower right.) Thus,

$$\begin{bmatrix} 5 & 0 \\ 0 & -1 \end{bmatrix} \quad \text{and} \quad \begin{bmatrix} 3 & 0 & 0 \\ 0 & 3 & 0 \\ 0 & 0 & 3 \end{bmatrix}$$

are both diagonal matrices of order 2×2 and 3×3 respectively. The zero matrix is the special diagonal matrix having all the elements on the main diagonal equal to zero.

The *identity* matrix is a diagonal matrix worthy of special consideration. Designated by $\mathbf{I}$, the identity is defined to be that diagonal matrix having all diagonal elements equal to one. Thus,

$$\begin{bmatrix} 1 & 0 \\ 0 & 1 \end{bmatrix} \quad \text{and} \quad \begin{bmatrix} 1 & 0 & 0 & 0 \\ 0 & 1 & 0 & 0 \\ 0 & 0 & 1 & 0 \\ 0 & 0 & 0 & 1 \end{bmatrix}$$

are the 2×2 and 4×4 identities respectively. The identity is perhaps the most important matrix of all. If the identity is of the appropriate order so that the following multiplication can be carried out, then for any arbitrary matrix $\mathbf{A}$,

$$\mathbf{AI} = \mathbf{A} \quad \text{and} \quad \mathbf{IA} = \mathbf{A}.$$

A *symmetric* matrix is a matrix that is equal to its transpose while a *skew symmetric* matrix is a matrix that is equal to the negative of its transpose. Thus, a matrix $\mathbf{A}$ is symmetric if $\mathbf{A} = \mathbf{A}'$ while it is skew symmetric if $\mathbf{A} = -\mathbf{A}'$. Examples of each are respectively

$$\begin{bmatrix} 1 & 2 & 3 \\ 2 & 4 & 5 \\ 3 & 5 & 6 \end{bmatrix} \quad \text{and} \quad \begin{bmatrix} 0 & 2 & -3 \\ -2 & 0 & 1 \\ 3 & -1 & 0 \end{bmatrix}.$$

Note that if a matrix is skew symmetric, then all its diagonal elements must be zero. Why?

A matrix $\mathbf{A} = [a_{ij}]$ is called *lower triangular* if $a_{ij} = 0$ for $j > i$ (that is, if all the elements above the main diagonal are zero) and *upper triangular* if $a_{ij} = 0$ for $i > j$ (that is, if all the elements below the main diagonal are zero). Examples of lower and upper triangular matrices are

$$\begin{bmatrix} 5 & 0 & 0 & 0 \\ -1 & 2 & 0 & 0 \\ 0 & 1 & 3 & 0 \\ 2 & 1 & 4 & 1 \end{bmatrix} \quad \text{and} \quad \begin{bmatrix} -1 & 2 & 4 & 1 \\ 0 & 1 & 3 & -1 \\ 0 & 0 & 2 & 5 \\ 0 & 0 & 0 & 5 \end{bmatrix}.$$

Finally, we define positive integral powers of a matrix in the obvious manner: $\mathbf{A}^2 = \mathbf{AA}$, $\mathbf{A}^3 = \mathbf{AAA}$ and, in general, if n is a positive integer,

$$\mathbf{A}^n = \underbrace{\mathbf{AA} \cdots \mathbf{A}}_{n \text{ times}}.$$

Thus, if

$$\mathbf{A} = \begin{bmatrix} 1 & -2 \\ 1 & 3 \end{bmatrix}, \quad \text{then } \mathbf{A}^2 = \begin{bmatrix} 1 & -2 \\ 1 & 3 \end{bmatrix}\begin{bmatrix} 1 & -2 \\ 1 & 3 \end{bmatrix} = \begin{bmatrix} -1 & -8 \\ 4 & 7 \end{bmatrix}.$$

PROBLEMS 1.5

1. Verify that $(\mathbf{A} + \mathbf{B})' = \mathbf{A}' + \mathbf{B}'$ where

$$\mathbf{A} = \begin{bmatrix} 1 & 5 & -1 \\ 2 & 1 & 3 \\ 0 & 7 & -8 \end{bmatrix} \quad \text{and} \quad \mathbf{B} = \begin{bmatrix} 6 & 1 & 3 \\ 2 & 0 & -1 \\ -1 & -7 & 2 \end{bmatrix}.$$

2. Verify that $(\mathbf{AB})' = \mathbf{B}'\mathbf{A}'$, where

$$\mathbf{A} = \begin{bmatrix} t & t^2 \\ 1 & 2t \\ 1 & 0 \end{bmatrix} \quad \text{and} \quad \mathbf{B} = \begin{bmatrix} 3 & t & t+1 & 0 \\ t & 2t & t^2 & t^3 \end{bmatrix}.$$

3. Prove that if $\mathbf{A}$ is any matrix, then $\mathbf{AA}'$ is symmetric.

4. Show that $\mathbf{AB} = \mathbf{BA}$, where

$$\mathbf{A} = \begin{bmatrix} -1 & 0 & 0 \\ 0 & 3 & 0 \\ 0 & 0 & 1 \end{bmatrix} \quad \text{and} \quad \mathbf{B} = \begin{bmatrix} 5 & 0 & 0 \\ 0 & 3 & 0 \\ 0 & 0 & 2 \end{bmatrix}.$$

5. Prove that if $\mathbf{A}$ and $\mathbf{B}$ are diagonal matrices of the same order then $\mathbf{AB} = \mathbf{BA}$.

6. If $\mathbf{A}$ is a square matrix and $\mathbf{B}$ is a diagonal matrix, describe what happens to $\mathbf{A}$ if it is postmultiplied by $\mathbf{B}$. What happens to $\mathbf{A}$ if it is premultiplied by $\mathbf{B}$?

7. Find $\mathbf{A}^3$ if

$$\mathbf{A} = \begin{bmatrix} 3 & 1 & 0 \\ 2 & -1 & 1 \\ 3 & 0 & 1 \end{bmatrix}.$$

8. Find $\mathbf{A}^3$ if

$$\mathbf{A} = \begin{bmatrix} 1 & 0 & 0 \\ 0 & 2 & 0 \\ 0 & 0 & 3 \end{bmatrix}.$$

9. Using problem 8 as a guide, what can be said about $\mathbf{A}^n$ if $\mathbf{A}$ is a diagonal matrix and n is a positive integer?

1.6 SUBMATRICES AND PARTITIONING

Given any matrix $\mathbf{A}$, a *submatrix* of $\mathbf{A}$ is a matrix obtained from $\mathbf{A}$ by the removal of any number of rows or columns. Thus, if

$$\mathbf{A} = \begin{bmatrix} 1 & 2 & 3 & 4 \\ 5 & 6 & 7 & 8 \\ 9 & 10 & 11 & 12 \\ 13 & 14 & 15 & 16 \end{bmatrix}, \qquad \mathbf{B} = \begin{bmatrix} 10 & 12 \\ 14 & 16 \end{bmatrix}, \qquad \text{and} \qquad \mathbf{C} = [2 \quad 3 \quad 4], \tag{11}$$

then $\mathbf{B}$ and $\mathbf{C}$ are both submatrices of $\mathbf{A}$. Here $\mathbf{B}$ was obtained by removing from $\mathbf{A}$ the first and second rows together with the first and third columns, while $\mathbf{C}$ was obtained by removing from $\mathbf{A}$ the second, third, and fourth rows together with the first column. By removing no rows and no columns from $\mathbf{A}$, it follows that $\mathbf{A}$ is a submatrix of itself.

A matrix is said to be partitioned if it is divided into submatrices by horizontal and vertical lines between the rows and columns. By varying the choices of where to put the horizontal and vertical lines, one can partition a matrix in many different ways. Thus,

$$\left[\begin{array}{cc|cc} 1 & 2 & 3 & 4 \\ 5 & 6 & 7 & 8 \\ \hline 9 & 10 & 11 & 12 \\ 13 & 14 & 15 & 16 \end{array}\right] \qquad \text{and} \qquad \left[\begin{array}{c|cc|c} 1 & 2 & 3 & 4 \\ \hline 5 & 6 & 7 & 8 \\ 9 & 10 & 11 & 12 \\ 13 & 14 & 15 & 16 \end{array}\right]$$

are examples of two different partitions of the matrix $\mathbf{A}$ given in (11).

If partitioning is carried out in a particularly judicious manner, it can be a great help in matrix multiplication. Consider the case where the two matrices **A** and **B** are to be multiplied together. If we partition both **A** and **B** into four submatrices, respectively, so that

$$\mathbf{A} = \left[\begin{array}{c|c} \mathbf{C} & \mathbf{D} \\ \hline \mathbf{E} & \mathbf{F} \end{array}\right] \quad \text{and} \quad \mathbf{B} = \left[\begin{array}{c|c} \mathbf{G} & \mathbf{H} \\ \hline \mathbf{J} & \mathbf{K} \end{array}\right]$$

where **C** through **K** represent submatrices, then the product **AB** may be obtained by simply carrying out the multiplication as if the submatrices were themselves elements. Thus,

$$\mathbf{AB} = \left[\begin{array}{c|c} \mathbf{CG} + \mathbf{DJ} & \mathbf{CH} + \mathbf{DK} \\ \hline \mathbf{EG} + \mathbf{FJ} & \mathbf{EH} + \mathbf{FK} \end{array}\right],$$

providing the partitioning was such that the indicated multiplications are defined.

EXAMPLE 1 Find **AB** if

$$\mathbf{A} = \begin{bmatrix} 3 & 1 & 2 \\ 1 & 4 & -1 \\ 3 & 1 & 2 \end{bmatrix} \quad \text{and} \quad \mathbf{B} = \begin{bmatrix} 1 & 3 & 2 \\ -1 & 0 & 1 \\ 0 & 1 & 1 \end{bmatrix}.$$

Solution We first partition **A** and **B** in the following manner

$$\mathbf{A} = \left[\begin{array}{cc|c} 3 & 1 & 2 \\ 1 & 4 & -1 \\ \hline 3 & 1 & 2 \end{array}\right] \quad \text{and} \quad \mathbf{B} = \left[\begin{array}{cc|c} 1 & 3 & 2 \\ -1 & 0 & 1 \\ \hline 0 & 1 & 1 \end{array}\right];$$

then,

$$\mathbf{AB} = \left[\begin{array}{c|c} \begin{bmatrix} 3 & 1 \\ 1 & 4 \end{bmatrix}\begin{bmatrix} 1 & 3 \\ -1 & 0 \end{bmatrix} + \begin{bmatrix} 2 \\ -1 \end{bmatrix}\begin{bmatrix} 0 & 1 \end{bmatrix} & \begin{bmatrix} 3 & 1 \\ 1 & 4 \end{bmatrix}\begin{bmatrix} 2 \\ 1 \end{bmatrix} + \begin{bmatrix} 2 \\ -1 \end{bmatrix}\begin{bmatrix} 1 \end{bmatrix} \\ \hline \begin{bmatrix} 3 & 1 \end{bmatrix}\begin{bmatrix} 1 & 3 \\ -1 & 0 \end{bmatrix} + \begin{bmatrix} 2 \end{bmatrix}\begin{bmatrix} 0 & 1 \end{bmatrix} & \begin{bmatrix} 3 & 1 \end{bmatrix}\begin{bmatrix} 2 \\ 1 \end{bmatrix} + \begin{bmatrix} 2 \end{bmatrix}\begin{bmatrix} 1 \end{bmatrix} \end{array}\right]$$

$$= \left[\begin{array}{c|c} \begin{bmatrix} 2 & 9 \\ -3 & 3 \end{bmatrix} + \begin{bmatrix} 0 & 2 \\ 0 & -1 \end{bmatrix} & \begin{bmatrix} 7 \\ 6 \end{bmatrix} + \begin{bmatrix} 2 \\ -1 \end{bmatrix} \\ \hline \begin{bmatrix} 2 & 9 \end{bmatrix} + \begin{bmatrix} 0 & 2 \end{bmatrix} & \begin{bmatrix} 7 \end{bmatrix} + \begin{bmatrix} 2 \end{bmatrix} \end{array}\right]$$

$$= \left[\begin{array}{cc|c} 2 & 11 & 9 \\ -3 & 2 & 5 \\ \hline 2 & 11 & 9 \end{array}\right] = \begin{bmatrix} 2 & 11 & 9 \\ -3 & 2 & 5 \\ 2 & 11 & 9 \end{bmatrix}.$$

EXAMPLE 2 Find **AB** if

$$\mathbf{A} = \left[\begin{array}{cc|c} 3 & 1 & 0 \\ 2 & 0 & 0 \\ \hline 0 & 0 & 3 \\ 0 & 0 & 1 \\ \hline 0 & 0 & 0 \end{array}\right] \quad \text{and} \quad \mathbf{B} = \left[\begin{array}{cc|ccc} 2 & 1 & 0 & 0 & 0 \\ -1 & 1 & 0 & 0 & 0 \\ \hline 0 & 1 & 0 & 0 & 1 \end{array}\right].$$

Solution From the indicated partitions, we find that

$$\mathbf{AB} = \left[\begin{array}{c|c} \begin{bmatrix} 3 & 1 \\ 2 & 0 \end{bmatrix}\begin{bmatrix} 2 & 1 \\ -1 & 1 \end{bmatrix} + \begin{bmatrix} 0 \\ 0 \end{bmatrix}[0 \quad 1] & \begin{bmatrix} 3 & 1 \\ 2 & 0 \end{bmatrix}\begin{bmatrix} 0 & 0 & 0 \\ 0 & 0 & 0 \end{bmatrix} + \begin{bmatrix} 0 \\ 0 \end{bmatrix}[0 \quad 0 \quad 1] \\ \hline \begin{bmatrix} 0 & 0 \\ 0 & 0 \end{bmatrix}\begin{bmatrix} 2 & 1 \\ -1 & 1 \end{bmatrix} + \begin{bmatrix} 3 \\ 1 \end{bmatrix}[0 \quad 1] & \begin{bmatrix} 0 & 0 \\ 0 & 0 \end{bmatrix}\begin{bmatrix} 0 & 0 & 0 \\ 0 & 0 & 0 \end{bmatrix} + \begin{bmatrix} 3 \\ 1 \end{bmatrix}[0 \quad 0 \quad 1] \\ \hline [0 \quad 0]\begin{bmatrix} 2 & 1 \\ -1 & 1 \end{bmatrix} + [0][0 \quad 1] & [0 \quad 0]\begin{bmatrix} 0 & 0 & 0 \\ 0 & 0 & 0 \end{bmatrix} + [0][0 \quad 0 \quad 1] \end{array}\right]$$

$$\mathbf{AB} = \left[\begin{array}{c|c} \begin{bmatrix} 5 & 4 \\ 4 & 2 \end{bmatrix} + \begin{bmatrix} 0 & 0 \\ 0 & 0 \end{bmatrix} & \begin{bmatrix} 0 & 0 & 0 \\ 0 & 0 & 0 \end{bmatrix} + \begin{bmatrix} 0 & 0 & 0 \\ 0 & 0 & 0 \end{bmatrix} \\ \hline \begin{bmatrix} 0 & 0 \\ 0 & 0 \end{bmatrix} + \begin{bmatrix} 0 & 3 \\ 0 & 1 \end{bmatrix} & \begin{bmatrix} 0 & 0 & 0 \\ 0 & 0 & 0 \end{bmatrix} + \begin{bmatrix} 0 & 0 & 3 \\ 0 & 0 & 1 \end{bmatrix} \\ \hline [0 \quad 0] + [0 \quad 0] & [0 \quad 0 \quad 0] + [0 \quad 0 \quad 0] \end{array}\right]$$

$$= \left[\begin{array}{cc|ccc} 5 & 4 & 0 & 0 & 0 \\ 4 & 2 & 0 & 0 & 0 \\ \hline 0 & 3 & 0 & 0 & 3 \\ 0 & 1 & 0 & 0 & 1 \\ \hline 0 & 0 & 0 & 0 & 0 \end{array}\right] = \begin{bmatrix} 5 & 4 & 0 & 0 & 0 \\ 4 & 2 & 0 & 0 & 0 \\ 0 & 3 & 0 & 0 & 3 \\ 0 & 1 & 0 & 0 & 1 \\ 0 & 0 & 0 & 0 & 0 \end{bmatrix}.$$

Note that we partitioned in order to make maximum use of the zero submatrices of both **A** and **B**.

Partitioning can be used to check matrix multiplication. The method is illustrated by the following example.

EXAMPLE 3 Find **AB** if

$$\mathbf{A} = \begin{bmatrix} 2 & 1 \\ -1 & 3 \end{bmatrix} \quad \text{and} \quad \mathbf{B} = \begin{bmatrix} 1 & 2 & 3 \\ 0 & 1 & 4 \end{bmatrix}.$$

Solution We first construct two new partitioned matrices **C** and **D** where **C** is obtained from **A** by adding to **A** another row consisting of the sums of the columns of **A**, and **D** is obtained from **B** by adding to **B** another column consisting of the sums of the rows of **B**. Thus,

$$\mathbf{C} = \left[\begin{array}{rr} 2 & 1 \\ -1 & 3 \\ \hline 1 & 4 \end{array}\right], \qquad \text{and} \qquad \mathbf{D} = \left[\begin{array}{ccc|c} 1 & 2 & 3 & 6 \\ 0 & 1 & 4 & 5 \end{array}\right].$$

We now calculate **CD**:

$$\mathbf{CD} = \left[\begin{array}{c|c} \begin{bmatrix} 2 & 1 \\ -1 & 3 \end{bmatrix}\begin{bmatrix} 1 & 2 & 3 \\ 0 & 1 & 4 \end{bmatrix} & \begin{bmatrix} 2 & 1 \\ -1 & 3 \end{bmatrix}\begin{bmatrix} 6 \\ 5 \end{bmatrix} \\ \hline [1 \quad 4]\begin{bmatrix} 1 & 2 & 3 \\ 0 & 1 & 4 \end{bmatrix} & [1 \quad 4]\begin{bmatrix} 6 \\ 5 \end{bmatrix} \end{array}\right]$$

$$= \left[\begin{array}{rrr|r} 2 & 5 & 10 & 17 \\ -1 & 1 & 9 & 9 \\ \hline 1 & 6 & 19 & 26 \end{array}\right].$$

The submatrix in the upper left hand corner should be **AB**, the answer to the problem; if it is correct, then the submatrix in the upper right hand corner will be the sum of the rows of **AB**, while the submatrix in the lower left-hand corner will be the sum of the columns of **AB**. The submatrix in the lower right-hand corner will be the sum of the column of the submatrix in the upper right corner and also the sum of the row of the submatrix in the lower left corner.

PROBLEMS 1.6

1. Which of the following are submatrices of the given **A** and why?

$$\mathbf{A} = \begin{bmatrix} 1 & 2 & 3 \\ 4 & 5 & 6 \\ 7 & 8 & 9 \end{bmatrix}$$

(a) $\begin{bmatrix} 1 & 3 \\ 7 & 9 \end{bmatrix}$ (b) $[1]$ (c) $\begin{bmatrix} 1 & 2 \\ 8 & 9 \end{bmatrix}$ (d) $\begin{bmatrix} 4 & 6 \\ 7 & 9 \end{bmatrix}$.

2. Given the matrices **A** and **B** (as shown), find **AB** using the partitionings indicated:

$$\mathbf{A} = \left[\begin{array}{rr|r} 1 & -1 & 2 \\ \hline 3 & 0 & 4 \\ 0 & 1 & 2 \end{array}\right], \qquad \mathbf{B} = \left[\begin{array}{rrr|r} 5 & 2 & 0 & 2 \\ 1 & -1 & 3 & 1 \\ \hline 0 & 1 & 1 & 4 \end{array}\right].$$

3. Partition the given matrices **A** and **B** and, using these results, find **AB**.

$$\mathbf{A} = \begin{bmatrix} 4 & 1 & 0 & 0 \\ 2 & 2 & 0 & 0 \\ 0 & 0 & 1 & 0 \\ 0 & 0 & 1 & 2 \end{bmatrix}, \qquad \mathbf{B} = \begin{bmatrix} 3 & 2 & 0 & 0 \\ -1 & 1 & 0 & 0 \\ 0 & 0 & 2 & 1 \\ 0 & 0 & 1 & -1 \end{bmatrix}.$$

4. Use the checking procedure to find **AB** if

$$\mathbf{A} = \begin{bmatrix} 2 & 1 \\ -3 & 2 \end{bmatrix} \quad \text{and} \quad \mathbf{B} = \begin{bmatrix} 11 & 9 \\ -2 & 4 \end{bmatrix}.$$

5. Use the checking procedure to find **AB** if

$$\mathbf{A} = \begin{bmatrix} 9 & -3 & 5 \\ 2 & 7 & 8 \\ 1 & -4 & 2 \end{bmatrix} \quad \text{and} \quad \mathbf{B} = \begin{bmatrix} 4 & -1 \\ 11 & -3 \\ 3 & 1 \end{bmatrix}.$$

1.7 VECTORS

Definition 1 A *vector* is a $1 \times n$ or $n \times 1$ matrix.

A $1 \times n$ matrix is called a *row vector* while an $n \times 1$ matrix is a *column vector*. The elements are called the *components* of the vector while the number of components in the vector, in this case n, is its *dimension*. Thus,

$$\begin{bmatrix} 1 \\ 2 \\ 3 \end{bmatrix}$$

is an example of a 3-dimensional column vector, while

$$[t \quad 2t \quad -t \quad 0]$$

is an example of a 4-dimensional row vector.

The reader who is already familiar with vectors will notice that we have not defined vectors as directed line segments. We have done this intentionally, first because in more than three dimensions this geometric interpretation loses its significance, and second, because in the general mathematical framework, vectors are not directed line segments. However, the idea of representing a finite dimensional vector by its components and hence as a matrix is one that is acceptable to the scientist, engineer, and mathematician. Also, as a bonus, since a vector is nothing more than a special matrix, we have already defined scalar multiplication, vector addition, and vector equality.

A vector $\mathbf{y}$ (vectors will be designated by boldface lowercase letters) has associated with it a nonnegative number called its *magnitude* or length designated by $\|\mathbf{y}\|$.

Definition 2 If $\mathbf{y} = [y_1 y_2 \cdots y_n]$ then $\|\mathbf{y}\| = \sqrt{(y_1)^2 + (y_2)^2 + \cdots + (y_n)^2}$.

EXAMPLE 1 Find $\|\mathbf{y}\|$ if $\mathbf{y} = [1 \quad 2 \quad 3 \quad 4]$.

Solution

$$\|\mathbf{y}\| = \sqrt{(1)^2 + (2)^2 + (3)^2 + (4)^2} = \sqrt{30}.$$

If $\mathbf{z}$ is a column vector, $\|\mathbf{z}\|$ is defined in a completely analogous manner.

EXAMPLE 2 Find $\|\mathbf{z}\|$ if

$$\mathbf{z} = \begin{bmatrix} -1 \\ 2 \\ -3 \end{bmatrix}$$

Solution $\|\mathbf{z}\| = \sqrt{(-1)^2 + (2)^2 + (-3)^2} = \sqrt{14}.$

A vector is called a *unit vector* if its magnitude is equal to one. A nonzero vector is said to be *normalized* if it is divided by its magnitude. Thus, a normalized vector is also a unit vector.

EXAMPLE 3 Normalize the vector $[1 \quad 0 \quad -3 \quad 2 \quad -1]$.

Solution The magnitude of this vector is

$$\sqrt{(1)^2 + (0)^2 + (-3)^2 + (2)^2 + (-1)^2} = \sqrt{15}.$$

Hence, the normalized vector is

$$\begin{bmatrix} \dfrac{1}{\sqrt{15}} & 0 & \dfrac{-3}{\sqrt{15}} & \dfrac{2}{\sqrt{15}} & \dfrac{-1}{\sqrt{15}} \end{bmatrix}.$$

In passing, we note that when a general vector is written $\mathbf{y} = [y_1 y_2 \cdots y_n]$ one of the subscripts of each element of the matrix is deleted. This is done solely for the sake of convenience. Since a row vector has only one row (a column vector has only one column), it is redundant and unnecessary to exhibit the row subscript (the column subscript).

PROBLEMS 1.7

1. Find p if $5\mathbf{x} - 2\mathbf{y} = \mathbf{b}$, where

$$\mathbf{x} = \begin{bmatrix} 1 \\ 3 \\ 0 \end{bmatrix}, \quad \mathbf{y} = \begin{bmatrix} 2 \\ p \\ 1 \end{bmatrix}, \quad \text{and} \quad \mathbf{b} = \begin{bmatrix} 1 \\ 13 \\ -2 \end{bmatrix}.$$

2. Find $\mathbf{x}$ if $3\mathbf{x} + 2\mathbf{y} = \mathbf{b}$, where

$$\mathbf{y} = \begin{bmatrix} 3 \\ 1 \\ 6 \\ 0 \end{bmatrix} \quad \text{and} \quad \mathbf{b} = \begin{bmatrix} 2 \\ -1 \\ 4 \\ 1 \end{bmatrix}.$$

3. Find $\|\mathbf{y}\|$ if

(a) $\mathbf{y} = [2 \quad 1 \quad -1 \quad 3]$,
(b) $\mathbf{y} = [0 \quad -1 \quad 5 \quad 3 \quad 2]$.

4. Normalize $\mathbf{y}$ if

(a) $\mathbf{y} = [3 \quad 0 \quad 1]$,
(b) $\mathbf{y} = [6 \quad -7 \quad 1 \quad 0 \quad -3]$.

5. Prove that a normalized vector must be a unit vector.

Chapter 2 DETERMINANTS

2.1 DETERMINANTS

Every square matrix has associated with it a scalar called its *determinant*. To be extremely rigorous we would have to define this scalar in terms of permutations on positive integers.[1] However, since in practice it is difficult to apply a definition of this sort, other procedures have been developed which yield the determinant in a more straightforward manner. In this chapter, therefore, we concern ourselves solely with those methods that can be applied easily. We note here for reference that determinants are only defined for square matrices.

[1] P. Lancaster, " Theory of Matrices," p. 32. Academic Press, New York, 1969.

Given a square matrix **A**, we use det(**A**) or |**A**| to designate its determinant. If the matrix can actually be exhibited, we then designate the determinant of **A** by replacing the brackets by vertical straight lines. For example, if

$$\mathbf{A} = \begin{bmatrix} 1 & 2 & 3 \\ 4 & 5 & 6 \\ 7 & 8 & 9 \end{bmatrix} \tag{1}$$

then,

$$\det(\mathbf{A}) = \begin{vmatrix} 1 & 2 & 3 \\ 4 & 5 & 6 \\ 7 & 8 & 9 \end{vmatrix}. \tag{2}$$

We cannot overemphasize the fact that (1) and (2) represent entirely different animals. (1) represents a matrix, a rectangular array, an entity unto itself while (2) represents a scalar, a number associated with the matrix in (1). There is absolutely no similarity between the two other than form!

We are now ready to calculate determinants.

Definition 1 The determinant of a 1×1 matrix $[a]$ is the scalar a.

Thus, the determinant of the matrix [5] is 5 and the determinant of the matrix $[-3]$ is -3.

Definition 2 The determinant of a 2×2 matrix

$$\begin{bmatrix} a & b \\ c & d \end{bmatrix}$$

is the scalar $ad-bc$.

EXAMPLE 1 Find det(**A**) if

$$\mathbf{A} = \begin{bmatrix} 1 & 2 \\ 4 & 3 \end{bmatrix}.$$

Solution

$$\det(\mathbf{A}) = \begin{vmatrix} 1 & 2 \\ 4 & 3 \end{vmatrix} = (1)(3) - (2)(4) = 3 - 8 = -5.$$

EXAMPLE 2 Find |**A**| if

$$\mathbf{A} = \begin{bmatrix} 2 & -1 \\ 4 & 3 \end{bmatrix}.$$

Solution

$$|\mathbf{A}| = \begin{vmatrix} 2 & -1 \\ 4 & 3 \end{vmatrix} = (2)(3) - (-1)(4) = 6 + 4 = 10.$$

We now could proceed to give separate rules which would enable one to compute determinants of 3×3, 4×4, and higher order matrices. This is unnecessary. In the next section, we will give a method that enables us to reduce all determinants of order n ($n > 2$) (if $\mathbf{A}$ has order $n \times n$ then $\det(\mathbf{A})$ is said to have order n) to a sum of determinants of order 2.

PROBLEMS 2.1

1. Find $|\mathbf{A}|$ if

$$\mathbf{A} = \begin{bmatrix} 1 & 0 \\ 2 & 1 \end{bmatrix}.$$

2. Find $|\mathbf{A}|$ if

$$\mathbf{A} = \begin{bmatrix} 1 & 3 \\ -1 & 2 \end{bmatrix}.$$

3. Find $|\mathbf{A}|$, $|\mathbf{B}|$, and $|\mathbf{AB}|$ if

$$\mathbf{A} = \begin{bmatrix} 1 & 3 \\ 2 & 1 \end{bmatrix} \quad \text{and} \quad \mathbf{B} = \begin{bmatrix} 4 & 2 \\ -1 & 2 \end{bmatrix}.$$

What is the relationship between these three determinants?

2.2 EXPANSION BY COFACTORS

Definition 1 Given a matrix $\mathbf{A}$, a *minor* is the determinant of any square submatrix of $\mathbf{A}$.

That is, given a square matrix $\mathbf{A}$, a minor is the determinant of any matrix formed from $\mathbf{A}$ by the removal of an equal number of rows and columns. As an example, if

$$\mathbf{A} = \begin{bmatrix} 1 & 2 & 3 \\ 4 & 5 & 6 \\ 7 & 8 & 9 \end{bmatrix},$$

then

$$\begin{vmatrix} 1 & 2 \\ 7 & 8 \end{vmatrix} \quad \text{and} \quad \begin{vmatrix} 5 & 6 \\ 8 & 9 \end{vmatrix}$$

are both minors since

$$\begin{bmatrix} 1 & 2 \\ 7 & 8 \end{bmatrix} \quad \text{and} \quad \begin{bmatrix} 5 & 6 \\ 8 & 9 \end{bmatrix}$$

are both submatrices of **A**, while

$$\begin{vmatrix} 1 & 2 \\ 8 & 9 \end{vmatrix} \quad \text{and} \quad |1 \quad 2|$$

are not minors since

$$\begin{bmatrix} 1 & 2 \\ 8 & 9 \end{bmatrix}$$

is not a submatrix of **A** and $[1 \quad 2]$, although a submatrix of **A**, is not square.

A more useful concept for our immediate purposes, since it will enable us to calculate determinants, is that of the cofactor of an element of a matrix.

Definition 2 Given a matrix $\mathbf{A} = [a_{ij}]$, the *cofactor of the element* a_{ij} is a scalar obtained by multiplying together the term $(-1)^{i+j}$ and the minor obtained from **A** by removing the ith row and jth column.

In other words, to compute the cofactor of the element a_{ij} we first form a submatrix of **A** by crossing out both the row and column in which the element a_{ij} appears. We then find the determinant of the submatrix and finally multiply it by the number $(-1)^{i+j}$.

EXAMPLE 1 Find the cofactor of the element 4 in the matrix

$$\mathbf{A} = \begin{bmatrix} 1 & 2 & 3 \\ 4 & 5 & 6 \\ 7 & 8 & 9 \end{bmatrix}.$$

Solution We first note that 4 appears in the (2, 1) position. The submatrix obtained by crossing out the second row and first column is

$$\begin{bmatrix} 1 & 2 & 3 \\ 4 & 5 & 6 \\ 7 & 8 & 9 \end{bmatrix} = \begin{bmatrix} 2 & 3 \\ 8 & 9 \end{bmatrix},$$

which has a determinant equal to $(2)(9) - (3)(8) = -6$. Since 4 appears in the (2, 1) position, $i = 2$ and $j = 1$. Thus, $(-1)^{i+j} = (-1)^{2+1} = (-1)^3 = (-1)$. The cofactor of 4 is $(-1)(-6) = 6$.

EXAMPLE 2 Using the same **A** as in Example 1, find the cofactor of the element 9.

Solution The element 9 appears in the (3, 3) position. Thus, crossing out the third row and third column, we obtain the submatrix

$$\begin{bmatrix} 1 & 2 & 3 \\ 4 & 5 & 6 \\ 7 & 8 & 9 \end{bmatrix} = \begin{bmatrix} 1 & 2 \\ 4 & 5 \end{bmatrix},$$

which has a determinant equal to $(1)(5) - (2)(4) = -3$. Since, in this case, $i = j = 3$, the cofactor of 9 is $(-1)^{3+3}(-3) = (-1)^6(-3) = -3$.

We now have enough tools at hand to find the determinant of any matrix.

EXPANSION BY COFACTORS To find the determinant of a matrix **A** of arbitrary order, (a) pick any one row or any one column of the matrix (dealer's choice), (b) for each element in the row or column chosen, find its cofactor, (c) multiply each element in the row or column chosen by its cofactor and sum the results. This sum is the determinant of the matrix.

EXAMPLE 3 Find det(**A**) if

$$\mathbf{A} = \begin{bmatrix} 3 & 5 & 0 \\ -1 & 2 & 1 \\ 3 & -6 & 4 \end{bmatrix}.$$

Solution In this example, we expand by the second column.

$$\begin{aligned} |\mathbf{A}| &= (5)\text{ (cofactor of 5)} + (2)\text{(cofactor of 2)} + (-6)\text{(cofactor of } -6) \\ &= (5)(-1)^{1+2}\begin{vmatrix} -1 & 1 \\ 3 & 4 \end{vmatrix} + (2)(-1)^{2+2}\begin{vmatrix} 3 & 0 \\ 3 & 4 \end{vmatrix} + (-6)(-1)^{3+2}\begin{vmatrix} 3 & 0 \\ -1 & 1 \end{vmatrix} \\ &= 5(-1)(-4-3) + (2)(1)(12-0) + (-6)(-1)(3-0) \\ &= (-5)(-7) + (2)(12) + (6)(3) = 35 + 24 + 18 = 77. \end{aligned}$$

EXAMPLE 4 Using the **A** of Example 3 and expanding by the first row, find det(**A**).

Solution

$$\begin{aligned} |\mathbf{A}| &= 3\text{(cofactor of 3)} + 5\text{(cofactor of 5)} + 0\text{(cofactor of 0)} \\ &= (3)(-1)^{1+1}\begin{vmatrix} 2 & 1 \\ -6 & 4 \end{vmatrix} + 5(-1)^{1+2}\begin{vmatrix} -1 & 1 \\ 3 & 4 \end{vmatrix} + 0 \\ &= (3)(1)(8+6) + (5)(-1)(-4-3) = (3)(14) + (-5)(-7) = 42 + 35 = 77. \end{aligned}$$

The previous examples illustrate two important properties of the method. First, the value of the determinant is the same regardless of which row or column we choose to expand by[2] and second, expanding by a row or column that contains zeros significantly reduces the number of computations involved.

EXAMPLE 5 Find det(**A**) if

$$\mathbf{A} = \begin{bmatrix} 1 & 0 & 5 & 2 \\ -1 & 4 & 1 & 0 \\ 3 & 0 & 4 & 1 \\ -2 & 1 & 1 & 3 \end{bmatrix}.$$

Solution We first check to see which row or column contains the most zeros and expand by it. Thus, expanding by the second column gives

$$|\mathbf{A}| = 0(\text{cofactor of } 0) + 4(\text{cofactor of } 4) + 0(\text{cofactor of } 0) + 1(\text{cofactor of } 1)$$

$$= 0 + 4(-1)^{2+2}\begin{vmatrix} 1 & 5 & 2 \\ 3 & 4 & 1 \\ -2 & 1 & 3 \end{vmatrix} + 0 + 1(-1)^{4+2}\begin{vmatrix} 1 & 5 & 2 \\ -1 & 1 & 0 \\ 3 & 4 & 1 \end{vmatrix}$$

$$= 4\begin{vmatrix} 1 & 5 & 2 \\ 3 & 4 & 1 \\ -2 & 1 & 3 \end{vmatrix} + \begin{vmatrix} 1 & 5 & 2 \\ -1 & 1 & 0 \\ 3 & 4 & 1 \end{vmatrix}.$$

Using expansion by cofactors on each of the determinants of order 3 yields

$$\begin{vmatrix} 1 & 5 & 2 \\ 3 & 4 & 1 \\ -2 & 1 & 3 \end{vmatrix} = 1(-1)^{1+1}\begin{vmatrix} 4 & 1 \\ 1 & 3 \end{vmatrix} + 5(-1)^{1+2}\begin{vmatrix} 3 & 1 \\ -2 & 3 \end{vmatrix} + 2(-1)^{1+3}\begin{vmatrix} 3 & 4 \\ -2 & 1 \end{vmatrix}$$

$$= -22$$

and

$$\begin{vmatrix} 1 & 5 & 2 \\ -1 & 1 & 0 \\ 3 & 4 & 1 \end{vmatrix} = 2(-1)^{1+3}\begin{vmatrix} -1 & 1 \\ 3 & 4 \end{vmatrix} + 0 + 1(-1)^{3+3}\begin{vmatrix} 1 & 5 \\ -1 & 1 \end{vmatrix} = -8.$$

Hence,

$$|\mathbf{A}| = 4(-22) - 8 = -88 - 8 = -96.$$

PROBLEMS 2.2

1. Find $|\mathbf{A}|$ if

$$\mathbf{A} = \begin{bmatrix} 1 & 3 & 2 \\ -1 & 4 & 1 \\ 5 & 3 & 8 \end{bmatrix}.$$

[2] F. M. Stein, "Intoduction to Matrices and Determinants" p. 46. Wadsworth, Belmont, California, 1967.

2. Find |A| if

$$\mathbf{A} = \begin{bmatrix} 3 & -2 & 0 \\ 1 & 1 & 2 \\ -3 & 4 & 1 \end{bmatrix}.$$

3. Find |A| if

$$\mathbf{A} = \begin{bmatrix} 1 & 2 & 1 & -1 \\ 4 & 0 & 3 & 0 \\ 1 & 1 & 0 & 5 \\ 2 & -2 & 1 & 1 \end{bmatrix}.$$

4. Find |A| if

$$\mathbf{A} = \begin{bmatrix} 11 & 1 & 0 & 9 & 0 \\ 2 & 1 & 1 & 0 & 0 \\ 4 & -1 & 1 & 0 & 0 \\ 3 & 2 & 2 & 1 & 0 \\ 0 & 0 & 1 & 2 & 0 \end{bmatrix}.$$

Can you generalize this result to any arbitrary matrix containing a zero column or zero row?

2.3 PROPERTIES OF DETERMINANTS

In this section, we give some useful properties of determinants. For the sake of expediency, we present most of these properties in terms of row operations; they are, however, equally valid for the analogous column operations. Similarly, we only prove each property for determinants of order three keeping in mind that, in general, these proofs may be extended, in a straightforward manner, to cover determinants of higher order.

Property 1 If one row of a matrix consists entirely of zeros, then the determinant is zero.

Proof Expanding by the zero row, we immediately obtain the desired result.

Property 2 If two rows of a matrix are interchanged, the determinant changes sign.

Proof Consider

$$\mathbf{A} = \begin{bmatrix} a_{11} & a_{12} & a_{13} \\ a_{21} & a_{22} & a_{23} \\ a_{31} & a_{32} & a_{33} \end{bmatrix}.$$

Expanding by the third row, we obtain

$$|\mathbf{A}| = a_{31}(a_{12}a_{23} - a_{13}a_{22}) - a_{32}(a_{11}a_{23} - a_{13}a_{21}) + a_{33}(a_{11}a_{22} - a_{12}a_{21}).$$

Now consider the matrix **B** obtained from **A** by interchanging the second and third rows:

$$\mathbf{B} = \begin{bmatrix} a_{11} & a_{12} & a_{13} \\ a_{31} & a_{32} & a_{33} \\ a_{21} & a_{22} & a_{23} \end{bmatrix}.$$

Expanding by the second row, we find that

$$\begin{aligned} |\mathbf{B}| = &-a_{31}(a_{12}a_{23} - a_{13}a_{22}) + a_{32}(a_{11}a_{23} - a_{13}a_{21}) \\ &- a_{33}(a_{11}a_{22} - a_{12}a_{21}). \end{aligned}$$

Thus, $|\mathbf{B}| = -|\mathbf{A}|$. Through similar reasoning, one can demonstrate that the result is valid regardless of which two rows are interchanged.

Property 3 If two rows of a determinant are identical, the determinant is zero.

Proof If we interchange the two identical rows of the matrix, the matrix remains unaltered; hence the determinant of the matrix remains constant. From Property 2, however, by interchanging two rows of a matrix, we change the sign of the determinant. Thus, the determinant must on one hand remain the same while on the other hand change sign. The only way both of these conditions can be met simultaneously is for the determinant to be zero.

Property 4 If the matrix **B** is obtained from the matrix **A** by multiplying every element in one row of **A** by the scalar λ, then $|\mathbf{B}| = \lambda|\mathbf{A}|$

Proof

$$\begin{vmatrix} \lambda a_{11} & \lambda a_{12} & \lambda a_{13} \\ a_{21} & a_{22} & a_{23} \\ a_{31} & a_{32} & a_{33} \end{vmatrix}$$

$$= \lambda a_{11} \begin{vmatrix} a_{22} & a_{23} \\ a_{32} & a_{33} \end{vmatrix} - \lambda a_{12} \begin{vmatrix} a_{21} & a_{23} \\ a_{31} & a_{33} \end{vmatrix} + \lambda a_{13} \begin{vmatrix} a_{21} & a_{22} \\ a_{31} & a_{32} \end{vmatrix}$$

$$= \lambda \left(a_{11} \begin{vmatrix} a_{22} & a_{23} \\ a_{32} & a_{33} \end{vmatrix} - a_{12} \begin{vmatrix} a_{21} & a_{23} \\ a_{31} & a_{33} \end{vmatrix} + a_{13} \begin{vmatrix} a_{21} & a_{22} \\ a_{31} & a_{32} \end{vmatrix} \right)$$

$$= \lambda \begin{vmatrix} a_{11} & a_{12} & a_{13} \\ a_{21} & a_{22} & a_{23} \\ a_{31} & a_{32} & a_{33} \end{vmatrix}.$$

In essence, Property 4 shows us how to multiply a scalar times a determinant. We know from Chapter 1 that multiplying a scalar times a matrix simply multiplies every element of the matrix by that scalar. Property 4, however, implies that multiplying a scalar times a determinant simply multiplies *one* row of the determinant (or, analogously, one column) by the scalar. Thus, while in matrices

$$8\begin{bmatrix} 1 & 2 \\ 3 & 4 \end{bmatrix} = \begin{bmatrix} 8 & 16 \\ 24 & 32 \end{bmatrix},$$

in determinants we have

$$8\begin{vmatrix} 1 & 2 \\ 3 & 4 \end{vmatrix} = \begin{vmatrix} 1 & 2 \\ 24 & 32 \end{vmatrix},$$

or alternatively

$$8\begin{vmatrix} 1 & 2 \\ 3 & 4 \end{vmatrix} = 4(2)\begin{vmatrix} 1 & 2 \\ 3 & 4 \end{vmatrix} = 4\begin{vmatrix} 2 & 4 \\ 3 & 4 \end{vmatrix} = \begin{vmatrix} 2 & 16 \\ 3 & 16 \end{vmatrix}.$$

Property 5 For an $n \times n$ matrix $\mathbf{A}$ and any scalar λ, $\det(\lambda\mathbf{A}) = \lambda^n \det(\mathbf{A})$.

Proof This proof makes continued use of Property 4.

$$\det(\lambda\mathbf{A}) = \det\left\{\lambda\begin{bmatrix} a_{11} & a_{12} & a_{13} \\ a_{21} & a_{22} & a_{23} \\ a_{31} & a_{32} & a_{33} \end{bmatrix}\right\} = \det\left\{\begin{bmatrix} \lambda a_{11} & \lambda a_{12} & \lambda a_{13} \\ \lambda a_{21} & \lambda a_{22} & \lambda a_{23} \\ \lambda a_{31} & \lambda a_{32} & \lambda a_{33} \end{bmatrix}\right\}$$

$$= \begin{vmatrix} \lambda a_{11} & \lambda a_{12} & \lambda a_{13} \\ \lambda a_{21} & \lambda a_{22} & \lambda a_{23} \\ \lambda a_{31} & \lambda a_{32} & \lambda a_{33} \end{vmatrix} = \lambda\begin{vmatrix} a_{11} & a_{12} & a_{13} \\ \lambda a_{21} & \lambda a_{22} & \lambda a_{23} \\ \lambda a_{31} & \lambda a_{32} & \lambda a_{33} \end{vmatrix}$$

$$= (\lambda)(\lambda)\begin{vmatrix} a_{11} & a_{12} & a_{13} \\ a_{21} & a_{22} & a_{23} \\ \lambda a_{31} & \lambda a_{32} & \lambda a_{33} \end{vmatrix} = \lambda(\lambda)(\lambda)\begin{vmatrix} a_{11} & a_{12} & a_{13} \\ a_{21} & a_{22} & a_{23} \\ a_{31} & a_{32} & a_{33} \end{vmatrix}$$

$$= \lambda^3 \det(\mathbf{A}).$$

Note that for a 3×3 matrix, $n = 3$.

Property 6 If a matrix $\mathbf{B}$ is obtained from a matrix $\mathbf{A}$ by adding to one row of $\mathbf{A}$, a scalar times another row of $\mathbf{A}$, then $|\mathbf{A}| = |\mathbf{B}|$.

Proof Let

$$\mathbf{A} = \begin{bmatrix} a_{11} & a_{12} & a_{13} \\ a_{21} & a_{22} & a_{23} \\ a_{31} & a_{32} & a_{33} \end{bmatrix}$$

and

$$\mathbf{B} = \begin{bmatrix} a_{11} & a_{12} & a_{13} \\ a_{21} & a_{22} & a_{23} \\ a_{31} + \lambda a_{11} & a_{32} + \lambda a_{12} & a_{33} + \lambda a_{13} \end{bmatrix},$$

where **B** has been obtained from **A** by adding λ times the first row of **A** to the third row of **A**. Expanding $|\mathbf{B}|$ by its third row, we obtain

$$\begin{aligned} |\mathbf{B}| &= (a_{31} + \lambda a_{11}) \begin{vmatrix} a_{12} & a_{13} \\ a_{22} & a_{23} \end{vmatrix} - (a_{32} + \lambda a_{12}) \begin{vmatrix} a_{11} & a_{13} \\ a_{21} & a_{23} \end{vmatrix} \\ &\quad + (a_{33} + \lambda a_{13}) \begin{vmatrix} a_{11} & a_{12} \\ a_{21} & a_{22} \end{vmatrix} \\ &= a_{31} \begin{vmatrix} a_{12} & a_{13} \\ a_{22} & a_{23} \end{vmatrix} - a_{32} \begin{vmatrix} a_{11} & a_{13} \\ a_{21} & a_{23} \end{vmatrix} + a_{33} \begin{vmatrix} a_{11} & a_{12} \\ a_{21} & a_{22} \end{vmatrix} \\ &\quad + \lambda \left\{ a_{11} \begin{vmatrix} a_{12} & a_{13} \\ a_{22} & a_{23} \end{vmatrix} - a_{12} \begin{vmatrix} a_{11} & a_{13} \\ a_{21} & a_{23} \end{vmatrix} + a_{13} \begin{vmatrix} a_{11} & a_{12} \\ a_{21} & a_{22} \end{vmatrix} \right\}. \end{aligned}$$

The first three terms of this sum are exactly $|\mathbf{A}|$ (expand $|\mathbf{A}|$ by its third row), while the last three terms of the sum are

$$\lambda \begin{vmatrix} a_{11} & a_{12} & a_{13} \\ a_{21} & a_{22} & a_{23} \\ a_{11} & a_{12} & a_{13} \end{vmatrix}$$

(expand this determinant by its third row). Thus, it follows that

$$|\mathbf{B}| = |\mathbf{A}| + \lambda \begin{vmatrix} a_{11} & a_{12} & a_{13} \\ a_{21} & a_{22} & a_{23} \\ a_{11} & a_{12} & a_{13} \end{vmatrix}.$$

From Property 3, however, this second determinant is zero since its first and third rows are identical, hence $|\mathbf{B}| = |\mathbf{A}|$.

The same type of argument will quickly show that this result is valid regardless of the two rows chosen.

EXAMPLE 1 Without expanding, show that

$$\begin{vmatrix} a & r & x \\ b & s & y \\ c & t & z \end{vmatrix} = \begin{vmatrix} a - r & r + 2x & x \\ b - s & s + 2y & y \\ c - t & t + 2z & z \end{vmatrix}.$$

Solution Using property 6, we have that

$$\begin{vmatrix} a & r & x \\ b & s & y \\ c & t & z \end{vmatrix} = \begin{vmatrix} a - r & r & x \\ b - s & s & y \\ c - t & t & z \end{vmatrix} \quad \left\{ \begin{array}{l} \text{by adding } (-1) \text{ times} \\ \text{the second column to} \\ \text{the first column.} \end{array} \right.$$

and that

$$\begin{vmatrix} a-r & r & x \\ b-s & s & y \\ c-t & t & z \end{vmatrix} = \begin{vmatrix} a-r & r+2x & x \\ b-s & s+2y & y \\ c-t & t+2z & z \end{vmatrix} \quad \left\{\begin{array}{l}\text{by adding twice} \\ \text{the third column} \\ \text{to the second} \\ \text{column.}\end{array}\right.$$

The result now follows by combining the two equalities.

To repeat, Properties 1–6 are equally valid if the word row is replaced by the word column.

Property 7 $\det(\mathbf{A}) = \det(\mathbf{A}')$.

Proof If

$$\mathbf{A} = \begin{bmatrix} a_{11} & a_{12} & a_{13} \\ a_{21} & a_{22} & a_{23} \\ a_{31} & a_{32} & a_{33} \end{bmatrix}, \qquad \text{then} \qquad \mathbf{A}' = \begin{bmatrix} a_{11} & a_{21} & a_{31} \\ a_{12} & a_{22} & a_{32} \\ a_{13} & a_{23} & a_{33} \end{bmatrix}.$$

Expanding $\det(\mathbf{A}')$ by the first column, it follows that

$$|\mathbf{A}'| = a_{11}\begin{vmatrix} a_{22} & a_{32} \\ a_{23} & a_{33} \end{vmatrix} - a_{12}\begin{vmatrix} a_{21} & a_{31} \\ a_{23} & a_{33} \end{vmatrix} + a_{13}\begin{vmatrix} a_{21} & a_{31} \\ a_{22} & a_{32} \end{vmatrix}$$

$$= a_{11}(a_{22}a_{33} - a_{32}a_{23}) - a_{12}(a_{21}a_{33} - a_{31}a_{23}) + a_{13}(a_{21}a_{32} - a_{31}a_{22}).$$

This, however, is exactly the expression we would obtain if we expand $\det(\mathbf{A})$ by the first row. Thus, $|\mathbf{A}'| = |\mathbf{A}|$.

Property 8 If **A** and **B** are of the same order, then $\det(\mathbf{A})\det(\mathbf{B}) = \det(\mathbf{AB})$. Because of its difficulty, the proof of Property 8 is omitted here.[3]

EXAMPLE 2 Show that Property 8 is valid for

$$\mathbf{A} = \begin{bmatrix} 2 & 3 \\ 1 & 4 \end{bmatrix} \qquad \text{and} \qquad \mathbf{B} = \begin{bmatrix} 6 & -1 \\ 7 & 4 \end{bmatrix}.$$

Solution $|\mathbf{A}| = 5$, $|\mathbf{B}| = 31$.

$$\mathbf{AB} = \begin{bmatrix} 33 & 10 \\ 34 & 15 \end{bmatrix} \qquad \text{thus} \qquad |\mathbf{AB}| = 155 = |\mathbf{A}||\mathbf{B}|.$$

[3] For a proof of Property 8, see E. D. Nearing, "Linear Algebra and Matrix Theory," p. 78. Wiley, New York, 1967.

PROBLEMS 2.3

1. Prove that the determinant of a diagonal matrix is the product of the elements on the main diagonal.
2. Prove that the determinant of an upper or lower triangular matrix is the product of the elements on the main diagonal.
3. Without expanding, show that

$$\begin{vmatrix} a+x & r-x & x \\ b+y & s-y & y \\ c+z & t-z & z \end{vmatrix} = \begin{vmatrix} a & r & x \\ b & s & y \\ c & t & z \end{vmatrix}.$$

4. Verify Property 5 for $\lambda = -3$ and

$$\mathbf{A} = \begin{bmatrix} 2 & 1 & 0 \\ 5 & -1 & 3 \\ 2 & 1 & 1 \end{bmatrix}.$$

5. Verify Property 8 for

$$\mathbf{A} = \begin{vmatrix} 6 & 1 \\ 1 & 2 \end{vmatrix} \quad \text{and} \quad \mathbf{B} = \begin{vmatrix} 3 & -1 \\ 2 & 1 \end{vmatrix}.$$

6. Without expanding, show that

$$\begin{vmatrix} 2a & 3r & x \\ 4b & 6s & 2y \\ -2c & -3t & -z \end{vmatrix} = -12 \begin{vmatrix} a & r & x \\ b & s & y \\ c & t & z \end{vmatrix}.$$

7. Without expanding, show that

$$\begin{vmatrix} a-3b & r-3s & x-3y \\ b-2c & s-2t & y-2z \\ 5c & 5t & 5z \end{vmatrix} = 5 \begin{vmatrix} a & r & x \\ b & s & y \\ c & t & z \end{vmatrix}.$$

2.4 PIVOTAL CONDENSATION

By making use of the properties of determinants given in the previous section, it is often possible to transform a specific determinant into another equivalent determinant which is easier to evaluate. One such procedure is the *method of pivotal condensation.* This method involves operating on a determinant so that one row or one column is transformed into a new row or column containing at most one nonzero element. The new determinant is then easily evaluated by expansion by cofactors. We illustrate the method with the following examples. Note, the only operations that are used are those that have been defined in the previous section.

EXAMPLE 1 Evaluate

$$\begin{vmatrix} 10 & -6 & -9 \\ 6 & -5 & -7 \\ -10 & 9 & 12 \end{vmatrix}.$$

Solution

$$\begin{vmatrix} 10 & -6 & -9 \\ 6 & -5 & -7 \\ -10 & 9 & 12 \end{vmatrix} = \begin{vmatrix} 10 & -6 & -9 \\ 6 & -5 & -7 \\ 0 & 3 & 3 \end{vmatrix} \quad \begin{cases} \text{by adding (1) times the} \\ \text{first row to the third row} \\ \text{(Property 6)} \end{cases}$$

$$= \begin{vmatrix} 10 & -6 & -3 \\ 6 & -5 & -2 \\ 0 & 3 & 0 \end{vmatrix} \quad \begin{cases} \text{by adding } (-1) \text{ times the} \\ \text{second column to the} \\ \text{third column (Property 6)} \end{cases}$$

$$= -3\begin{vmatrix} 10 & -3 \\ 6 & -2 \end{vmatrix} \quad \{\text{by expansion by cofactors}$$

$$= -3(-20 + 18) = 6.$$

EXAMPLE 2 Evaluate

$$\begin{vmatrix} 4 & 2 & -1 \\ 1 & 9 & 2 \\ 2 & -16 & -5 \end{vmatrix}.$$

Solution

$$\begin{vmatrix} 4 & 2 & -1 \\ 1 & 9 & 2 \\ 2 & -16 & -5 \end{vmatrix} = \begin{vmatrix} 4 & 2 & -1 \\ 1 & 9 & 2 \\ 0 & -34 & -9 \end{vmatrix} \quad \begin{cases} \text{by adding } (-2) \text{ times the} \\ \text{second row to the third} \\ \text{row (Property 6)} \end{cases}$$

$$= \begin{vmatrix} 0 & -34 & -9 \\ 1 & 9 & 2 \\ 0 & -34 & -9 \end{vmatrix} \quad \begin{cases} \text{by adding } (-4) \text{ times the} \\ \text{second row to the first} \\ \text{row (Property 6)} \end{cases}$$

$$= (-1)\begin{vmatrix} -34 & -9 \\ -34 & -9 \end{vmatrix} \quad \begin{cases} \text{by expansion by} \\ \text{cofactors} \end{cases}$$

$$= 0 \quad \{\text{by Property 3.}$$

In this example we decided to transform the first column into a column containing at most one nonzero element. We made this choice solely because of the unit element appearing in the (2, 1) position which would tend to simplify calculations. There was, however, nothing magical in this choice. We could have chosen some other column or row to transform as we do in the next example.

EXAMPLE 3 Evaluate the determinant of Example 2 by operating on another row or column.

Solution

$$\begin{vmatrix} 4 & 2 & -1 \\ 1 & 9 & 2 \\ 2 & -16 & -5 \end{vmatrix} = \begin{vmatrix} 0 & 2 & -1 \\ -17 & 9 & 2 \\ 34 & -16 & -5 \end{vmatrix} \quad \left\{ \begin{array}{l} \text{by adding } (-2) \text{ times} \\ \text{the second column to} \\ \text{the first column} \end{array} \right.$$

$$= \begin{vmatrix} 0 & 2 & 0 \\ -17 & 9 & \frac{13}{2} \\ 34 & -16 & -13 \end{vmatrix} \quad \left\{ \begin{array}{l} \text{by adding } (\frac{1}{2}) \text{ times the} \\ \text{second column to the} \\ \text{third column} \end{array} \right.$$

$$= -2 \begin{vmatrix} -17 & \frac{13}{2} \\ 34 & -13 \end{vmatrix} \quad \left\{ \begin{array}{l} \text{by expansion by cofac-} \\ \text{tors} \end{array} \right.$$

$$= (-2)(-\tfrac{1}{2}) \begin{vmatrix} 34 & -13 \\ 34 & -13 \end{vmatrix} \quad \{\text{by Property 4}$$

$$= 0 \quad \{\text{by Property 3.}$$

EXAMPLE 4 Evaluate

$$\begin{vmatrix} 3 & -1 & 0 & 2 \\ 0 & 1 & 4 & 1 \\ 3 & -2 & 3 & 5 \\ 9 & 7 & 0 & 2 \end{vmatrix}.$$

Solution Since the third column already contains two zeros, it would seem advisable to work on that one.

$$\begin{vmatrix} 3 & -1 & 0 & 2 \\ 0 & 1 & 4 & 1 \\ 3 & -2 & 3 & 5 \\ 9 & 7 & 0 & 2 \end{vmatrix} = \begin{vmatrix} 3 & -1 & 0 & 2 \\ 0 & 1 & 4 & 1 \\ 3 & -\frac{11}{4} & 0 & \frac{17}{4} \\ 9 & 7 & 0 & 2 \end{vmatrix} \quad \left\{ \begin{array}{l} \text{by adding } (-\frac{3}{4}) \text{ times} \\ \text{the second row to} \\ \text{the third row.} \end{array} \right.$$

$$= -4 \begin{vmatrix} 3 & -1 & 2 \\ 3 & -\frac{11}{4} & \frac{17}{4} \\ 9 & 7 & 2 \end{vmatrix} \quad \left\{ \begin{array}{l} \text{by expansion by} \\ \text{cofactors} \end{array} \right.$$

$$= -4(\tfrac{1}{4}) \begin{vmatrix} 3 & -1 & 2 \\ 12 & -11 & 17 \\ 9 & 7 & 2 \end{vmatrix} \quad \{\text{by Property 4}$$

$$= (-1) \begin{vmatrix} 3 & -1 & 2 \\ 0 & -7 & 9 \\ 9 & 7 & 2 \end{vmatrix} \quad \left\{ \begin{array}{l} \text{by adding } (-4) \text{ times} \\ \text{the first row to the} \\ \text{second row} \end{array} \right.$$

$$= (-1)\begin{vmatrix} 3 & -1 & 2 \\ 0 & -7 & 9 \\ 0 & 10 & -4 \end{vmatrix} \quad \begin{cases}\text{by adding } (-3) \text{ times} \\ \text{the first row to the} \\ \text{third row}\end{cases}$$

$$= (-1)(3)\begin{vmatrix} -7 & 9 \\ 10 & -4 \end{vmatrix} \quad \begin{cases}\text{by expansion by} \\ \text{cofactors}\end{cases}$$

$$= (-3)(28–90) = 186.$$

PROBLEMS 2.4

In all the following problems, find the determinants by the method of pivotal condensation:

1. $\begin{vmatrix} 3 & -4 & 2 \\ -1 & 5 & 7 \\ 1 & 9 & -6 \end{vmatrix}$,

2. $\begin{vmatrix} 1 & 2 & 3 \\ 4 & 5 & 6 \\ 7 & 8 & 9 \end{vmatrix}$,

3. $\begin{vmatrix} -2 & 0 & 1 & 3 \\ 4 & 0 & 2 & -2 \\ -3 & 1 & 0 & 1 \\ 5 & 4 & 1 & 7 \end{vmatrix}$,

4. $\begin{vmatrix} 3 & 5 & 4 & 6 \\ -2 & 1 & 0 & 7 \\ -5 & 4 & 7 & 2 \\ 8 & -3 & 1 & 1 \end{vmatrix}$.

2.5 CRAMER'S RULE

Using the properties of determinants, one may easily derive Cramer's rule which, as students of algebra probably know, provides a method for solving systems of simultaneous linear equations.[4] In this section we shall first state the rule, then illustrate its usage by an example and finally use the results of Section 2.3 to prove its validity.

Cramer's rule states that given a system of simultaneous linear equations in the matrix form $\mathbf{Ax} = \mathbf{b}$ (see Section 1.4), the ith component of $\mathbf{x}$ (or, equivalently, the ith unknown) is the quotient of two determinants. The

[4] For limitations of this method, see the last paragraph of this section.

determinant in the numerator is the determinant of a matrix obtained from **A** by replacing the ith column of **A** by the vector **b**, while the determinant in the denominator is just $|\mathbf{A}|$. Thus, if we are considering the system

$$\begin{aligned} a_{11}x_1 + a_{12}x_2 + a_{13}x_3 &= b_1, \\ a_{21}x_1 + a_{22}x_2 + a_{23}x_3 &= b_2, \\ a_{31}x_1 + a_{32}x_2 + a_{33}x_3 &= b_3, \end{aligned}$$

where x_1, x_2, and x_3 represent the unknowns, then Cramer's rule states that

$$x_1 = \frac{\begin{vmatrix} b_1 & a_{12} & a_{13} \\ b_2 & a_{22} & a_{23} \\ b_3 & a_{32} & a_{33} \end{vmatrix}}{|\mathbf{A}|}, \qquad x_2 = \frac{\begin{vmatrix} a_{11} & b_1 & a_{13} \\ a_{21} & b_2 & a_{23} \\ a_{31} & b_3 & a_{33} \end{vmatrix}}{|\mathbf{A}|},$$

$$x_3 = \frac{\begin{vmatrix} a_{11} & a_{12} & b_1 \\ a_{21} & a_{22} & b_2 \\ a_{31} & a_{32} & b_3 \end{vmatrix}}{|\mathbf{A}|}, \qquad \text{where } |\mathbf{A}| = \begin{vmatrix} a_{11} & a_{12} & a_{13} \\ a_{21} & a_{22} & a_{23} \\ a_{31} & a_{32} & a_{33} \end{vmatrix}.$$

Two restrictions on the application of Cramer's rule are immediate. First, the systems under consideration must have exactly the same number of equations as unknowns to insure that all matrices involved are square and hence have determinants. Second, the determinant of the coefficient matrix must not be zero since it appears in the denominator. If $|\mathbf{A}| = 0$, then Cramer's rule can not be applied.

EXAMPLE 1 Solve the system

$$\begin{aligned} x + 2y - 3z + w &= -5, \\ y + 3z + w &= 6, \\ 2x + 3y + z + w &= 4, \\ x + z + w &= 1. \end{aligned}$$

Solution

$$\mathbf{A} = \begin{bmatrix} 1 & 2 & -3 & 1 \\ 0 & 1 & 3 & 1 \\ 2 & 3 & 1 & 1 \\ 1 & 0 & 1 & 1 \end{bmatrix}, \qquad \mathbf{x} = \begin{bmatrix} x \\ y \\ z \\ w \end{bmatrix}, \qquad \text{and} \qquad \mathbf{b} = \begin{bmatrix} -5 \\ 6 \\ 4 \\ 1 \end{bmatrix}.$$

Since $|\mathbf{A}| = 20$, Cramer's rule can be applied, and

$$x = \frac{\begin{vmatrix} -5 & 2 & -3 & 1 \\ 6 & 1 & 3 & 1 \\ 4 & 3 & 1 & 1 \\ 1 & 0 & 1 & 1 \end{vmatrix}}{20} = \frac{0}{20} = 0, \qquad y = \frac{\begin{vmatrix} 1 & -5 & -3 & 1 \\ 0 & 6 & 3 & 1 \\ 2 & 4 & 1 & 1 \\ 1 & 1 & 1 & 1 \end{vmatrix}}{20} = \frac{20}{20} = 1,$$

$$z = \frac{\begin{vmatrix} 1 & 2 & -5 & 1 \\ 0 & 1 & 6 & 1 \\ 2 & 3 & 4 & 1 \\ 1 & 0 & 1 & 1 \end{vmatrix}}{20} = \frac{40}{20} = 2, \qquad w = \frac{\begin{vmatrix} 1 & 2 & -3 & -5 \\ 0 & 1 & 3 & 6 \\ 2 & 3 & 1 & 4 \\ 1 & 0 & 1 & 1 \end{vmatrix}}{20} = \frac{-20}{20} = -1.$$

We now derive Cramer's rule using only those properties of determinants given in Section 2.3. We consider the general system $\mathbf{Ax} = \mathbf{b}$ where

$$\mathbf{A} = \begin{bmatrix} a_{11} & a_{12} & a_{13} & \cdots & a_{1n} \\ a_{21} & a_{22} & a_{23} & \cdots & a_{2n} \\ a_{31} & a_{32} & a_{33} & \cdots & a_{3n} \\ \vdots & \vdots & \vdots & & \vdots \\ a_{n1} & a_{n2} & a_{n3} & \cdots & a_{nn} \end{bmatrix}, \qquad \mathbf{x} = \begin{bmatrix} x_1 \\ x_2 \\ x_3 \\ \vdots \\ x_n \end{bmatrix}, \qquad \text{and} \qquad \mathbf{b} = \begin{bmatrix} b_1 \\ b_2 \\ b_3 \\ \vdots \\ b_n \end{bmatrix}.$$

Then

$$x_1|\mathbf{A}| = \begin{vmatrix} a_{11}x_1 & a_{12} & a_{13} & \cdots & a_{1n} \\ a_{21}x_1 & a_{22} & a_{23} & \cdots & a_{2n} \\ a_{31}x_1 & a_{32} & a_{33} & \cdots & a_{3n} \\ \vdots & \vdots & \vdots & & \vdots \\ a_{n1}x_1 & a_{n2} & a_{n3} & \cdots & a_{nn} \end{vmatrix} \quad \{\text{by Property 4}$$

$$= \begin{vmatrix} a_{11}x_1 + a_{12}x_2 & a_{12} & a_{13} & \cdots & a_{1n} \\ a_{21}x_1 + a_{22}x_2 & a_{22} & a_{23} & \cdots & a_{2n} \\ a_{31}x_1 + a_{32}x_2 & a_{32} & a_{33} & \cdots & a_{3n} \\ \vdots & \vdots & \vdots & & \vdots \\ a_{n1}x_1 + a_{n2}x_2 & a_{n2} & a_{n3} & \cdots & a_{nn} \end{vmatrix} \quad \begin{cases} \text{by adding } (x_2) \text{ times} \\ \text{the second column to} \\ \text{the first column} \end{cases}$$

$$= \begin{vmatrix} a_{11}x_1 + a_{12}x_2 + a_{13}x_3 & a_{12} & a_{13} & \cdots & a_{1n} \\ a_{21}x_1 + a_{22}x_2 + a_{23}x_3 & a_{22} & a_{23} & \cdots & a_{2n} \\ a_{31}x_1 + a_{32}x_2 + a_{33}x_3 & a_{32} & a_{33} & \cdots & a_{3n} \\ \vdots & \vdots & \vdots & & \vdots \\ a_{n1}x_1 + a_{n2}x_2 + a_{n3}x_3 & a_{n2} & a_{n3} & \cdots & a_{nn} \end{vmatrix} \quad \begin{cases} \text{by adding } (x_3) \\ \text{times the third} \\ \text{column to the} \\ \text{first column} \end{cases}$$

$$= \begin{vmatrix} a_{11}x_1 + a_{12}x_2 + a_{13}x_3 + \cdots + a_{1n}x_n & a_{12} & a_{13} & \cdots & a_{1n} \\ a_{21}x_1 + a_{22}x_2 + a_{23}x_3 + \cdots + a_{2n}x_n & a_{22} & a_{23} & \cdots & a_{2n} \\ a_{31}x_1 + a_{32}x_2 + a_{33}x_3 + \cdots + a_{3n}x_n & a_{32} & a_{33} & \cdots & a_{3n} \\ \vdots & \vdots & \vdots & & \vdots \\ a_{n1}x_1 + a_{n2}x_2 + a_{n3}x_3 + \cdots + a_{nn}x_n & a_{n2} & a_{n3} & \cdots & a_{nn} \end{vmatrix}$$

by making continued use of Property 6 in the obvious manner. We now note that the first column of the new determinant is nothing more than **Ax**. Hence, since $\mathbf{Ax} = \mathbf{b}$, the first column is simply **b**.
Thus,

$$x_1|\mathbf{A}| = \begin{vmatrix} b_1 & a_{12} & a_{13} & \cdots & a_{1n} \\ b_2 & a_{22} & a_{23} & \cdots & a_{2n} \\ b_3 & a_{32} & a_{33} & \cdots & a_{3n} \\ \vdots & \vdots & \vdots & & \vdots \\ b_n & a_{n2} & a_{n3} & \cdots & a_{nn} \end{vmatrix}$$

or

$$x_1 = \frac{\begin{vmatrix} b_1 & a_{12} & \cdots & a_{1n} \\ b_2 & a_{22} & \cdots & a_{2n} \\ \vdots & \vdots & & \vdots \\ b_n & a_{n2} & \cdots & a_{nn} \end{vmatrix}}{|\mathbf{A}|}$$

providing $|\mathbf{A}| \neq 0$. This expression is Cramer's rule for obtaining x_1. A similar argument applied to the jth column, instead of the first column, quickly shows that Cramer's rule is valid for every x_j, $j = 1, 2, \ldots, n$.

Although Cramer's rule gives a systematic method for the solution of simultaneous linear equations, the number of computations involved can become awesome if the order of the determinant is large. Thus, for large systems, Cramer's rule is almost never used and other methods of solution, some of which we shall discuss later, are employed.

PROBLEMS 2.5

Solve the following systems of equations by Cramer's rule.

1.
$$\begin{aligned} 3x + y + z &= 4, \\ x - y + 2z &= 15, \\ 2x - 2y - z &= 5. \end{aligned}$$

2.
$$\begin{aligned} 3x + 3y + 2z &= 0, \\ x - y + z &= 0, \\ 5x + 2y - 3z &= 0. \end{aligned}$$

3.
$$\begin{aligned} x + 2y + z + w &= 7, \\ 3x + 4y - 2z - 4w &= 13, \\ 2x + y - z + w &= -4, \\ x - 3y + 4z + 5w &= 0. \end{aligned}$$

Chapter 3 THE INVERSE

3.1 THE INVERSE

Definition 1 The *inverse* of an $n \times n$ matrix $\mathbf{A}$ is an $n \times n$ matrix $\mathbf{B}$ having the property that

$$\mathbf{AB} = \mathbf{BA} = \mathbf{I}.$$

Here, $\mathbf{B}$ is called the inverse of $\mathbf{A}$ and is usually denoted by $\mathbf{A}^{-1}$. If a square matrix $\mathbf{A}$ has an inverse, it is said to be *invertible* or *nonsingular*. If $\mathbf{A}$ does not possess an inverse, it is said to be *singular*. In particular, the identity matrix is invertible and is its own inverse since

$$\mathbf{II} = \mathbf{I}$$

while the zero matrix is singular. *Note* that inverses are only defined for square matrices.

One method for finding inverses requires calculating the cofactor and adjoint matrices.

Definition 2 The *cofactor matrix* associated with an $n \times n$ matrix $\mathbf{A}$ is an $n \times n$ matrix $\mathbf{A}^c$ obtained from $\mathbf{A}$ by replacing each element of $\mathbf{A}$ by its cofactor.

EXAMPLE 1 Find $\mathbf{A}^c$ if

$$\mathbf{A} = \begin{bmatrix} 3 & 1 & 2 \\ -2 & 5 & 4 \\ 1 & 3 & 6 \end{bmatrix}.$$

Solution

$$\mathbf{A}^c = \begin{bmatrix} (-1)^{1+1}\begin{vmatrix} 5 & 4 \\ 3 & 6 \end{vmatrix} & (-1)^{1+2}\begin{vmatrix} -2 & 4 \\ 1 & 6 \end{vmatrix} & (-1)^{1+3}\begin{vmatrix} -2 & 5 \\ 1 & 3 \end{vmatrix} \\ (-1)^{2+1}\begin{vmatrix} 1 & 2 \\ 3 & 6 \end{vmatrix} & (-1)^{2+2}\begin{vmatrix} 3 & 2 \\ 1 & 6 \end{vmatrix} & (-1)^{2+3}\begin{vmatrix} 3 & 1 \\ 1 & 3 \end{vmatrix} \\ (-1)^{3+1}\begin{vmatrix} 1 & 2 \\ 5 & 4 \end{vmatrix} & (-1)^{3+2}\begin{vmatrix} 3 & 2 \\ -2 & 4 \end{vmatrix} & (-1)^{3+3}\begin{vmatrix} 3 & 1 \\ -2 & 5 \end{vmatrix} \end{bmatrix},$$

$$\mathbf{A}^c = \begin{bmatrix} 18 & 16 & -11 \\ 0 & 16 & -8 \\ -6 & -16 & 17 \end{bmatrix}.$$

If $\mathbf{A} = [a_{ij}]$, we will use the notation $\mathbf{A}^c = [a_{ij}^c]$ to represent the cofactor matrix. Thus a_{ij}^c represents the cofactor of a_{ij}.

Definition 3 The *adjoint* of an $n \times n$ matrix $\mathbf{A}$ is the transpose of the cofactor matrix of $\mathbf{A}$.

Thus, if we designate the adjoint of $\mathbf{A}$ by $\mathbf{A}^a$, we have that $\mathbf{A}^a = (\mathbf{A}^c)'$.

EXAMPLE 2 Find $\mathbf{A}^a$ for the $\mathbf{A}$ given in Example 1.

Solution

$$\mathbf{A}^a = \begin{bmatrix} 18 & 0 & -6 \\ 16 & 16 & -16 \\ -11 & -8 & 17 \end{bmatrix}.$$

The importance of the adjoint is given in the following theorem, which is proved in the appendix to this chapter.

Theorem 1 $AA^a = A^aA = |A|I.$

If $|A| \neq 0$, we may divide by it in Theorem 1 and obtain

$$A\left(\frac{A^a}{|A|}\right) = \left(\frac{A^a}{|A|}\right)A = I.$$

Thus, using the definition of the inverse, we have that

$$\boxed{A^{-1} = \frac{1}{|A|}A^a} \qquad \text{if } |A| \neq 0. \tag{1}$$

That is, if $|A| \neq 0$, then A^{-1} may be obtained by dividing the adjoint of A by the determinant of A.

EXAMPLE 3 Find A^{-1} for the A given in Example 1.

Solution The determinant of A is found to be 48. From Example 2, we have that

$$A^a = \begin{bmatrix} 18 & 0 & -6 \\ 16 & 16 & -16 \\ -11 & -8 & 17 \end{bmatrix}.$$

Thus,

$$A^{-1} = \left(\frac{A^a}{|A|}\right) = 1/48\begin{bmatrix} 18 & 0 & -6 \\ 16 & 16 & -16 \\ -11 & -8 & 17 \end{bmatrix} = \begin{bmatrix} 3/8 & 0 & -1/8 \\ 1/3 & 1/3 & -1/3 \\ -11/48 & -1/6 & 17/48 \end{bmatrix}.$$

EXAMPLE 4 Find A^{-1} if

$$A = \begin{bmatrix} 5 & 8 & 1 \\ 0 & 2 & 1 \\ 4 & 3 & -1 \end{bmatrix}.$$

Solution $\det(A) = -1 \neq 0$, therefore A^{-1} exists.

$$A^c = \begin{bmatrix} -5 & 4 & -8 \\ 11 & -9 & 17 \\ 6 & -5 & 10 \end{bmatrix}, \qquad A^a = (A^c)' = \begin{bmatrix} -5 & 11 & 6 \\ 4 & -9 & -5 \\ -8 & 17 & 10 \end{bmatrix},$$

$$A^{-1} = \frac{A^a}{|A|} = \begin{bmatrix} 5 & -11 & -6 \\ -4 & 9 & 5 \\ 8 & -17 & -10 \end{bmatrix},$$

EXAMPLE 5 Find $\mathbf{A}^{-1}$ if

$$\mathbf{A} = \begin{bmatrix} 1 & 2 \\ 3 & 4 \end{bmatrix}.$$

Solution $|\mathbf{A}| = -2$, therefore $\mathbf{A}^{-1}$ exists.

$$\mathbf{A}^c = \begin{bmatrix} 4 & -3 \\ -2 & 1 \end{bmatrix}, \qquad \mathbf{A}^a = (\mathbf{A}^c)' = \begin{bmatrix} 4 & -2 \\ -3 & 1 \end{bmatrix},$$

$$\mathbf{A}^{-1} = \frac{\mathbf{A}^a}{|\mathbf{A}|} = (-\tfrac{1}{2})\begin{bmatrix} 4 & -2 \\ -3 & 1 \end{bmatrix} = \begin{bmatrix} -2 & 1 \\ \frac{3}{2} & -\frac{1}{2} \end{bmatrix}.$$

One note of warning—students are apt to make arithmetic errors in using the above method for finding inverses. This is especially true in computing the cofactor matrix. The student is well advised, therefore, to check his results every time he calculates an inverse by multiplying the calculated inverse by the original matrix and observing whether or not the identity matrix is obtained.

The reader might suspect that since the method given above depends on calculating cofactors, hence, involves evaluating determinants, the amount of computations involved can become awesome if the matrix is of large order. This being the case, one generally applies other methods to invert such matrices. See, for example, Section 3.4.

PROBLEMS 3.1

1. Find the inverse of

$$\begin{bmatrix} 1 & 1 \\ 3 & 4 \end{bmatrix}.$$

2. Find the inverse of

$$\begin{bmatrix} 1 & \frac{1}{2} \\ \frac{1}{2} & \frac{1}{3} \end{bmatrix}.$$

3. Find the inverse of

$$\begin{bmatrix} 3 & 2 & 1 \\ 4 & 0 & 1 \\ 3 & 9 & 2 \end{bmatrix}.$$

4. Find $\mathbf{A}^{-1}$ if $\mathbf{A}$ is the upper triangular matrix

$$\begin{bmatrix} 4 & 1 & 5 \\ 0 & 3 & -1 \\ 0 & 0 & 2 \end{bmatrix}.$$

5. Find $\mathbf{A}^{-1}$ if $\mathbf{A}$ is the lower triangular matrix

$$\begin{bmatrix} 1 & 0 & 0 & 0 \\ 2 & -1 & 0 & 0 \\ 4 & 6 & 2 & 0 \\ 3 & 2 & 4 & -1 \end{bmatrix}.$$

6. Find $\mathbf{A}^{-1}$ if $\mathbf{A}$ is the symmetric matrix

$$\begin{bmatrix} 1 & 2 & -1 \\ 2 & 0 & 1 \\ -1 & 1 & 3 \end{bmatrix}.$$

7. Find $\mathbf{A}^{-1}$ if $\mathbf{A}$ is the diagonal matrix

$$\begin{bmatrix} \lambda_1 & 0 & 0 & 0 \\ 0 & \lambda_2 & 0 & 0 \\ 0 & 0 & \lambda_3 & 0 \\ 0 & 0 & 0 & \lambda_4 \end{bmatrix}$$

where $\lambda_k \neq 0$, $k = 1, 2, 3, 4$. Can you generalize this to higher orders?

3.2 SIMULTANEOUS EQUATIONS

One of the major uses of the inverse is in the solution of systems of simultaneous linear equations. Recall, from Section 1.4, that any such system may be written in the form

$$\mathbf{Ax} = \mathbf{b}, \tag{2}$$

where $\mathbf{A}$ is the coefficient matrix, $\mathbf{b}$ is a known vector, and $\mathbf{x}$ is the unknown vector we wish to find. If $\mathbf{A}$ is invertible, then we can premultiply (2) by $\mathbf{A}^{-1}$ and obtain

$$\mathbf{A}^{-1}\mathbf{Ax} = \mathbf{A}^{-1}\mathbf{b}.$$

But $\mathbf{A}^{-1}\mathbf{A} = \mathbf{I}$, therefore

$$\mathbf{Ix} = \mathbf{A}^{-1}\mathbf{b}$$

or

$$\mathbf{x} = \mathbf{A}^{-1}\mathbf{b}. \tag{3}$$

Hence, (3) shows that if $\mathbf{A}$ is invertible, then $\mathbf{x}$ can be obtained by premultiplying $\mathbf{b}$ by the inverse of $\mathbf{A}$.

EXAMPLE 1 Solve the following system for x and y:

$$\begin{aligned} x - 2y &= -9, \\ -3x + y &= 2. \end{aligned}$$

Solution Define

$$\mathbf{A} = \begin{bmatrix} 1 & -2 \\ -3 & 1 \end{bmatrix}, \qquad \mathbf{x} = \begin{bmatrix} x \\ y \end{bmatrix}, \qquad \mathbf{b} = \begin{bmatrix} -9 \\ 2 \end{bmatrix};$$

then the system can be written as $\mathbf{Ax} = \mathbf{b}$, hence $\mathbf{x} = \mathbf{A}^{-1}\mathbf{b}$. Using the method given in Section 3.1 we find that

$$\mathbf{A}^{-1} = (-\tfrac{1}{5})\begin{bmatrix} 1 & 2 \\ 3 & 1 \end{bmatrix}.$$

Thus,

$$\begin{bmatrix} x \\ y \end{bmatrix} = \mathbf{x} = \mathbf{A}^{-1}\mathbf{b} = (-\tfrac{1}{5})\begin{bmatrix} 1 & 2 \\ 3 & 1 \end{bmatrix}\begin{bmatrix} -9 \\ 2 \end{bmatrix}$$

$$= (-\tfrac{1}{5})\begin{bmatrix} -5 \\ -25 \end{bmatrix} = \begin{bmatrix} 1 \\ 5 \end{bmatrix}.$$

Using the definition of matrix equality (two matrices are equal if and only if their corresponding elements are equal), we have that $x = 1$ and $y = 5$.

EXAMPLE 2 Solve the following system for x, y and z:

$$\begin{aligned} 5x + 8y + z &= 2, \\ 2y + z &= -1, \\ 4x + 3y - z &= 3. \end{aligned}$$

Solution

$$\mathbf{A} = \begin{bmatrix} 5 & 8 & 1 \\ 0 & 2 & 1 \\ 4 & 3 & -1 \end{bmatrix}, \qquad \mathbf{x} = \begin{bmatrix} x \\ y \\ z \end{bmatrix}, \qquad \mathbf{b} = \begin{bmatrix} 2 \\ -1 \\ 3 \end{bmatrix}.$$

$\mathbf{A}^{-1}$ is found to be

$$\begin{bmatrix} 5 & -11 & -6 \\ -4 & 9 & 5 \\ 8 & -17 & -10 \end{bmatrix}.$$

Thus,

$$\begin{bmatrix} x \\ y \\ z \end{bmatrix} = \mathbf{x} = \mathbf{A}^{-1}\mathbf{b} = \begin{bmatrix} 5 & -11 & -6 \\ -4 & 9 & 5 \\ 8 & -17 & -10 \end{bmatrix}\begin{bmatrix} 2 \\ -1 \\ 3 \end{bmatrix} = \begin{bmatrix} 3 \\ -2 \\ 3 \end{bmatrix},$$

hence $x = 3$, $y = -2$, and $z = 3$.

Not only does the invertibility of **A** provide us with a solution of the system $\mathbf{Ax} = \mathbf{b}$, it also provides us with a means of showing that this solution is unique (that is, there is no other solution to the system).

Theorem 1 If **A** is invertible, then the system of simultaneous linear equations given by $\mathbf{Ax} = \mathbf{b}$ has one and only one solution.

Proof Define $\mathbf{w} = \mathbf{A}^{-1}\mathbf{b}$. Since we have already shown that **w** is a solution to $\mathbf{Ax} = \mathbf{b}$, it follows that

$$\mathbf{Aw} = \mathbf{b}. \tag{4}$$

Assume that there exists another solution **y**. Since **y** is a solution, we have that

$$\mathbf{Ay} = \mathbf{b}. \tag{5}$$

Equations (4) and (5) imply that

$$\mathbf{Aw} = \mathbf{Ay}. \tag{6}$$

Premultiply both sides of (6) by $\mathbf{A}^{-1}$. Then

$$\mathbf{A}^{-1}\mathbf{Aw} = \mathbf{A}^{-1}\mathbf{Ay},$$
$$\mathbf{Iw} = \mathbf{Iy},$$

or

$$\mathbf{w} = \mathbf{y}.$$

Thus, we see that if **y** is assumed to be a solution of $\mathbf{Ax} = \mathbf{b}$, it must, in fact, equal **w**. Therefore, $\mathbf{w} = \mathbf{A}^{-1}\mathbf{b}$ is the only solution to the problem.

PROBLEMS 3.2

1. Use inversion to solve the following system for x and y:

$$\begin{aligned} 4x + 2y &= 6, \\ 2x - 3y &= 7. \end{aligned}$$

2. Use inversion to solve the following system for l and p:

$$\begin{aligned} 4l - p &= 1, \\ 5l - 2p &= -1. \end{aligned}$$

3. Use inversion to solve the following system for x, y, and z:

$$\begin{aligned} 2x + 3y - z &= 4, \\ -x - 2y + z &= -2, \\ 3x - y \quad &= 2. \end{aligned}$$

4. Use inversion to solve the following system for l, m, and n:

$$\begin{aligned} 60l + 30m + 20n &= 0, \\ 30l + 20m + 15n &= -10, \\ 20l + 15m + 12n &= -10. \end{aligned}$$

3.3 PROPERTIES OF THE INVERSE

Once the previous sections are understood, two questions are immediate! First, is the inverse unique? Second, if $\det(\mathbf{A}) = 0$, is it still possible for $\mathbf{A}$ to be invertible? That is, does there exist another method (different from the one developed in Section 3.1) that can be used to find inverses for matrices having zero determinants? Fortunately, we can answer both of these questions with the following theorems:

Theorem 1 The inverse of a matrix is unique.

Proof Suppose that both $\mathbf{B}$ and $\mathbf{C}$ are inverses of $\mathbf{A}$. Then, by the definition of an inverse, we have that

$$\mathbf{AB} = \mathbf{I}, \tag{7}$$

$$\mathbf{BA} = \mathbf{I}, \tag{8}$$

$$\mathbf{AC} = \mathbf{I}, \tag{9}$$

$$\mathbf{CA} = \mathbf{I}. \tag{10}$$

Equations (7) and (9) imply that

$$\mathbf{AB} = \mathbf{AC}.$$

Premultiplying both sides of this equation by $\mathbf{B}$, we obtain

$$\mathbf{B}(\mathbf{AB}) = \mathbf{B}(\mathbf{AC})$$

or

$$(\mathbf{BA})\mathbf{B} = (\mathbf{BA})\mathbf{C}$$

Using (8), we obtain

$$\mathbf{IB} = \mathbf{IC}$$

or

$$\mathbf{B} = \mathbf{C}.$$

Thus, if **B** and **C** are both inverses of **A**, they must in fact be equal. Hence, the inverse is unique.

Theorem 2 If the determinant of a matrix is zero, the matrix does not have an inverse.

Proof Let $\det(\mathbf{A}) = 0$. Our method of proof is the following: we will assume that $\mathbf{A}^{-1}$ exists and will derive the result $1 = 0$. Since the conclusion $1 = 0$ is absurd, something in our argument must have been wrong. A quick check will show that the only possible error is the assumption that $\mathbf{A}^{-1}$ exists. Thus, it will follow that $\mathbf{A}^{-1}$ does not exist which is what we want to prove.

Now assume $\mathbf{A}^{-1}$ exists. Then $\mathbf{A}\mathbf{A}^{-1} = \mathbf{I}$. Taking the determinant of both sides, using the property of determinants that $\det(\mathbf{A}\mathbf{A}^{-1}) = \det(\mathbf{A})\det(\mathbf{A}^{-1})$, and observing that $\det(\mathbf{I}) = 1$, we have that

$$\det(\mathbf{A})\det(\mathbf{A}^{-1}) = 1;$$

but $\det(\mathbf{A}) = 0$

$$0 \cdot \det(\mathbf{A}^{-1}) = 1$$

or

$$0 = 1.$$

This conclusion is absurd, hence our assumption is incorrect, and $\mathbf{A}^{-1}$ does not exist.

Using Theorem 1, we are now in a position to prove some useful properties of the inverse of a matrix **A**.

Property 1 $(\mathbf{A}^{-1})^{-1} = \mathbf{A}$.

Proof See Problem 1.

Property 2 $(\mathbf{AB})^{-1} = \mathbf{B}^{-1}\mathbf{A}^{-1}$.

Proof $(\mathbf{AB})^{-1}$ denotes the inverse of **AB**. However, $(\mathbf{B}^{-1}\mathbf{A}^{-1})(\mathbf{AB}) = \mathbf{B}^{-1}(\mathbf{A}^{-1}\mathbf{A})\mathbf{B} = \mathbf{B}^{-1}\mathbf{I}\mathbf{B} = \mathbf{B}^{-1}\mathbf{B} = \mathbf{I}$. Thus, $\mathbf{B}^{-1}\mathbf{A}^{-1}$ is also an inverse for **AB**, and, by uniqueness of the inverse, $\mathbf{B}^{-1}\mathbf{A}^{-1} = (\mathbf{AB})^{-1}$.

Property 3 $(\mathbf{A}_1\mathbf{A}_2 \cdots \mathbf{A}_n)^{-1} = \mathbf{A}_n^{-1}\mathbf{A}_{n-1}^{-1} \cdots \mathbf{A}_2^{-1}\mathbf{A}_1^{-1}$.

Proof This is an extension of Property 2 and, as such, is proved in a similar manner.

Caution: Note that Property 3 states that the inverse of a product is *not* the product of the inverses but rather the product of the inverses commuted.

Property 4 $(\mathbf{A}')^{-1} = (\mathbf{A}^{-1})'$

Proof $(\mathbf{A}')^{-1}$ denotes the inverse of $\mathbf{A}'$. However, using the property of the transpose that $(\mathbf{AB})' = \mathbf{B}'\mathbf{A}'$, we have that

$$(\mathbf{A}')(\mathbf{A}^{-1})' = (\mathbf{A}^{-1}\mathbf{A})' = \mathbf{I}' = \mathbf{I}.$$

Thus, $(\mathbf{A}^{-1})'$ is an inverse of $\mathbf{A}'$, and by uniqueness of the inverse, $(\mathbf{A}^{-1})' = (\mathbf{A}')^{-1}$.

Property 5 $(\lambda\mathbf{A})^{-1} = (1/\lambda)(\mathbf{A})^{-1}$ if λ is a nonzero scalar.

Proof $(\lambda\mathbf{A})^{-1}$ denotes the inverse of $\lambda\mathbf{A}$. However,

$$(\lambda\mathbf{A})(1/\lambda)\mathbf{A}^{-1} = \lambda(1/\lambda)\mathbf{A}\mathbf{A}^{-1} = 1 \cdot \mathbf{I} = \mathbf{I}.$$

Thus, $(1/\lambda)\mathbf{A}^{-1}$ is an inverse of $\lambda\mathbf{A}$, and by uniqueness of the inverse, $(1/\lambda)\mathbf{A}^{-1} = (\lambda\mathbf{A})^{-1}$.

Property 6 The inverse of a nonsingular diagonal matrix is diagonal.

Proof One can readily check that the inverse of

$$\mathbf{A} = \begin{bmatrix} \lambda_1 & 0 & \cdots & 0 \\ 0 & \lambda_2 & \cdots & 0 \\ \vdots & \vdots & & \vdots \\ 0 & 0 & \cdots & \lambda_n \end{bmatrix}$$

is

$$\mathbf{A}^{-1} = \begin{bmatrix} 1/\lambda_1 & 0 & \cdots & 0 \\ 0 & 1/\lambda_2 & \cdots & 0 \\ \vdots & \vdots & & \vdots \\ 0 & 0 & \cdots & 1/\lambda_n \end{bmatrix}.$$

Note that *no* λ_i, $i = 1, 2, \ldots, n$, can equal zero, otherwise $\det(\mathbf{A}) = 0$, which implies, by Theorem 2, that $\mathbf{A}^{-1}$ does not exist.

Property 7 The inverse of a nonsingular symmetric matrix is symmetric.

Proof See Problem 6.

Property 8 The inverse of a nonsingular upper or lower triangular matrix is again an upper or lower triangular matrix respectively.

Proof We omit the proof of Property 8.

Finally, the inverse provides us with a straightforward way of defining square matrices raised to negative integral powers. If $\mathbf{A}$ is nonsingular then we define $\mathbf{A}^{-n} = (\mathbf{A}^{-1})^n$.

EXAMPLE 1 Find $\mathbf{A}^{-2}$ if

$$\mathbf{A} = \begin{bmatrix} \frac{1}{3} & \frac{1}{2} \\ \frac{1}{2} & 1 \end{bmatrix}.$$

Solution

$$\mathbf{A}^{-2} = (\mathbf{A}^{-1})^2$$

$$= \begin{bmatrix} 12 & -6 \\ -6 & 4 \end{bmatrix}^2 = \begin{bmatrix} 12 & -6 \\ -6 & 4 \end{bmatrix}\begin{bmatrix} 12 & -6 \\ -6 & 4 \end{bmatrix} = \begin{bmatrix} 180 & -96 \\ -96 & 52 \end{bmatrix}.$$

PROBLEMS 3.3

1. Prove Property 1.
2. Prove that $(\mathbf{ABC})^{-1} = \mathbf{C}^{-1}\mathbf{B}^{-1}\mathbf{A}^{-1}$.
3. Verify the result of Problem 2 if

$$\mathbf{A} = \begin{bmatrix} 1 & 3 \\ 0 & 2 \end{bmatrix}, \quad \mathbf{B} = \begin{bmatrix} 4 & 0 \\ 0 & 2 \end{bmatrix}, \quad \text{and} \quad \mathbf{C} = \begin{bmatrix} -1 & 0 \\ 2 & 2 \end{bmatrix}.$$

4. Verify Property 4 for

$$\mathbf{A} = \begin{bmatrix} 2 & -1 \\ 4 & 1 \end{bmatrix}.$$

5. Verify Property 5 for $\lambda = 2$ and

$$\mathbf{A} = \begin{bmatrix} 1 & 0 & 2 \\ 2 & 3 & -1 \\ -1 & 0 & 3 \end{bmatrix}.$$

6. Prove Property 7 (Hint: use Property 4).
7. Verify Property 7 for

$$\mathbf{A} = \begin{bmatrix} 1 & 2 \\ 2 & 3 \end{bmatrix}.$$

8. Find $\mathbf{A}^{-3}$ if

$$\mathbf{A} = \begin{bmatrix} 1 & -2 \\ 2 & 1 \end{bmatrix}.$$

9. If $\mathbf{A}$ is symmetric, prove the identity

$$(\mathbf{BA}^{-1})'(\mathbf{A}^{-1}\mathbf{B}')^{-1} = \mathbf{I}.$$

10. Prove that $\det(\mathbf{A}^{-1}) = 1/\det(\mathbf{A})$. (Hint: use Property 8 of Section 2.3)

3.4 ANOTHER METHOD FOR INVERSION

Although the method of cofactors provides a systematic procedure for obtaining inverses, it can become involved in cumbersome computations. In this section, we give another method for inversion which, for a particular problem, may be simpler to use.

We define the following as elementary row operations on a matrix:

(1) interchanging any two rows;
(2) multiplying any row by a nonzero scalar;
(3) multiplying one row by a scalar and adding it to another row.

Elementary column operations are defined analogously. It can be shown that a matrix is invertible if and only if it can be transformed into the identity by successive row (or, similarly, column) operations.[5]

EXAMPLE 1 Using elementary row operations, determine whether or not

$$\mathbf{A} = \begin{bmatrix} 1 & 2 \\ 3 & 4 \end{bmatrix}$$

is invertible.

Solution

$$\begin{bmatrix} 1 & 2 \\ 3 & 4 \end{bmatrix} \sim \begin{bmatrix} 1 & 2 \\ 0 & -2 \end{bmatrix} \quad \left\{\begin{array}{l}\text{by adding } (-3) \text{ times} \\ \text{the first row to the} \\ \text{second row}\end{array}\right.$$

$$\sim \begin{bmatrix} 1 & 2 \\ 0 & 1 \end{bmatrix} \quad \left\{\begin{array}{l}\text{by multiplying the} \\ \text{second row by } (-\tfrac{1}{2})\end{array}\right.$$

$$\sim \begin{bmatrix} 1 & 0 \\ 0 & 1 \end{bmatrix} \quad \left\{\begin{array}{l}\text{by adding } (-2) \text{ times} \\ \text{the second row to the} \\ \text{first row.}\end{array}\right.$$

Thus, since $\mathbf{A}$ can be transformed into the identity by elementary row operations, it is invertible. The notation ($\sim$) should be read "is transformed into." Note that an equality sign can not be used here. Why?

[5] See K. Hoffman and R. Kunze, "Linear Algebra," p. 24. Prentice-Hall, Englewood Cliffs, New Jersey, 1961.

We are now able to give an alternative method for inversion. Let $\mathbf{A}$ be the $n \times n$ matrix we wish to invert. Place next to it another $n \times n$ matrix $\mathbf{B}$ which is initially the identity. Using elementary row operations on $\mathbf{A}$, transform it into the identity. Each time an operation is performed on $\mathbf{A}$, repeat the exact same operation on $\mathbf{B}$. After $\mathbf{A}$ is transformed into the identity, the matrix obtained from transforming $\mathbf{B}$ will be $\mathbf{A}^{-1}$.

EXAMPLE 2 Invert

$$\mathbf{A} = \begin{bmatrix} 2 & 1 \\ 3 & 4 \end{bmatrix}.$$

Solution

$$\begin{bmatrix} 2 & 1 \\ 3 & 4 \end{bmatrix}\begin{bmatrix} 1 & 0 \\ 0 & 1 \end{bmatrix} \sim \begin{bmatrix} 2 & 1 \\ -5 & 0 \end{bmatrix}\begin{bmatrix} 1 & 0 \\ -4 & 1 \end{bmatrix} \quad \left\{\begin{array}{l}\text{by adding } (-4) \text{ times} \\ \text{the first row to the} \\ \text{second row}\end{array}\right.$$

$$\sim \begin{bmatrix} 2 & 1 \\ 1 & 0 \end{bmatrix}\begin{bmatrix} 1 & 0 \\ \frac{4}{5} & -\frac{1}{5} \end{bmatrix} \quad \left\{\begin{array}{l}\text{by multiplying the} \\ \text{second row by } (-\frac{1}{5})\end{array}\right.$$

$$\sim \begin{bmatrix} 0 & 1 \\ 1 & 0 \end{bmatrix}\begin{bmatrix} -\frac{3}{5} & \frac{2}{5} \\ \frac{4}{5} & -\frac{1}{5} \end{bmatrix} \quad \left\{\begin{array}{l}\text{by adding } (-2) \text{ times} \\ \text{the second row to the} \\ \text{first row}\end{array}\right.$$

$$\sim \begin{bmatrix} 1 & 0 \\ 0 & 1 \end{bmatrix}\begin{bmatrix} \frac{4}{5} & -\frac{1}{5} \\ -\frac{3}{5} & \frac{2}{5} \end{bmatrix} \quad \left\{\begin{array}{l}\text{by interchanging the} \\ \text{first and second rows.}\end{array}\right.$$

Thus,

$$\mathbf{A}^{-1} = \begin{bmatrix} \frac{4}{5} & -\frac{1}{5} \\ -\frac{3}{5} & \frac{2}{5} \end{bmatrix}.$$

If one prefers to work with columns instead of rows, an analogous procedure is used.

EXAMPLE 3 Invert

$$\mathbf{A} = \begin{bmatrix} 3 & 1 & 0 \\ 2 & 0 & 0 \\ 4 & 0 & 1 \end{bmatrix}.$$

Solution

$$\begin{bmatrix} 3 & 1 & 0 \\ 2 & 0 & 0 \\ 4 & 0 & 1 \end{bmatrix}\begin{bmatrix} 1 & 0 & 0 \\ 0 & 1 & 0 \\ 0 & 0 & 1 \end{bmatrix} \sim \begin{bmatrix} 3 & 1 & 0 \\ 2 & 0 & 0 \\ 0 & 0 & 1 \end{bmatrix}\begin{bmatrix} 1 & 0 & 0 \\ 0 & 1 & 0 \\ -4 & 0 & 1 \end{bmatrix} \quad \left\{\begin{array}{l}\text{by adding } (-4) \text{ times} \\ \text{the third column to} \\ \text{the first column}\end{array}\right.$$

$$\sim \begin{bmatrix} 0 & 1 & 0 \\ 2 & 0 & 0 \\ 0 & 0 & 1 \end{bmatrix} \begin{bmatrix} 1 & 0 & 0 \\ -3 & 1 & 0 \\ -4 & 0 & 1 \end{bmatrix} \quad \left\{\begin{array}{l}\text{by adding } (-3) \text{ times} \\ \text{the second column} \\ \text{to the first column}\end{array}\right.$$

$$\sim \begin{bmatrix} 0 & 1 & 0 \\ 1 & 0 & 0 \\ 0 & 0 & 1 \end{bmatrix} \begin{bmatrix} \frac{1}{2} & 0 & 0 \\ -\frac{3}{2} & 1 & 0 \\ -2 & 0 & 1 \end{bmatrix} \quad \left\{\begin{array}{l}\text{by multiplying the} \\ \text{first column by } (\frac{1}{2})\end{array}\right.$$

$$\sim \begin{bmatrix} 1 & 0 & 0 \\ 0 & 1 & 0 \\ 0 & 0 & 1 \end{bmatrix} \begin{bmatrix} 0 & \frac{1}{2} & 0 \\ 1 & -\frac{3}{2} & 0 \\ 0 & -2 & 1 \end{bmatrix} \quad \left\{\begin{array}{l}\text{by interchanging the} \\ \text{first and second} \\ \text{columns.}\end{array}\right.$$

Thus,

$$\mathbf{A}^{-1} = \begin{bmatrix} 0 & \frac{1}{2} & 0 \\ 1 & -\frac{3}{2} & 0 \\ 0 & -2 & 1 \end{bmatrix}.$$

Caution: Row and column operations can not be used together. Once an inversion is started by elementary row (or column) operations, it must be continued by elementary row (or column) operations only. Interchanging row and column operations in the same problem will generally lead to an incorrect answer. Furthermore, although the matrix **B** is placed next to the matrix **A**, the student should be aware that no multiplication of the matrices is intended; such positioning is for convenience only.

PROBLEMS 3.4

Invert the following matrices by elementary row operations and then by elementary column operations.

1. $\begin{bmatrix} 7 & 1 \\ 2 & -1 \end{bmatrix}$

2. $\begin{bmatrix} 1 & 0 & 1 \\ 3 & 1 & 2 \\ 1 & 2 & 2 \end{bmatrix}$

3. $\begin{bmatrix} 2 & 2 & 3 \\ 1 & 0 & 1 \\ 1 & 1 & 1 \end{bmatrix}$

4. $\begin{bmatrix} 2 & 1 & -1 & 3 \\ 1 & 4 & 2 & 1 \\ 0 & 0 & -1 & 1 \\ 0 & 1 & 0 & 1 \end{bmatrix}$

APPENDIX TO CHAPTER 3

We shall now prove Theorem 1 of Section 3.1 dealing with the product of a matrix and its adjoint. For this proof, we will need the following lemma:

Lemma 1 If each element of one row of a matrix is multiplied by the cofactor of the corresponding element of a different row, the sum is zero.

Proof We prove this lemma only for an arbitrary 3×3 matrix **A** where

$$\mathbf{A} = \begin{bmatrix} a_{11} & a_{12} & a_{13} \\ a_{21} & a_{22} & a_{23} \\ a_{31} & a_{32} & a_{33} \end{bmatrix}.$$

Consider the case in which we multiply every element of the third row by the cofactor of the corresponding element in the second row and then sum the results. Thus,

a_{31}(cofactor of a_{21}) + a_{32}(cofactor of a_{22}) + a_{33}(cofactor of a_{23})

$$= a_{31}(-1)^3 \begin{vmatrix} a_{12} & a_{13} \\ a_{32} & a_{33} \end{vmatrix} + a_{32}(-1)^4 \begin{vmatrix} a_{11} & a_{13} \\ a_{31} & a_{33} \end{vmatrix} + a_{33}(-1)^5 \begin{vmatrix} a_{11} & a_{12} \\ a_{31} & a_{32} \end{vmatrix}$$

$$= \begin{vmatrix} a_{11} & a_{12} & a_{13} \\ a_{31} & a_{32} & a_{33} \\ a_{31} & a_{32} & a_{33} \end{vmatrix} = 0 \text{ \{from Property 3, Section 2.3.}$$

Note that this property is equally valid if we replace the word row by the word column.

Theorem 1 $\mathbf{AA}^a = |\mathbf{A}|\mathbf{I}$.

Proof We prove this theorem only for matrices of order 3×3. The proof easily may be extended to cover matrices of any arbitrary order. This extension is left as an exercise for the student.

$$\mathbf{AA}^a = \begin{bmatrix} a_{11} & a_{12} & a_{13} \\ a_{21} & a_{22} & a_{23} \\ a_{31} & a_{32} & a_{33} \end{bmatrix} \begin{bmatrix} a^c_{11} & a^c_{21} & a^c_{31} \\ a^c_{12} & a^c_{22} & a^c_{32} \\ a^c_{13} & a^c_{23} & a^c_{33} \end{bmatrix}.$$

If we denote this product matrix by $[b_{ij}]$, then

$$b_{11} = a_{11}a^c_{11} + a_{12}a^c_{12} + a_{13}a^c_{13},$$
$$b_{12} = a_{11}a^c_{21} + a_{12}a^c_{22} + a_{13}a^c_{23},$$
$$b_{23} = a_{21}a^c_{31} + a_{22}a^c_{32} + a_{23}a^c_{33},$$
$$b_{22} = a_{21}a^c_{21} + a_{22}a^c_{22} + a_{23}a^c_{23},$$

etc.

We now note that $b_{11} = |\mathbf{A}|$ since it is precisely the term obtained when one computes det(**A**) by cofactors, expanding by the first row. Similarly, $b_{22} = |\mathbf{A}|$ since it is precisely the term obtained by computing det(**A**) by cofactors after expanding by the second row. It follows from the above lemma that $b_{12} = 0$ and $b_{23} = 0$ since b_{12} is the term obtained by multiplying each element in the first row of **A** by the cofactor of the corresponding element in the second row and adding, while b_{23} is the term obtained by multiplying each element in the second row of **A** by the cofactor of the corresponding element in the third row and adding. Continuing this analysis for each b_{ij}, we find that

$$\mathbf{AA}^a = \begin{bmatrix} |\mathbf{A}| & 0 & 0 \\ 0 & |\mathbf{A}| & 0 \\ 0 & 0 & |\mathbf{A}| \end{bmatrix} = |\mathbf{A}| \begin{bmatrix} 1 & 0 & 0 \\ 0 & 1 & 0 \\ 0 & 0 & 1 \end{bmatrix},$$

$$\mathbf{AA}^a = |\mathbf{A}|\mathbf{I}.$$

Theorem 2 $\mathbf{A}^a\mathbf{A} = |\mathbf{A}|\mathbf{I}.$

Proof This proof is completely analogous to the previous one and is left as an exercise for the student.

Chapter 4 SIMULTANEOUS LINEAR EQUATIONS

4.1 LINEAR SYSTEMS

Systems of simultaneous equations appear frequently in engineering and scientific problems. Because of their importance and because they lend themselves to matrix analysis, we devote this entire chapter to their solutions.

We are interested in systems of the form

$$\begin{aligned} a_{11}x_1 + a_{12}x_2 + \cdots + a_{1n}x_n &= b_1, \\ a_{21}x_1 + a_{22}x_2 + \cdots + a_{2n}x_n &= b_2, \\ &\vdots \\ a_{m1}x_1 + a_{m2}x_2 + \cdots + a_{mn}x_n &= b_m. \end{aligned} \tag{1}$$

We assume that the coefficients a_{ij} $(i = 1, 2, \ldots, m; j = 1, 2, \ldots, n)$ and quantities b_i $(i = 1, 2, \ldots, m)$ are all known scalars. The quantities $x_1, x_2, \ldots, x_n$ represent unknowns.

Definition 1 A *solution* to Eq. (1) is a set of n scalars $x_1, x_2, \ldots, x_n$ that when substituted into Eq. (1) satisfies the given equations (that is, the equalities are valid).

Equation (1) is a generalization of systems considered earlier in that m can differ from n. If $m > n$, the system has more equations than unknowns. If $m < n$, the system has more unknowns than equations. If $m = n$, the system has as many unknowns as equations. In any case, the methods of Section 1.4 may be used to convert Eq. (1) into the matrix form

$$\mathbf{Ax} = \mathbf{b}, \tag{2}$$

where

$$\mathbf{A} = \begin{bmatrix} a_{11} & a_{12} & \cdots & a_{1n} \\ a_{21} & a_{22} & \cdots & a_{2n} \\ \vdots & \vdots & & \vdots \\ a_{m1} & a_{m2} & \cdots & a_{mn} \end{bmatrix}, \qquad \mathbf{x} = \begin{bmatrix} x_1 \\ x_2 \\ \vdots \\ x_n \end{bmatrix}, \qquad \mathbf{b} = \begin{bmatrix} b_1 \\ b_2 \\ \vdots \\ b_m \end{bmatrix}.$$

Thus, if $m \neq n$, $\mathbf{A}$ will be rectangular and the dimensions of $\mathbf{x}$ and $\mathbf{b}$ will be different.

EXAMPLE 1 Convert the following system to matrix form:

$$\begin{aligned} x + 2y - z + w &= 4, \\ x + 3y + 2z + 4w &= 9. \end{aligned}$$

Solution

$$\mathbf{A} = \begin{bmatrix} 1 & 2 & -1 & 1 \\ 1 & 3 & 2 & 4 \end{bmatrix}, \qquad \mathbf{x} = \begin{bmatrix} x \\ y \\ z \\ w \end{bmatrix}, \qquad \mathbf{b} = \begin{bmatrix} 4 \\ 9 \end{bmatrix}.$$

EXAMPLE 2 Convert the following system to matrix form:

$$\begin{aligned} x - 2y &= -9, \\ 4x + y &= 9, \\ 2x + y &= 7, \\ x - y &= -1. \end{aligned}$$

Solution

$$\mathbf{A} = \begin{bmatrix} 1 & -2 \\ 4 & 1 \\ 2 & 1 \\ 1 & -1 \end{bmatrix}, \qquad \mathbf{x} = \begin{bmatrix} x \\ y \end{bmatrix}, \qquad \mathbf{b} = \begin{bmatrix} -9 \\ 9 \\ 7 \\ -1 \end{bmatrix}.$$

A system of equations given by (1) or (2) can possess no solutions, exactly one solution, or more than one solution (note that by a solution to (2) we mean a vector $\mathbf{x}$ which satisfies the matrix equality (2)). Examples of such systems are

$$\begin{aligned} x + y &= 1, \\ x + y &= 2, \end{aligned} \tag{3}$$

$$\begin{aligned} x + y &= 1, \\ x - y &= 0, \end{aligned} \tag{4}$$

$$\begin{aligned} x + y &= 0, \\ 2x + 2y &= 0. \end{aligned} \tag{5}$$

Equation (3) has no solutions, (4) admits only the solution $x = y = \frac{1}{2}$, while (5) has solutions $x = -y$ for any value of y.

Definition 2 A system of simultaneous linear equations is *consistent* if it possesses at least one solution. If no solution exists, the system is *inconsistent*.

Equation (3) is an example of an inconsistent system, while (4) and (5) represent examples of consistent systems.

Definition 3 A system given by (2) is *homogeneous* if $\mathbf{b} = \mathbf{0}$ (the zero vector). If $\mathbf{b} \neq \mathbf{0}$ (at least one component of $\mathbf{b}$ differs from zero) the system is *nonhomogeneous*.

Equation (5) is an example of a homogeneous system.

PROBLEMS 4.1

In problems 1 and 2, determine whether or not the proposed values of x, y and z are solutions of the given systems.

1. $x + y + 2z = 2,$ (a) $x = 1, y = -3, z = 2$
$x - y - 2z = 0,$ (b) $x = 1, y = -1, z = 1$
$x + 2y + 2z = 1.$

2. $x + 2y + 3z = 6,$ (a) $x = 1, y = 1, z = 1$
$x - 3y + 2z = 0,$ (b) $x = 2, y = 2, z = 0$
$3x - 4y + 7z = 6.$ (c) $x = 14, y = 2, z = -4$

4.2 SOLUTIONS BY INVERSION

If Eq. (1) has as many equations as unknowns (that is, $m = n$), then some results regarding its solutions have been obtained in Sections 3.2 and 3.3. For completeness, we review those results here.

Consider the system $\mathbf{Ax} = \mathbf{b}$ ($\mathbf{A}$ is now a square matrix). If the determinant of $\mathbf{A}$ is not equal to zero, then $\mathbf{A}^{-1}$ exists and the unique solution to the system is

$$\mathbf{x} = \mathbf{A}^{-1}\mathbf{b}. \tag{6}$$

If the determinant of $\mathbf{A}$ equals zero, then $\mathbf{A}^{-1}$ does not exist and other methods must be developed to obtain the solution.

Recall from Section 2.5 that Cramer's rule also provides a method of solution for Eq. (1). Once again however, this method is applicable only when the determinant of $\mathbf{A}$ is not equal to zero.

4.3 GAUSSIAN ELIMINATION

Most readers have probably encountered simultaneous equations in high school algebra. At that time, matrices were not available; hence other methods were developed to solve these systems, in particular, the method of Gaussian elimination or substitution. The reader should recall that this method, although relatively simple, was extremely powerful since it allowed one to solve all systems; it had no restrictions such as the ones placed on the methods given in the previous section. Due to the utility of this method, we review it in this section. The matrix equivalent of Gaussian elimination, and thus the matrix technique which can be used to solve all systems, will be discussed in Section 4.7.

Consider the system given by Eq. (1):

$$\begin{aligned}
a_{11}x_1 + a_{12}x_2 + \cdots + a_{1n}x_n &= b_1,\\
a_{21}x_1 + a_{22}x_2 + \cdots + a_{2n}x_n &= b_2,\\
&\ \,\vdots\\
a_{m1}x_1 + a_{m2}x_2 + \cdots + a_{mn}x_n &= b_m.
\end{aligned}$$

The method of Gaussian elimination is the following: take the first equation and solve for x_1 in terms of $x_2, x_3, \ldots, x_n$ and then substitute this value of x_1 into all the other equations, thus eliminating it from those equations. (If x_1 does not appear in the first equation, rearrange the equations so that it does. For example, one might have to interchange the order of the first and second equations.) This new set of equations is called the *first derived set*. Working with the first derived set, solve the second equation for x_2 in terms of $x_3, x_4, \ldots, x_n$ and then substitute this value of x_2 into the third, fourth, etc. equations, thus eliminating it. This new set is the second derived set. This process is kept up until the following set of equations is obtained:

$$\begin{aligned}
x_1 &= c_{12}x_2 + c_{13}x_3 + c_{14}x_4 + \cdots + c_{1n}x_n + d_1, \\
x_2 &= \qquad\quad c_{23}x_3 + c_{24}x_4 + \cdots + c_{2n}x_n + d_2, \\
x_3 &= \qquad\qquad\qquad c_{34}x_4 + \cdots + c_{3n}x_n + d_3, \\
&\vdots \\
x_m &= \qquad\qquad c_{m,m+1}x_{m+1} + \cdots + c_{m,n}x_n + d_m,
\end{aligned} \tag{7}$$

where the c_{ij}'s and the d_i's are some combination of the original a_{ij}'s and b_i's. System (7) can be quickly solved by inspection.

EXAMPLE 1 Use Gaussian elimination to solve the system

$$\begin{aligned}
r + 2s + t &= 3, \\
3r - 2s - 4t &= -2, \\
2r + 3s - t &= -6.
\end{aligned}$$

Solution By solving the first equation for r and then substituting it into the second and third equations, we obtain the first derived set

$$\begin{aligned}
r &= 3 - 2s - t, \\
-8s - 7t &= -11, \\
-s - 3t &= -12.
\end{aligned}$$

For convenience, we interchange the second and third equation. Note that this in no way alters the system. Thus, we obtain the system

$$\begin{aligned}
r &= 3 - 2s - t, \\
-s - 3t &= -12, \\
-8s - 7t &= -11.
\end{aligned}$$

By solving the second equation for s and then substituting it into the third equation, we obtain the second derived set

$$\begin{aligned}
r &= 3 - 2s - t, \\
s &= 12 - 3t, \\
17t &= 85.
\end{aligned}$$

By solving for t in the third equation and then substituting it into the remaining equations (of which there are none), we obtain the third derived set

$$\begin{aligned} r &= 3 - 2s - t, \\ s &= 12 - 3t, \\ t &= 5. \end{aligned}$$

Thus, the solution is $t = 5$, $s = -3$, $r = 4$.

Example 1 could have been solved also by the methods given in Section 4.2. The following examples show how to handle systems that can not be solved by inversion or Cramer's rule.

EXAMPLE 2 Use Gaussian elimination to solve the system

$$\begin{aligned} x + y + 3z &= -1, \\ 2x - 2y - z &= 1, \\ 5x + y + 8z &= -2. \end{aligned}$$

Solution The first derived set is

$$\begin{aligned} x &= -1 - y - 3z, \\ -4y - 7z &= 3, \\ -4y - 7z &= 3. \end{aligned}$$

The second derived set is

$$\begin{aligned} x &= -1 - y - 3z, \\ y &= -\tfrac{3}{4} - \tfrac{7}{4}z, \\ 0 &= 0. \end{aligned}$$

Since the third equation can not be solved for z, this is as far as we can go. Thus, since we can not obtain a unique value for z, the first and second equations will not yield a unique value for x and y. *Caution*: The third equation does *not* imply that $z = 0$. On the contrary, this equation says nothing at all about z, consequently z is completely arbitrary. The second equation gives y in terms of z. Substituting this value into the first equation, we obtain x in terms of z. The solution therefore is $x = -\frac{1}{4} - \frac{5}{4}z$ and $y = -\frac{3}{4} - \frac{7}{4}z$, z is arbitrary. Thus there are infinitely many solutions to the above system. However, once z is chosen, x and y are determined. If z is chosen to be -1, then $x = y = 1$, while if z is chosen to be 3, then $x = -4$, $y = -6$. The solutions can be expressed in the vector form

$$\begin{bmatrix} x \\ y \\ z \end{bmatrix} = \begin{bmatrix} -\frac{1}{4} & -\frac{5}{4}z \\ -\frac{3}{4} & -\frac{7}{4}z \\ & z \end{bmatrix} = \begin{bmatrix} -\frac{1}{4} \\ -\frac{3}{4} \\ 0 \end{bmatrix} + z\begin{bmatrix} -\frac{5}{4} \\ -\frac{7}{4} \\ 1 \end{bmatrix}.$$

EXAMPLE 3 Use Gaussian elimination to solve

$$\begin{aligned} a + 2b - 3c + \ \ d &= 1, \\ 2a + 6b + 4c + 2d &= 8. \end{aligned}$$

Solution The first derived set is

$$\begin{aligned} a &= 1 - 2b + 3c - d, \\ 2b + 10c &= 6. \end{aligned}$$

The second derived set is

$$\begin{aligned} a &= 1 - 2b + 3c - d \\ b &= 3 - 5c \end{aligned}$$

Again, since there are no more equations, this is as far as we can go, and since there are no defining equations for c and d, these two unknowns must be arbitrary. Solving for a and b in terms of c and d, we obtain the solution $a = -5 + 13c - d$, $b = 3 - 5c$; c and d are arbitrary. The solutions can be expressed in the vector form

$$\begin{bmatrix} a \\ b \\ c \\ d \end{bmatrix} = \begin{bmatrix} -5 + 13c - d \\ 3 - 5c \\ c \\ d \end{bmatrix} = \begin{bmatrix} -5 \\ 3 \\ 0 \\ 0 \end{bmatrix} + c\begin{bmatrix} 13 \\ -5 \\ 1 \\ 0 \end{bmatrix} + d\begin{bmatrix} -1 \\ 0 \\ 0 \\ 1 \end{bmatrix}.$$

Note that while c and d are arbitrary, once they are given a particular value, a and b are automatically determined. For example, if c is chosen as -1 and d as 4, a solution is $a = -22$, $b = 8$, $c = -1$, $d = 4$, while if c is chosen as 0 and d as -3, a solution is $a = -2$, $b = 3$, $c = 0$, $d = -3$.

EXAMPLE 4 Use Gaussian elimination to solve the following system:

$$\begin{aligned} x + \ \ 3y &= 4, \\ 2x - \ \ \ y &= 1, \\ 3x + \ \ 2y &= 5, \\ 5x + 15y &= 20. \end{aligned}$$

Solution The first derived set is

$$\begin{aligned} x &= 4 - 3y, \\ -7y &= -7, \\ -7y &= -7, \\ 0 &= 0. \end{aligned}$$

The second derived set is

$$\begin{aligned} x &= 4 - 3y, \\ y &= 1, \\ 0 &= 0, \\ 0 &= 0. \end{aligned}$$

Thus, the solution is $y = 1$, $x = 1$, or in vector form

$$\begin{bmatrix} x \\ y \end{bmatrix} = \begin{bmatrix} 1 \\ 1 \end{bmatrix}.$$

PROBLEMS 4.3

Use Gaussian elimination to solve the following systems:

1. $$\begin{aligned} x + 2y - 2z &= -1, \\ 2x + y + z &= 5, \\ -x + y - z &= -2. \end{aligned}$$

2. $$\begin{aligned} x + y - z &= 0, \\ 3x + 2y + 4z &= 0. \end{aligned}$$

3. $$\begin{aligned} x + 3y &= 4, \\ 2x - y &= 1, \\ -2x - 6y &= -8, \\ 4x - 9y &= -5, \\ -6x + 3y &= -3. \end{aligned}$$

4. $$\begin{aligned} 4r - 3s + 2t &= 1, \\ r + s - 3t &= 4, \\ 5r - 2s - t &= 5. \end{aligned}$$

5. $$\begin{aligned} 2l - m + n - p &= 1, \\ l + 2m - n + 2p &= -1, \\ l - 3m + 2n - 3p &= 2. \end{aligned}$$

4.4 LINEAR INDEPENDENCE

We momentarily digress from our discussion of simultaneous equations to develop the concepts of linearly independent vectors and rank of a matrix, both of which will prove indispensable to us in the ensuing sections.

Definition 1 A vector $\mathbf{V}_1$ is a *linear combination* of the vectors $\mathbf{V}_2, \mathbf{V}_3, \ldots, \mathbf{V}_n$ if there exist scalars $d_2, d_3, \ldots, d_n$ such that

$$\mathbf{V}_1 = d_2\mathbf{V}_2 + d_3\mathbf{V}_3 + \cdots + d_n\mathbf{V}_n.$$

EXAMPLE 1 Show that (1 2 3) is a linear combination of (2 4 0) and (0 0 1).

Solution (1 2 3) = $\frac{1}{2}$(2 4 0) + 3(0 0 1).

Referring to Example 1, we could say that the row vector (1 2 3) depends linearly on the other two vectors or, more generally, that the set of vectors {(1 2 3), (2 4 0), (0 0 1)} is *linearly dependent*. Another way of expressing this dependence would be to say that there exist constants c_1, c_2, c_3 not all zero such that $c_1(1\ 2\ 3) + c_2(2\ 4\ 0) + c_3(0\ 0\ 1) = (0\ 0\ 0)$. Such a set would be $c_1 = -1$, $c_2 = \frac{1}{2}$, $c_3 = 3$. Note that the set $c_1 = c_2 = c_3 = 0$ is also a suitable set. The important fact about dependent sets, however, is that there exists a set of constants, *not all equal to zero*, that satisfies the equality.

Now consider the set given by $\mathbf{V}_1 = (1\ 0\ 0)$, $\mathbf{V}_2 = (0\ 1\ 0)$, $\mathbf{V}_3 = (0\ 0\ 1)$. It is easy to verify that no vector in this set is a linear combination of the other two. Thus, each vector is linearly independent of the other two or, more generally, the set of vectors is *linearly independent*. Another way of expressing this independence would be to say the only scalars that satisfy the equation $c_1(1\ 0\ 0) + c_2(0\ 1\ 0) + c_3(0\ 0\ 1) = (0\ 0\ 0)$ are $c_1 = c_2 = c_3 = 0$.

Definition 2 A set of vectors $\{\mathbf{V}_1, \mathbf{V}_2, \ldots, \mathbf{V}_n\}$, of the same dimension, is *linearly dependent* if there exist scalars $c_1, c_2, \ldots, c_n$, not all zero, such that

$$c_1\mathbf{V}_1 + c_2\mathbf{V}_2 + c_3\mathbf{V}_3 + \cdots + c_n\mathbf{V}_n = \mathbf{0}. \tag{8}$$

The vectors are *linearly independent* if the only set of scalars that satisfies (8) is the set $c_1 = c_2 = \cdots = c_n = 0$.

Therefore, to test whether or not a given set of vectors is linearly independent, first form the vector equation (8) and ask "What values for the c's satisfy this equation?" Clearly $c_1 = c_2 = \cdots = c_n = 0$ is a suitable set. If this is the only set of values that satisfies (8) then the vectors are linearly independent. If there exists a set of values that is not all zero, then the vectors are linearly dependent.

Note that it is not necessary for all the c's to be different from zero for a set of vectors to be linearly dependent. Consider the vectors $\mathbf{V}_1 = (1, 2)$, $\mathbf{V}_2 = (1, 4)$, $\mathbf{V}_3 = (2, 4)$. $c_1 = 2$, $c_2 = 0$, $c_3 = -1$ is a set of scalars, *not all zero*, such that $c_1\mathbf{V}_1 + c_2\mathbf{V}_2 + c_3\mathbf{V}_3 = \mathbf{0}$. Thus, this set is linearly dependent.

EXAMPLE 2 Is the set {(1, 2), (3, 4)} linearly independent?

Solution The vector equation is

$$c_1(1\ 2) + c_2(3\ 4) = (0\ 0).$$

This equation can be rewritten as

$$(c_1\ 2c_1) + (3c_2\ 4c_2) = (0\ 0)$$

or as

$$(c_1 + 3c_2 \qquad 2c_1 + 4c_2) = (0\ 0).$$

Equating components, we see that this vector equation is equivalent to the system

$$\begin{aligned} c_1 + 3c_2 &= 0, \\ 2c_1 + 4c_2 &= 0. \end{aligned}$$

Using Gaussian elimination, we find that the only solution to this system is $c_1 = c_2 = 0$, hence the original set of vectors is linearly independent.

Although we have worked exclusively with row vectors, the above definitions are equally applicable to column vectors.

EXAMPLE 3 Is the set

$$\left\{ \begin{bmatrix} 2 \\ 6 \\ -2 \end{bmatrix}, \begin{bmatrix} 3 \\ 1 \\ 2 \end{bmatrix}, \begin{bmatrix} 8 \\ 16 \\ -3 \end{bmatrix} \right\}$$

linearly independent?

Solution Consider the vector equation

$$c_1 \begin{bmatrix} 2 \\ 6 \\ -2 \end{bmatrix} + c_2 \begin{bmatrix} 3 \\ 1 \\ 2 \end{bmatrix} + c_3 \begin{bmatrix} 8 \\ 16 \\ -3 \end{bmatrix} = \begin{bmatrix} 0 \\ 0 \\ 0 \end{bmatrix}. \tag{9}$$

This equation can be rewritten as

$$\begin{bmatrix} 2c_1 \\ 6c_1 \\ -2c_1 \end{bmatrix} + \begin{bmatrix} 3c_2 \\ c_2 \\ 2c_2 \end{bmatrix} + \begin{bmatrix} 8c_3 \\ 16c_3 \\ -3c_3 \end{bmatrix} = \begin{bmatrix} 0 \\ 0 \\ 0 \end{bmatrix}$$

or as

$$\begin{bmatrix} 2c_1 + 3c_2 + 8c_3 \\ 6c_1 + c_2 + 16c_3 \\ -2c_1 + 2c_2 - 3c_3 \end{bmatrix} = \begin{bmatrix} 0 \\ 0 \\ 0 \end{bmatrix}.$$

By equating components, we see that this vector equation is equivalent to the system

$$\begin{aligned} 2c_1 + 3c_2 + 8c_3 &= 0, \\ 6c_1 + c_2 + 16c_3 &= 0, \\ -2c_1 + 2c_2 - 3c_3 &= 0. \end{aligned}$$

By using Gaussian elimination, we find that the solution to this system is $c_1 = (-\frac{5}{2})c_3$, $c_2 = -c_3$, c_3 arbitrary. Thus, choosing $c_3 = 2$, we obtain

$c_1 = -5$, $c_2 = -2$, $c_3 = 2$ as a particular nonzero set of constants that satisfies (9); hence, the original vectors are linearly dependent.

EXAMPLE 4 Is the set

$$\left\{\begin{bmatrix}1\\2\end{bmatrix}, \begin{bmatrix}5\\7\end{bmatrix}, \begin{bmatrix}-3\\1\end{bmatrix}\right\}$$

linearly independent?

Solution Consider the vector equation

$$c_1\begin{bmatrix}1\\2\end{bmatrix} + c_2\begin{bmatrix}5\\7\end{bmatrix} + c_3\begin{bmatrix}-3\\1\end{bmatrix} = \begin{bmatrix}0\\0\end{bmatrix}. \tag{10}$$

This is equivalent to the system

$$\begin{aligned} c_1 + 5c_2 - 3c_3 &= 0,\\ 2c_1 + 7c_2 + c_3 &= 0. \end{aligned}$$

By using Gaussian elimination, we find that the solution to this system is $c_1 = (-26/3)c_3$, $c_2 = (7/3)c_3$, c_3 arbitrary. Hence a particular nonzero solution is found by choosing $c_3 = 3$; then $c_1 = -26$, $c_2 = 7$, and, therefore, the vectors are linearly dependent.

We conclude this section with a few important theorems on linear independence and dependence.

Theorem 1 A set of vectors is linearly dependent if and only if one of the vectors is a linear combination of the others.

Proof Let $\{\mathbf{V}_1, \mathbf{V}_2, \ldots, \mathbf{V}_n\}$ be a linearly dependent set. Then there exist scalars $c_1, c_2, \ldots, c_n$, not all zero, such that (8) is satisfied. Assume $c_1 \neq 0$. (Since at least one of the c's must differ from zero, we lose no generality in assuming it is c_1.) Equation (8) can be rewritten as

$$c_1\mathbf{V}_1 = -c_2\mathbf{V}_2 - c_3\mathbf{V}_3 - \cdots - c_n\mathbf{V}_n,$$

or as

$$\mathbf{V}_1 = -\frac{c_2}{c_1}\mathbf{V}_2 - \frac{c_3}{c_1}\mathbf{V}_3 - \cdots - \frac{c_n}{c_1}\mathbf{V}_n.$$

Thus, $\mathbf{V}_1$ is a linear combination of $\mathbf{V}_2, \mathbf{V}_3, \ldots, \mathbf{V}_n$. To complete the proof, we must show that if one vector is a linear combination of the others, then the set is linearly dependent. We leave this as an exercise for the student (see Problem 7).

OBSERVATION In order for a set of vectors to be linearly dependent, it is not necessary for *every* vector to be a linear combination of the others, only that there exists *one* vector that is a linear combination of the others. For example, consider the vectors (1 0), (2 0), (0 1). Here, (0, 1) cannot be written as a linear combination of the other two vectors; however, (2 0) can be written as a linear combination of (1 0) and (0 1), namely, (2 0) = 2(1 0) + 0(0 1); hence, the vectors are linearly dependent.

Theorem 2 The set consisting of the single vector $\mathbf{V}_1$ is a linearly independent set if and only if $\mathbf{V}_1 \neq 0$.

Proof Consider the equation $c_1\mathbf{V}_1 = \mathbf{0}$. If $\mathbf{V}_1 \neq \mathbf{0}$, then the only way this equation can be valid is if $c_1 = 0$; hence, the set is linearly independent. If $\mathbf{V}_1 = \mathbf{0}$, then any $c_1 \neq 0$ will satisfy the equation; hence, the set is linearly dependent.

Theorem 3 Any set of vectors that contains the zero vector is linearly dependent.

Proof Consider the set $\{\mathbf{V}_1, \mathbf{V}_2, \ldots, \mathbf{V}_n, \mathbf{0}\}$. Pick $c_1 = c_2 = \cdots = c_n = 0$, $c_{n+1} = 5$ (any other number will do). Then this is a set of scalars, not all zero, such that

$$c_1\mathbf{V}_1 + c_2\mathbf{V}_2 + \cdots + c_n\mathbf{V}_n + c_{n+1}\mathbf{0} = \mathbf{0};$$

hence, the set of vectors is linearly dependent.

Theorem 4 If a set of vectors is linearly independent, any subset of these vectors is also linearly independent.

Proof See Problem 8.

Theorem 5 If a set of vectors is linearly dependent, then any larger set, containing this set, is also linearly dependent.

Proof See Problem 9.

PROBLEMS 4.4

In Problems 1–4, determine whether or not the given set is linearly independent.

1. $\{(1\ 3), (2\ -1), (1\ 1)\}$

2. $\left\{\begin{bmatrix}4\\5\\1\end{bmatrix}, \begin{bmatrix}3\\0\\2\end{bmatrix}, \begin{bmatrix}1\\1\\1\end{bmatrix}\right\}$

3. $\left\{\begin{bmatrix}2\\1\\1\\3\end{bmatrix}, \begin{bmatrix}4\\-1\\2\\-1\end{bmatrix}, \begin{bmatrix}8\\1\\4\\5\end{bmatrix}\right\}$

4. $\{(2\ 1\ 1), (3\ -1\ 4), (1\ 3\ -2)\}$.

5. Express the vector

$$\begin{bmatrix}2\\1\\2\end{bmatrix}$$

as a linear combination of

$$\begin{bmatrix}1\\1\\0\end{bmatrix}, \quad \begin{bmatrix}1\\0\\-1\end{bmatrix}, \quad \begin{bmatrix}1\\1\\1\end{bmatrix}.$$

6. Can the vector $(3\ 1\ 1)$ be expressed as a linear combination of the vectors $(1\ 2\ 1)$ and $(1\ 3\ -1)$?

7. Finish the proof of Theorem 1. (Hint: Assume that $\mathbf{V}_1$ can be written as a linear combination of the other vectors.)

8. Prove Theorem 4.

9. Prove Theorem 5.

4.5 RANK

Recall from Section 2.2 that a minor of a matrix $\mathbf{A}$ is defined to be the determinant of a square submatrix of $\mathbf{A}$ (submatrices were defined in Section 1.6). For example, if

$$\mathbf{A} = \begin{bmatrix}1 & 4 & 7 & 10\\2 & 5 & 8 & 11\\3 & 6 & 9 & 12\end{bmatrix}$$

then

$$\begin{vmatrix}1 & 7 & 10\\2 & 8 & 11\\3 & 9 & 12\end{vmatrix}, \quad \begin{vmatrix}4 & 10\\6 & 12\end{vmatrix}, \quad \text{and } |\,8\,|$$

are all minors of $\mathbf{A}$, while

$$\begin{vmatrix}1 & 4\\9 & 12\end{vmatrix}$$

is not since the matrix

$$\begin{bmatrix} 1 & 4 \\ 9 & 12 \end{bmatrix}$$

is not a submatrix of **A**.

Consider the matrix **A** given above. It has four different minors of order 3 associated with it (delete a different column each time to obtain them). When these minors are evaluated, they may or may not equal zero. If at least one of them is different from zero, we say that the rank of **A** is 3. If all the minors of order 3 equal zero, we then consider the 18 minors of order 2. If any one of these minors differs from zero, we say that the rank of **A** is 2. If all the minors of order 2 equal zero, we then consider the 12 minors of order 1. If any one of these minors differ from zero, we say that the rank of **A** is 1. If all the minors of order 1 equal zero, we then say that the rank of **A** is zero. In particular, for the **A** just given,

$$\begin{vmatrix} 4 & 7 & 10 \\ 5 & 8 & 11 \\ 6 & 9 & 12 \end{vmatrix} = \begin{vmatrix} 1 & 7 & 10 \\ 2 & 8 & 11 \\ 3 & 9 & 12 \end{vmatrix} = \begin{vmatrix} 1 & 4 & 10 \\ 2 & 5 & 11 \\ 3 & 6 & 12 \end{vmatrix} = \begin{vmatrix} 1 & 4 & 7 \\ 2 & 5 & 8 \\ 3 & 6 & 9 \end{vmatrix} = 0.$$

Thus, all minors of order 3 are zero. However,

$$\begin{vmatrix} 1 & 4 \\ 2 & 5 \end{vmatrix} = -3$$

is a minor of order 2 not equal to zero, hence the rank of **A** is 2.

Definition 1 The rank of a matrix **A**, designated $r(\mathbf{A})$, is the order of the largest nonzero minor of **A**.

As used here, "largest" refers to the orders of the minors involved and not to their respective values. For example, consider the matrix

$$\begin{bmatrix} 2 & 1 & 3 & 1 \\ 3 & 4 & 7 & 1 \\ 5 & 6 & 11 & 1 \end{bmatrix}.$$

The minor

$$\begin{vmatrix} 2 & 1 & 3 \\ 3 & 4 & 7 \\ 5 & 6 & 11 \end{vmatrix} \quad \text{of order 3 is larger than the minor} \quad \begin{vmatrix} 2 & 1 \\ 3 & 4 \end{vmatrix}$$

of order 2, even though

$$\begin{vmatrix} 2 & 1 & 3 \\ 3 & 4 & 7 \\ 5 & 6 & 11 \end{vmatrix} = 0 \quad \text{and} \quad \begin{vmatrix} 2 & 1 \\ 3 & 4 \end{vmatrix} = 5.$$

EXAMPLE 1 Find the rank of

$$\mathbf{A} = \begin{bmatrix} 1 & 2 & 3 & 2 & 5 \\ 3 & 2 & 5 & 1 & 6 \\ -1 & 4 & 3 & 0 & 3 \end{bmatrix}.$$

Solution The largest minor that can be formed from **A** is of order 3; there are 10 such minors. Although some of these minors equal zero, for instance,

$$\begin{vmatrix} 1 & 2 & 3 \\ 3 & 2 & 5 \\ -1 & 4 & 3 \end{vmatrix},$$

there is at least one minor, namely

$$\begin{vmatrix} 1 & 2 & 5 \\ 3 & 2 & 6 \\ -1 & 4 & 3 \end{vmatrix} = 22,$$

which does not. Thus, the rank of **A** is 3.

EXAMPLE 2 Find the rank of

$$\mathbf{A} = \begin{bmatrix} 1 & 2 & 5 & -4 & 3 \\ 3 & 2 & 11 & -4 & 5 \\ 1 & -4 & -1 & 8 & -3 \end{bmatrix}.$$

Solution All 10 minors of order 3 equal zero, so the rank of **A** will be 2 or less. Checking all minors of order 2, we find one of them, namely

$$\begin{vmatrix} 3 & 2 \\ 1 & -4 \end{vmatrix}$$

differs from zero, so $r(\mathbf{A}) = 2$.

EXAMPLE 3 Find the rank of

$$\mathbf{A} = \begin{bmatrix} 1 & 3 & -4 \\ -1 & -3 & 4 \\ 2 & 6 & -8 \end{bmatrix}.$$

Solution The largest minor that can be formed from **A** is of order 3; there is only one such minor, namely det(**A**), and it is equal to zero. Thus, the rank of **A** will be 2 or less. Checking all the 9 minors of order 2, we find that each of them is also equal to zero; hence, the rank of **A** will be 1 or zero. Checking minors of order 1, we find one which is not zero (in fact all are nonzero); therefore, $r(\mathbf{A}) = 1$.

Since a minor is the determinant of a square matrix, it follows that the rank of a matrix can be no greater than the number of rows or columns of the matrix whichever is smaller. Thus, for example, if **A** has 5 rows and 2 columns, it is impossible to form a minor of **A** larger than 2, from which it follows that $r(\mathbf{A}) \leq 2$.

By employing the elementary row and column operations introduced in Section 3.4, we can give an alternative method for determining rank which, for matrices of large orders, will be the most expedient method to use. Let **A** be a rectangular matrix whose rank we wish to determine. Use elementary row *and* column operations to transform **A** into a matrix **B** that has the property that every row and column of **B** contains at most one nonzero element. The number of nonzero rows (or, equivalently, nonzero columns) is the rank of **A**.[6]

EXAMPLE 4 Find the rank of the matrix given in Example 1.

Solution By using both elementary row and column operations, we can transform the matrix **A** into the matrix

$$\mathbf{B} = \begin{bmatrix} 0 & -2 & 0 & 0 & 0 \\ 0 & 0 & 0 & 1 & 0 \\ -11 & 0 & 0 & 0 & 0 \end{bmatrix}$$

which has the property that every row and column contains at most one nonzero element (note that some rows and columns contain only zero elements). Since **B** has 3 nonzero rows (and columns) the rank of **A** is 3.

EXAMPLE 5 Find the rank of the matrix given in Example 2.

Solution By using both elementary row and column operations, we can transform the matrix **A** into the matrix

$$\mathbf{B} = \begin{bmatrix} 1 & 0 & 0 & 0 & 0 \\ 0 & -4 & 0 & 0 & 0 \\ 0 & 0 & 0 & 0 & 0 \end{bmatrix}.$$

Here, **B** has 2 nonzero rows, hence the rank of **A** is 2.

Note that **B** is *not* unique. That is, since elementary row (column) operations include interchanging rows (columns) and multiplying rows (columns) by scalars, there are many acceptable **B** matrices. Thus, the matrix given in

[6] For a proof of this statement see F. M. Stein, "Introduction to Matrices and Determinants," p. 68. Wadsworth, Belmont, California, 1967.

Example 2 could have been transformed into either

$$\begin{bmatrix} 2 & 0 & 0 & 0 & 0 \\ 0 & 0 & 0 & 0 & 0 \\ 0 & 0 & 0 & 0 & -107 \end{bmatrix} \quad \text{or} \quad \begin{bmatrix} 0 & 0 & 0 & 0 & 0 \\ 0 & -4 & 0 & 0 & 0 \\ 0 & 0 & 0 & 7 & 0 \end{bmatrix}$$

rather than the **B** given in Example 5. However, regardless of which matrix we choose, we still obtain the same answer for the rank of **A**, namely $r(\mathbf{A}) = 2$.

EXAMPLE 6 Find the rank of the matrix given in Example 3.

Solution This matrix can be transformed into the matrix

$$\mathbf{B} = \begin{bmatrix} 1 & 0 & 0 \\ 0 & 0 & 0 \\ 0 & 0 & 0 \end{bmatrix};$$

hence, the rank of **A** is 1.

One final comment: Recall that when finding inverses with elementary row and column operations, the two could not be used together. This is not true in determining rank. Here elementary row and column operations can be mixed freely.

The concepts of rank and linear independence are closely related.

Definition 2 The *row rank* of a matrix is the maximum number of linearly independent rows. Analogously, the *column rank* of a matrix is the maximum number of linearly independent columns.

Theorem 1 The row rank, column rank, and rank of a matrix are all equal.[7]

EXAMPLE 7 Is the set $\{(1, 3), (2, -1), (1, 1)\}$ linearly independent?

Solution Consider the matrix

$$\mathbf{A} = \begin{bmatrix} 1 & 2 & 1 \\ 3 & -1 & 1 \end{bmatrix}.$$

The rank of **A** is 2, hence by Theorem 1, the column rank is also 2 and the three columns of **A** are linearly dependent. (If the three columns of **A** were linearly independent, then the maximum number of linearly independent columns would be three, which would imply that the column rank of **A** is 3).

[7] For a proof of Theorem 1, see D. Zelinsky, "A First Course in Linear Algebra," p. 155. Academic Press, New York, 1968.

Since the columns of **A** are precisely the vectors under consideration, it follows that the vectors are linearly dependent.

EXAMPLE 8 Can the vector

$$\begin{bmatrix} 1 \\ 1 \end{bmatrix}$$

be written as a linear combination of the vectors

$$\begin{bmatrix} 3 \\ 6 \end{bmatrix} \quad \text{and} \quad \begin{bmatrix} 2 \\ 4 \end{bmatrix}?$$

Solution Consider the matrix

$$\begin{bmatrix} 3 & 2 \\ 6 & 4 \end{bmatrix}.$$

It has rank 1; hence, there exists just one linearly independent column vector. Now consider the matrix

$$\begin{bmatrix} 3 & 2 & 1 \\ 6 & 4 & 1 \end{bmatrix}.$$

It has rank 2; hence this matrix has two linearly independent column vectors. Since the second matrix is precisely the first with the one additional column, it follows that the additional column

$$\begin{bmatrix} 1 \\ 1 \end{bmatrix}$$

is independent of the other two and, therefore, cannot be written as a linear combination of the other two vectors.

PROBLEMS 4.5

In Problems 1–5, find the rank of the given matrix.

1. $\begin{bmatrix} 1 & 2 & 0 \\ 3 & 1 & -5 \end{bmatrix}$ **2.** $\begin{bmatrix} 4 & 1 \\ 2 & 3 \\ 2 & 2 \end{bmatrix}$

3. $\begin{bmatrix} 1 & 4 & -2 \\ 2 & 8 & -4 \\ -1 & -4 & 2 \end{bmatrix}$ **4.** $\begin{bmatrix} 1 & 2 & 4 & 2 \\ 1 & 1 & 3 & 2 \\ 1 & 2 & 4 & 2 \end{bmatrix}$

5. $\begin{bmatrix} 1 & 7 & 0 \\ 0 & 1 & 1 \\ 1 & 1 & 0 \end{bmatrix}.$

6. What is the rank of the zero matrix?
7. Can (3 7) be written as a linear combination of the vectors (1 2) and (3 2)?
8. Can (3 7) be written as a linear combination of the vectors (1 2) and (4 8)?
9. Prove that an $n \times n$ matrix is invertible if and only if its rank is n.

4.6 THEORY OF SOLUTIONS

Definition 1 Given the system $\mathbf{Ax} = \mathbf{b}$, the *augmented matrix*, designated by $\mathbf{A}^{\mathbf{b}}$, is a matrix obtained from $\mathbf{A}$ by adding to it one extra column, namely $\mathbf{b}$.

Thus, if

$$\mathbf{A} = \begin{bmatrix} 1 & 2 & 3 \\ 4 & 5 & 6 \end{bmatrix} \quad \text{and} \quad \mathbf{b} = \begin{bmatrix} 7 \\ 8 \end{bmatrix},$$

then

$$\mathbf{A}^{\mathbf{b}} = \begin{bmatrix} 1 & 2 & 3 & 7 \\ 4 & 5 & 6 & 8 \end{bmatrix},$$

while if

$$\mathbf{A} = \begin{bmatrix} 1 & 2 & 3 \\ 4 & 5 & 6 \\ 7 & 8 & 9 \end{bmatrix} \quad \text{and} \quad \mathbf{b} = \begin{bmatrix} -1 \\ -2 \\ -3 \end{bmatrix},$$

then

$$\mathbf{A}^{\mathbf{b}} = \begin{bmatrix} 1 & 2 & 3 & -1 \\ 4 & 5 & 6 & -2 \\ 7 & 8 & 9 & -3 \end{bmatrix}.$$

Consider once again the system $\mathbf{Ax} = \mathbf{b}$ of m equations and n unknowns given in Eq. (2). Designate the n columns of $\mathbf{A}$ by the vectors $\mathbf{V}_1, \mathbf{V}_2, \ldots, \mathbf{V}_n$. Then Eq. (2) can be rewritten in the vector form

$$x_1\mathbf{V}_1 + x_2\mathbf{V}_2 + \cdots + x_n\mathbf{V}_n = \mathbf{b}. \tag{11}$$

EXAMPLE 1 Rewrite the following system in the vector form (11):

$$x - 2y + 3z = 7,$$
$$4x + 5y - 6z = 8.$$

Solution

$$x\begin{bmatrix} 1 \\ 4 \end{bmatrix} + y\begin{bmatrix} -2 \\ 5 \end{bmatrix} + z\begin{bmatrix} 3 \\ -6 \end{bmatrix} = \begin{bmatrix} 7 \\ 8 \end{bmatrix}.$$

Thus, finding solutions to (1) and (2) is equivalent to finding scalars $x_1, x_2, \ldots, x_n$ that satisfy (11). This, however, is asking precisely the question "Is the vector $\mathbf{b}$ a linear combination of $\mathbf{V}_1, \mathbf{V}_2, \ldots, \mathbf{V}_n$?" If $\mathbf{b}$ is a linear combination of $\mathbf{V}_1, \mathbf{V}_2, \ldots, \mathbf{V}_n$, then there will exist scalars $x_1, x_2, \ldots, x_n$ that satisfy (11) and the system is consistent. If $\mathbf{b}$ is not a linear combination of these vectors, that is, if $\mathbf{b}$ is linearly independent of the vectors $\mathbf{V}_1, \mathbf{V}_2, \ldots, \mathbf{V}_n$, then no scalars $x_1, x_2, \ldots, x_n$ will exist that satisfy (11) and the system is inconsistent.

Taking a hint from Example 8, and Problems 7 and 8 of Section 4.5, we have the following theorem.

Theorem 1 The system $\mathbf{Ax} = \mathbf{b}$ is consistent if and only if $r(\mathbf{A}) = r(\mathbf{A}^{\mathbf{b}})$.

Once a system is found to be consistent, the following theorem can be used to ascertain the number of solutions it will have.

Theorem 2 If the system $\mathbf{Ax} = \mathbf{b}$ is consistent and the $r(\mathbf{A}) = k$ then the solutions are expressible in terms of $n - k$ arbitrary unknowns (n represents the number of unknowns in the system).[8]

EXAMPLE 2 Discuss the solutions of the system

$$x + y - z = 1,$$
$$x + y - z = 0.$$

Solution

$$\mathbf{A} = \begin{bmatrix} 1 & 1 & -1 \\ 1 & 1 & -1 \end{bmatrix}, \qquad \mathbf{b} = \begin{bmatrix} 1 \\ 0 \end{bmatrix}, \qquad \mathbf{A}^{\mathbf{b}} = \begin{bmatrix} 1 & 1 & -1 & 1 \\ 1 & 1 & -1 & 0 \end{bmatrix}.$$

Here, $r(\mathbf{A}) = 1$, $r(\mathbf{A}^{\mathbf{b}}) = 2$. Thus, $r(\mathbf{A}) \neq r(\mathbf{A}^{\mathbf{b}})$ and no solution exists.

EXAMPLE 3 Discuss the solutions of the system

$$\begin{aligned} x + y + w &= 3, \\ 2x + 2y + 2w &= 6, \\ -x - y - w &= -3. \end{aligned}$$

Solution

$$\mathbf{A} = \begin{bmatrix} 1 & 1 & 1 \\ 2 & 2 & 2 \\ -1 & -1 & -1 \end{bmatrix}, \qquad \mathbf{b} = \begin{bmatrix} 3 \\ 6 \\ -3 \end{bmatrix}, \qquad \mathbf{A}^{\mathbf{b}} = \begin{bmatrix} 1 & 1 & 1 & 3 \\ 2 & 2 & 2 & 6 \\ -1 & -1 & -1 & -3 \end{bmatrix}.$$

[8] For a proof of Theorem 2, see F. M. Stein, "Introduction to Matrices and Determinants," p. 126. Wadsworth, Belmont, California, 1967.

Here, $r(\mathbf{A}) = r(\mathbf{A^b}) = 1$; hence, the system is consistent. In this case, $n = 3$ and $k = 1$; thus, the solutions are expressible in terms of $3 - 1 = 2$ arbitrary unknowns. Using Gaussian elimination, we find that the solution is $x = 3 - y - w$ where y and w are both arbitrary.

EXAMPLE 4 Discuss the solutions of the system

$$\begin{aligned} 2x - 3y + z &= -1, \\ x - y + 2z &= 2, \\ 2x + y - 3z &= 3. \end{aligned}$$

Solution

$$\mathbf{A} = \begin{bmatrix} 2 & -3 & 1 \\ 1 & -1 & 2 \\ 2 & 1 & -3 \end{bmatrix}, \qquad \mathbf{b} = \begin{bmatrix} -1 \\ 2 \\ 3 \end{bmatrix}, \qquad \mathbf{A^b} = \begin{bmatrix} 2 & -3 & 1 & -1 \\ 1 & -1 & 2 & 2 \\ 2 & 1 & -3 & 3 \end{bmatrix}.$$

Here $r(\mathbf{A}) = r(\mathbf{A^b}) = 3$, hence the system is consistent. Since $n = 3$ and $k = 3$, the solution will be in $n - k = 0$ arbitrary unknowns. Thus, the solution is unique (none of the unknowns are arbitrary) and can be obtained by matrix inversion to be $x = y = 2$, $z = 1$.

EXAMPLE 5 Discuss the solutions of the system

$$\begin{aligned} x + y - 2z &= 1, \\ 2x + y + z &= 2, \\ 3x + 2y - z &= 3, \\ 4x + 2y + 2z &= 4. \end{aligned}$$

Solution

$$\mathbf{A} = \begin{bmatrix} 1 & 1 & -2 \\ 2 & 1 & 1 \\ 3 & 2 & -1 \\ 4 & 2 & 2 \end{bmatrix}, \qquad \mathbf{b} = \begin{bmatrix} 1 \\ 2 \\ 3 \\ 4 \end{bmatrix}, \qquad \mathbf{A^b} = \begin{bmatrix} 1 & 1 & -2 & 1 \\ 2 & 1 & 1 & 2 \\ 3 & 2 & -1 & 3 \\ 4 & 2 & 2 & 4 \end{bmatrix}.$$

Here $r(\mathbf{A}) = r(\mathbf{A^b}) = 2$. Thus, the system is consistent and the solutions will be in terms of $3 - 2 = 1$ arbitrary unknowns. Using Gaussian elimination, we find that the solution is $x = 1 - 3z$, $y = 5z$, and z is arbitrary.

In a consistent system, the solution is unique if $k = n$. If $k \neq n$, the solution will be in terms of arbitrary unknowns. Since these arbitrary unknowns can be chosen to be any constants whatsoever, it follows that there will be an infinite number of solutions. Thus, a consistent system will possess exactly one solution or an infinite number of solutions; there is no inbetween.

PROBLEMS 4.6

In Problems 1–5, discuss the solutions of the given system in terms of consistency and number of solutions. Check your answers by solving the systems wherever possible.

1.
$$\begin{aligned} x - 2y &= 0, \\ x + y &= 1, \\ 2x - y &= 1. \end{aligned}$$

2.
$$\begin{aligned} x + y &= 0, \\ 2x - 2y &= 1, \\ x - y &= 0. \end{aligned}$$

3.
$$\begin{aligned} x + y + z &= 1, \\ x - y + z &= 2, \\ 3x + y + 3z &= 4. \end{aligned}$$

4.
$$\begin{aligned} x + 3y + 2z - w &= 2, \\ 2x - y + z + w &= 3. \end{aligned}$$

5.
$$\begin{aligned} 2x - y + z &= 0, \\ x + 2y - z &= 4, \\ x + y + z &= 1. \end{aligned}$$

6. Prove that $r(\mathbf{A}) \leq r(\mathbf{A}^{\mathbf{b}})$.

7. Prove that for a homogeneous system $r(\mathbf{A}) = r(\mathbf{A}^{\mathbf{b}})$, and that such a system is always consistent.

4.7 MATRIX SOLUTIONS

Once a system is deemed consistent, the next step is to find the solutions. In this section, we give a matrix method for obtaining solutions to the system $\mathbf{Ax} = \mathbf{b}$. (Recall that Gaussian elimination gives us a method that does not use matrices.) We assume that all systems under consideration have already been checked and found consistent.

METHOD Assume that the rank of the coefficient matrix $\mathbf{A}$ is equal to k. Then there must exist at least one $k \times k$ submatrix of $\mathbf{A}$ having a nonzero determinant. Consider the k equations of the system that contributed coefficients to this submatrix and disregard all other equations (all of the disregarded equations are linear combinations of the ones kept and, as such, add nothing to the system). Using only these k equations, pick out the k unknowns whose coefficients contributed to the nonzero determinant; for the sake of argument, let us say that these unknowns are $x_1, x_2, \ldots, x_k$. Rewrite the k equations that are being kept so that the k unknowns $x_1, x_2, \ldots, x_k$ along with their coefficients appear on the left-hand side of the equations and everything else appears on the right-hand side. Solve for $x_1, x_2, \ldots, x_k$ in terms of the other unknowns by matrix inversion.

EXAMPLE 1 Find all solutions of the system:

$$\begin{aligned} x + 4y + 7z &= -1, \\ 2x + 5y + 8z &= -2, \\ 3x + 6y + 9z &= -3. \end{aligned}$$

Solution

$$\mathbf{A} = \begin{bmatrix} 1 & 4 & 7 \\ 2 & 5 & 8 \\ 3 & 6 & 9 \end{bmatrix}.$$

Here, $r(\mathbf{A}) = r(\mathbf{A}^{\mathbf{b}}) = 2$, which can be obtained by considering the submatrix

$$\begin{bmatrix} 1 & 4 \\ 2 & 5 \end{bmatrix}.$$

Since the coefficients of this submatrix are taken from the first two equations, we consider only these equations and disregard the third. Since the coefficients are associated with the unknowns x and y, we rewrite the equation as,

$$\begin{aligned} x + 4y &= -1 - 7z, \\ 2x + 5y &= -2 - 8z, \end{aligned}$$

or, in matrix form,

$$\begin{bmatrix} 1 & 4 \\ 2 & 5 \end{bmatrix} \begin{bmatrix} x \\ y \end{bmatrix} = \begin{bmatrix} (-1 - 7z) \\ (-2 - 8z) \end{bmatrix}.$$

Solving for x and y by inversion, we obtain

$$\begin{bmatrix} x \\ y \end{bmatrix} = \begin{bmatrix} -1 \\ 0 \end{bmatrix} + z \begin{bmatrix} 1 \\ -2 \end{bmatrix}.$$

Hence, the solution is $x = -1 + z$, $y = -2z$, z is arbitrary. Note that there were other 2×2 submatrices of $\mathbf{A}$ that could have been used, for example

$$\begin{bmatrix} 5 & 8 \\ 6 & 9 \end{bmatrix}.$$

If we had used this one instead of the submatrix originally chosen, we would have solved for y and z in terms of x, and x would have become the arbitrary unknown. This would have been as valid a solution as the one we obtained. In fact, the reader can check that the two solutions are equivalent.

EXAMPLE 2 Find all solutions of the system

$$\begin{aligned} 2x - y - 3z + w &= 0, \\ x + y + z + w &= 1, \\ 2x - 7y - 13z - w &= -4. \end{aligned}$$

Solution

$$\mathbf{A} = \begin{bmatrix} 2 & -1 & 3 & 1 \\ 1 & 1 & 1 & 1 \\ 2 & -7 & -13 & -1 \end{bmatrix}.$$

Here, $r(\mathbf{A}) = r(\mathbf{A}^{\mathbf{b}}) = 2$, which can be obtained by considering the submatrix

$$\begin{bmatrix} 2 & 1 \\ 1 & 1 \end{bmatrix}$$

determined by deleting the third row and the second and third columns of $\mathbf{A}$. Therefore, we consider only the first two equations and rewrite them as

$$\begin{aligned} 2x + w &= 0 + y + 3z, \\ x + w &= 1 - y - \ z, \end{aligned}$$

or in matrix form as

$$\begin{bmatrix} 2 & 1 \\ 1 & 1 \end{bmatrix} \begin{bmatrix} x \\ w \end{bmatrix} = \begin{bmatrix} (y + 3z) \\ (1 - y - z) \end{bmatrix}.$$

Solving for x and w by inversion, we obtain

$$\begin{bmatrix} x \\ w \end{bmatrix} = \begin{bmatrix} -1 \\ 2 \end{bmatrix} + y \begin{bmatrix} 2 \\ -3 \end{bmatrix} + z \begin{bmatrix} 4 \\ -5 \end{bmatrix}.$$

The solution is, therefore, $x = -1 + 2y + 4z$, $w = 2 - 3y - 5z$, y and z are arbitrary.

EXAMPLE 3 Find all solutions of the system

$$\begin{aligned} x + \ y + \ w &= 2, \\ 2x + 2y + 2w &= 4, \\ x - \ y + 2w &= 0, \\ 3x + \ y + 4w &= 4. \end{aligned}$$

Solution

$$\mathbf{A} = \begin{bmatrix} 1 & 1 & 1 \\ 2 & 2 & 2 \\ 1 & -1 & 2 \\ 3 & 1 & 4 \end{bmatrix}.$$

Here $r(\mathbf{A}) = r(\mathbf{A}^{\mathbf{b}}) = 2$, which can be obtained by considering the submatrix

$$\begin{bmatrix} 1 & 1 \\ 1 & 2 \end{bmatrix}$$

determined by deleting the second and fourth rows and the second column. Therefore, we consider only the first and third equations and rewrite them as

$$\begin{aligned} x + \quad w &= 2 - y, \\ x + 2w &= 0 + y, \end{aligned}$$

or, in matrix form, as

$$\begin{bmatrix} 1 & 1 \\ 1 & 2 \end{bmatrix} \begin{bmatrix} x \\ w \end{bmatrix} = \begin{bmatrix} (2 - y) \\ y \end{bmatrix}.$$

Solving for x and w by inversion, we obtain

$$\begin{bmatrix} x \\ w \end{bmatrix} = \begin{bmatrix} 4 \\ -2 \end{bmatrix} + y \begin{bmatrix} -3 \\ 2 \end{bmatrix}.$$

Thus, $x = 4 - 3y$, $w = -2 + 2y$, y is arbitrary.

PROBLEMS 4.7

Use matrix methods to find the solutions of the following consistent systems. Do not check for consistency.

1. Problem 1, Section 4.6

2. Problem 3, Section 4.6

3. Problem 4, Section 4.6

4. Problem 5, Section 4.6

5.
$$\begin{aligned} x + \quad y - \quad z &= 2, \\ 2x + 2y - 2z &= 4, \\ x - \quad y + \quad z &= 3. \end{aligned}$$

6.
$$\begin{aligned} x + \quad y + 2z - \quad w &= 4, \\ 2x + 2y + 4z - 2w &= 8, \\ 2x - 2y + \quad z - \quad w &= 1, \\ 4x - 4y + 2z - 2w &= 2. \end{aligned}$$

4.8 HOMOGENEOUS SYSTEMS

A homogeneous system of simultaneous linear equations has the form

$$\begin{aligned} a_{11}x_1 + a_{12}x_2 + \cdots + a_{1n}x_n &= 0, \\ a_{21}x_1 + a_{22}x_2 + \cdots + a_{2n}x_n &= 0, \\ &\ \vdots \\ a_{m1}x_1 + a_{m2}x_2 + \cdots + a_{mn}x_n &= 0, \end{aligned} \tag{12}$$

or the matrix form

$$\mathbf{Ax} = \mathbf{0}. \tag{13}$$

Since Eq. (13) is a special case of Eq. (2), (that is, $\mathbf{b} = \mathbf{0}$), the theory developed in Sections 4.6 and 4.7 remains valid. Because of the simplified structure of a homogeneous system, however, one can draw conclusions about it that are not valid for a nonhomogeneous system. For instance, a homogeneous system is always consistent. To verify this statement, note that $x_1 = x_2 = \cdots = x_n = 0$ is always a solution to Eq. (12). Such a solution is called the *trivial solution*. It is, in general, the *nontrivial solutions* (solutions in which one or more of the unknowns is different from zero) that are of the greatest interest.

Consider Theorem 2 of Section 4.6 for the system $\mathbf{Ax} = \mathbf{0}$. If the rank of $\mathbf{A}$ is less than n (n being the number of unknowns), then the solution will be in terms of arbitrary unknowns. Since these arbitrary unknowns can be assigned nonzero values, it follows that nontrivial solutions exist. On the other hand, if the rank of $\mathbf{A}$ equals n, then the solution will be unique, and, hence, must be the trivial solution (why?). Thus, it follows that:

Theorem 1 The homogeneous system (12) will admit nontrivial solutions if and only if $r(\mathbf{A}) \neq n$. In particular, if the homogeneous system is square (that is, if $m = n$), then it will admit nontrivial solutions if and only if $\det(\mathbf{A}) = 0$.

Theorem 1 provides us with a new method for solving some old problems.

EXAMPLE 1 Is the set

$$\left\{ \begin{bmatrix} 1 \\ 2 \end{bmatrix}, \begin{bmatrix} 5 \\ 7 \end{bmatrix}, \begin{bmatrix} -3 \\ 1 \end{bmatrix} \right\}$$

linearly independent?

Solution The vector equation

$$c_1 \begin{bmatrix} 1 \\ 2 \end{bmatrix} + c_2 \begin{bmatrix} 5 \\ 7 \end{bmatrix} + c_3 \begin{bmatrix} -3 \\ 1 \end{bmatrix} = \begin{bmatrix} 0 \\ 0 \end{bmatrix}$$

is equivalent to the homogeneous system

$$\begin{aligned} c_1 + 5c_2 - 3c_3 &= 0, \\ 2c_1 + 7c_2 + c_3 &= 0. \end{aligned}$$

We are seeking the values c_1, c_2, c_3 that satisfy this system. Since the rank of the coefficient matrix is less than 3, Theorem 1 implies that nontrivial solutions exist. This, in turn, implies that the vectors are linearly dependent. Note that although we did not specifically solve for the c's, we are still able to deduce that a nontrivial set exists, hence that the vectors are linearly dependent.

EXAMPLE 2 Find a value for x such that the vectors $(1\ 3\ 5)$, $(2\ -1\ 3)$, and $(4\ x\ 1)$ are linearly dependent.

Solution The vector equation

$$c_1(1\ 3\ 5) + c_2(2\ -1\ 3) + c_3(4\ x\ 1) = (0\ 0\ 0)$$

is equivalent to the system

$$\begin{aligned} c_1 + 2c_2 + 4c_3 &= 0, \\ 3c_1 - c_2 + xc_3 &= 0, \\ 5c_1 + 3c_2 + c_3 &= 0. \end{aligned} \tag{14}$$

The vectors will be linearly dependent if there exists a nontrivial set of c's such that (14) is satisfied. Since (14) is a square system, we have from Theorem 1 that a nontrivial set of c's will exist if and only if the determinant of the coefficient matrix is zero. The coefficient matrix is

$$\mathbf{A} = \begin{bmatrix} 1 & 2 & 4 \\ 3 & -1 & x \\ 5 & 3 & 1 \end{bmatrix},$$

which has a determinant of $7x + 49$. Thus, the determinant will be zero, and hence, the vectors will be linearly dependent, if and only if $x = -7$.

We conclude this chapter with one of the more important theorems of linear algebra.

Theorem 2 In an n-dimensional vector space, every set of $n + 1$ vectors is linearly dependent.

Proof This proof is a generalization of the solution of Example 1 and is left as an exercise for the student (see Problem 6).

PROBLEMS 4.8

Without specifically solving for the constants c_1, c_2, c_3, determine whether or not the vectors given in Problems 1 and 2 are linearly dependent.

1. $(1\ 3\ 2)$, $(-2\ 1\ -1)$, $(0\ 7\ 3)$

2. $(1\ 4\ 1\ 1)$, $(1\ 3\ -1\ 1)$, $(1\ 0\ 1\ 1)$.

In Problems 3–5, find values for x such that the given vectors will be linearly dependent.

3. $(1\ 2\ 3)$, $(2\ x\ 2)$, $(3\ 6\ 9)$

4. $(1\ 1\ 1)$, $(2\ 0\ 1)$, $(x\ 1\ 0)$

5. $(2\ 1\ 1\ 1)$, $(3\ -2\ 1\ 0)$, $(x\ -1\ 2\ 0)$.

6. Prove Theorem 2.

Chapter 5 EIGENVALUES AND EIGENVECTORS

5.1 DEFINITIONS

Consider the matrix **A** and the vectors $\mathbf{x}_1$, $\mathbf{x}_2$, $\mathbf{x}_3$ that are given by

$$\mathbf{A} = \begin{bmatrix} 1 & 4 & -1 \\ 0 & 2 & 1 \\ 0 & 0 & 3 \end{bmatrix}, \qquad \mathbf{x}_1 = \begin{bmatrix} 4 \\ 1 \\ 0 \end{bmatrix}, \qquad \mathbf{x}_2 = \begin{bmatrix} 3 \\ 2 \\ 2 \end{bmatrix}, \qquad \mathbf{x}_3 = \begin{bmatrix} 3 \\ 0 \\ 0 \end{bmatrix}.$$

If we form the products $\mathbf{Ax}_1$, $\mathbf{Ax}_2$, and $\mathbf{Ax}_3$, we obtain

$$\mathbf{Ax}_1 = \begin{bmatrix} 8 \\ 2 \\ 0 \end{bmatrix}, \qquad \mathbf{Ax}_2 = \begin{bmatrix} 9 \\ 6 \\ 6 \end{bmatrix}, \qquad \mathbf{Ax}_3 = \begin{bmatrix} 3 \\ 0 \\ 0 \end{bmatrix}.$$

But

$$\begin{bmatrix} 8 \\ 2 \\ 0 \end{bmatrix} = 2\mathbf{x}_1, \quad \begin{bmatrix} 9 \\ 6 \\ 6 \end{bmatrix} = 3\mathbf{x}_2, \quad \text{and} \begin{bmatrix} 3 \\ 0 \\ 0 \end{bmatrix} = 1\mathbf{x}_3;$$

hence,

$$\begin{aligned} \mathbf{Ax}_1 &= 2\mathbf{x}_1, \\ \mathbf{Ax}_2 &= 3\mathbf{x}_2, \\ \mathbf{Ax}_3 &= 1\mathbf{x}_3. \end{aligned}$$

That is, multiplying $\mathbf{A}$ by any one of the vectors $\mathbf{x}_1$, $\mathbf{x}_2$, or $\mathbf{x}_3$ is equivalent to simply multiplying the vector by a suitable scalar.

Definition 1 A nonzero vector $\mathbf{x}$ is an *eigenvector* (or characteristic vector) of a square matrix $\mathbf{A}$ if there exists a scalar λ such that $\mathbf{Ax} = \lambda\mathbf{x}$. Then λ is an *eigenvalue* (or characteristic value) of $\mathbf{A}$.

Thus, in the above example, $\mathbf{x}_1$, $\mathbf{x}_2$, and $\mathbf{x}_3$ are eigenvectors of $\mathbf{A}$ and 2, 3, 1 are eigenvalues of $\mathbf{A}$.

Note that eigenvectors and eigenvalues are only defined for square matrices. Furthermore, note that the zero vector can *not* be an eigenvector even though $\mathbf{A} \cdot \mathbf{0} = \lambda \cdot \mathbf{0}$ for every scalar λ. An eigenvalue, however, can be zero.

EXAMPLE 1 Show that

$$\mathbf{x} = \begin{bmatrix} 5 \\ 0 \\ 0 \end{bmatrix}$$

is an eigenvector of

$$\mathbf{A} = \begin{bmatrix} 0 & 5 & 7 \\ 0 & -1 & 2 \\ 0 & 3 & 1 \end{bmatrix}.$$

Solution

$$\mathbf{Ax} = \begin{bmatrix} 0 & 5 & 7 \\ 0 & -1 & 2 \\ 0 & 3 & 1 \end{bmatrix} \begin{bmatrix} 5 \\ 0 \\ 0 \end{bmatrix} = \begin{bmatrix} 0 \\ 0 \\ 0 \end{bmatrix} = 0 \begin{bmatrix} 5 \\ 0 \\ 0 \end{bmatrix}.$$

Thus, $\mathbf{x}$ is an eigenvector of $\mathbf{A}$ and $\lambda = 0$ is an eigenvalue.

EXAMPLE 2 Is

$$\mathbf{x} = \begin{bmatrix} 1 \\ 1 \end{bmatrix}$$

an eigenvector of

$$\mathbf{A} = \begin{bmatrix} 1 & 2 \\ 3 & 4 \end{bmatrix}?$$

Solution

$$\mathbf{Ax} = \begin{bmatrix} 1 & 2 \\ 3 & 4 \end{bmatrix}\begin{bmatrix} 1 \\ 1 \end{bmatrix} = \begin{bmatrix} 3 \\ 7 \end{bmatrix}.$$

Thus, if $\mathbf{x}$ is to be an eigenvector of $\mathbf{A}$, there must exist a scalar λ such that $\mathbf{Ax} = \lambda\mathbf{x}$, or such that

$$\begin{bmatrix} 3 \\ 7 \end{bmatrix} = \lambda\begin{bmatrix} 1 \\ 1 \end{bmatrix} = \begin{bmatrix} \lambda \\ \lambda \end{bmatrix}.$$

It is quickly verified that no such λ exists, hence $\mathbf{x}$ is not an eigenvector of $\mathbf{A}$.

PROBLEMS 5.1

1. Determine which of the following vectors are eigenvectors for

$$\mathbf{A} = \begin{bmatrix} 1 & 3 & 0 & 0 \\ 1 & -1 & 0 & 0 \\ 0 & 0 & 1 & 2 \\ 0 & 0 & 4 & 3 \end{bmatrix}.$$

(a) $\begin{bmatrix} 1 \\ -1 \\ 0 \\ 0 \end{bmatrix}$ (b) $\begin{bmatrix} 0 \\ 0 \\ 1 \\ -1 \end{bmatrix}$ (c) $\begin{bmatrix} 1 \\ 0 \\ 0 \\ -1 \end{bmatrix}$ (d) $\begin{bmatrix} 3 \\ 1 \\ 0 \\ 0 \end{bmatrix}$

(e) $\begin{bmatrix} 0 \\ 0 \\ 0 \\ 0 \end{bmatrix}$ (f) $\begin{bmatrix} 1 \\ 1 \\ 0 \\ 0 \end{bmatrix}$.

2. What are the eigenvalues that correspond to the eigenvectors found in Problem 1?

5.2 EIGENVALUES

Let $\mathbf{x}$ be an eigenvector of the matrix $\mathbf{A}$. Then there must exist an eigenvalue λ such that

$$\mathbf{Ax} = \lambda\mathbf{x} \tag{1}$$

or, equivalently,

$$\mathbf{Ax} - \lambda\mathbf{x} = \mathbf{0}$$

or

$$(\mathbf{A} - \lambda\mathbf{I})\mathbf{x} = \mathbf{0}. \tag{2}$$

Caution: We could not have written (2) as $(\mathbf{A} - \lambda)\mathbf{x} = \mathbf{0}$ since the term $\mathbf{A} - \lambda$ would require subtracting a scalar from a matrix, an operation which is not defined. The quantity $\mathbf{A} - \lambda\mathbf{I}$, however, is defined since we are now subtracting one matrix from another.

Define a new matrix

$$\mathbf{B} = \mathbf{A} - \lambda\mathbf{I}. \tag{3}$$

Then (2) may be rewritten as

$$\mathbf{Bx} = \mathbf{0}, \tag{4}$$

which is a linear homogeneous system of equations for the unknown $\mathbf{x}$. If $\det(\mathbf{B}) \neq 0$, then we have that $\mathbf{x} = \mathbf{B}^{-1}\mathbf{0}$; consequently, $\mathbf{x} = \mathbf{0}$. This result, however, is absurd since $\mathbf{x}$ is an eigenvector and can not be zero. Thus, it follows (see Section 4.8) that $\mathbf{x}$ will be an eigenvector of $\mathbf{A}$ if and only if $\det(\mathbf{B}) = 0$, or, by (3), if and only if

$$\det(\mathbf{A} - \lambda\mathbf{I}) = 0. \tag{5}$$

Equation (5) is called the *characteristic equation of* $\mathbf{A}$. The roots of (5) determine the eigenvalues of $\mathbf{A}$.

EXAMPLE 1 Find the eigenvalues of

$$\mathbf{A} = \begin{bmatrix} 1 & 2 \\ 4 & 3 \end{bmatrix}.$$

Solution

$$\mathbf{A} - \lambda\mathbf{I} = \begin{bmatrix} 1 & 2 \\ 4 & 3 \end{bmatrix} - \lambda\begin{bmatrix} 1 & 0 \\ 0 & 1 \end{bmatrix} = \begin{bmatrix} 1 & 2 \\ 4 & 3 \end{bmatrix} - \begin{bmatrix} \lambda & 0 \\ 0 & \lambda \end{bmatrix}$$
$$= \begin{bmatrix} 1-\lambda & 2 \\ 4 & 3-\lambda \end{bmatrix}.$$

$\det(\mathbf{A} - \lambda\mathbf{I}) = (1 - \lambda)(3 - \lambda) - 8 = \lambda^2 - 4\lambda - 5$. The characteristic equation of $\mathbf{A}$ is $\det(\mathbf{A} - \lambda\mathbf{I}) = 0$, or $\lambda^2 - 4\lambda - 5 = 0$. Solving for λ, we have that $\lambda = -1, 5$; hence the eigenvalues of $\mathbf{A}$ are $\lambda_1 = -1$, $\lambda_2 = 5$.

EXAMPLE 2 Find the eigenvalues of

$$\mathbf{A} = \begin{bmatrix} 1 & -2 \\ 1 & 1 \end{bmatrix}.$$

Solution

$$\mathbf{A} - \lambda\mathbf{I} = \begin{bmatrix} 1 & -2 \\ 1 & 1 \end{bmatrix} - \lambda \begin{bmatrix} 1 & 0 \\ 0 & 1 \end{bmatrix} = \begin{bmatrix} 1-\lambda & -2 \\ 1 & 1-\lambda \end{bmatrix},$$

$$\det(\mathbf{A} - \lambda\mathbf{I}) = (1-\lambda)(1-\lambda) + 2 = \lambda^2 - 2\lambda + 3.$$

The characteristic equation is $\lambda^2 - 2\lambda + 3 = 0$; hence, solving for λ by the quadratic formula, we have that $\lambda_1 = 1 + \sqrt{2}i$, $\lambda_2 = 1 - \sqrt{2}i$ which are eigenvalues of **A**.

Note: Even if the elements of a matrix are real, the eigenvalues may be complex.

EXAMPLE 3 Find the eigenvalues of

$$\mathbf{A} = \begin{bmatrix} t & 2t \\ 2t & -t \end{bmatrix}.$$

Solution

$$\mathbf{A} - \lambda\mathbf{I} = \begin{bmatrix} t & 2t \\ 2t & -t \end{bmatrix} - \lambda \begin{bmatrix} 1 & 0 \\ 0 & 1 \end{bmatrix} = \begin{bmatrix} t-\lambda & 2t \\ 2t & -t-\lambda \end{bmatrix}$$

$$\det(\mathbf{A} - \lambda\mathbf{I}) = (t-\lambda)(-t-\lambda) - 4t^2 = \lambda^2 - 5t^2.$$

The characteristic equation is $\lambda^2 - 5t^2 = 0$, hence, the eigenvalues are $\lambda_1 = \sqrt{5}t$, $\lambda_2 = -\sqrt{5}t$.

Note: If the matrix **A** depends on a parameter (in this case the parameter is t), then the eigenvalues may also depend on the parameter.

EXAMPLE 4 Find the eigenvalues for

$$\mathbf{A} = \begin{bmatrix} 2 & -1 & 1 \\ 3 & -2 & 1 \\ 0 & 0 & 1 \end{bmatrix}.$$

Solution

$$\mathbf{A} - \lambda\mathbf{I} = \begin{bmatrix} 2 & -1 & 1 \\ 3 & -2 & 1 \\ 0 & 0 & 1 \end{bmatrix} - \lambda \begin{bmatrix} 1 & 0 & 0 \\ 0 & 1 & 0 \\ 0 & 0 & 1 \end{bmatrix} = \begin{bmatrix} 2-\lambda & -1 & 1 \\ 3 & -2-\lambda & 1 \\ 0 & 0 & 1-\lambda \end{bmatrix}.$$

$$\det(\mathbf{A} - \lambda\mathbf{I}) = (1-\lambda)[(2-\lambda)(-2-\lambda) + 3] = (1-\lambda)(\lambda^2 - 1).$$

The characteristic equation is $(1-\lambda)(\lambda^2-1)=0$; hence, the eigenvalues are $\lambda_1=\lambda_2=1$, $\lambda_3=-1$.

Note: The roots of the characteristic equation can be repeated. That is, $\lambda_1=\lambda_2=\lambda_3=\cdots=\lambda_k$. When this happens, the eigenvalue is said to be of *multiplicity k*. Thus, in Example 4, $\lambda=1$ is an eigenvalue of multiplicity 2 while, $\lambda=-1$ is an eigenvalue of multiplicity 1.

From the definition of the characteristic equation (5), it can be shown that if **A** is an $n \times n$ matrix then the characteristic equation of **A** is an nth degree polynomial in λ.[9] It follows from the fundamental theorem of algebra, that the characteristic equation has n roots, counting multiplicity. Hence, **A** has exactly n eigenvalues, counting multiplicity. (See Examples 1 and 4).

Finally, the student should be aware of the fact that, in general, it is very difficult to find the eigenvalues of a matrix. First the characteristic equation must be obtained. For matrices of high order this in itself is a lengthy task. Once the characteristic equation is determined, it must be solved for its roots. If the equation is of high order, this can be an impossibility in practice. For example, the reader is invited to find the eigenvalues of

$$\mathbf{A}=\begin{bmatrix}1 & 3 & 2 & -1\\ 1 & 1 & 2 & -3\\ 3 & 1 & 1 & -1\\ 2 & -2 & 1 & 2\end{bmatrix}.$$

For these reasons, eigenvalues are rarely found by the method just given, and numerical techniques are used to obtain approximate values.

PROBLEMS 5.2

Find the eigenvalues of the following matrices:

1. $$\mathbf{A}=\begin{bmatrix}1 & 0 & 0\\ 0 & 1 & 0\\ 0 & 0 & 0\end{bmatrix}$$

2. $$\mathbf{A}=\begin{bmatrix}3 & 5\\ -5 & -3\end{bmatrix}$$

3. $$\mathbf{A}=\begin{bmatrix}2 & 5\\ -1 & -2\end{bmatrix}$$

4. $$\mathbf{A}=\begin{bmatrix}7t & 4t\\ 2t & 5t\end{bmatrix}$$

5. $$\mathbf{A}=\begin{bmatrix}1 & 1 & -1\\ 0 & 0 & 0\\ 1 & 2 & 3\end{bmatrix}$$

6. $$\mathbf{A}=\begin{bmatrix}1 & -1 & 0 & 0\\ 3 & 5 & 0 & 0\\ 0 & 0 & 1 & 5\\ 0 & 0 & -1 & 1\end{bmatrix}.$$

[9] Henceforth, we will agree to denote the characteristic equation of an $n \times n$ matrix **A** as either $\det(\mathbf{A}-\lambda\mathbf{I})=0$, or $-\det(\mathbf{A}-\lambda\mathbf{I})=0$, depending whether n is even or odd. In this way, we will guarantee that the coefficient of λ^n in the characteristic equation is always $+1$.

5.3 EIGENVECTORS

To each distinct eigenvalue of a matrix **A** there will correspond at least one eigenvector which can be found by solving the appropriate set of homogeneous equations. If an eigenvalue λ_i is substituted in (2), the corresponding eigenvector $\mathbf{x}_i$ is the solution of

$$(\mathbf{A} - \lambda_i \mathbf{I})\mathbf{x}_i = \mathbf{0}. \tag{6}$$

EXAMPLE 1 Find the eigenvectors of

$$\mathbf{A} = \begin{bmatrix} 1 & 2 \\ 4 & 3 \end{bmatrix}.$$

Solution The eigenvalues of **A** have already been found to be $\lambda_1 = -1$, $\lambda_2 = 5$ (see Example 1 of Section 5.2). We first calculate the eigenvector corresponding to λ_1. From (6),

$$(\mathbf{A} - (-1)\mathbf{I})\mathbf{x}_1 = \mathbf{0}. \tag{7}$$

If we designate the unknown vector $\mathbf{x}_1$ by

$$\begin{bmatrix} x_1 \\ y_1 \end{bmatrix},$$

Eq. (7) becomes

$$\left\{\begin{bmatrix} 1 & 2 \\ 4 & 3 \end{bmatrix} + \begin{bmatrix} 1 & 0 \\ 0 & 1 \end{bmatrix}\right\}\begin{bmatrix} x_1 \\ y_1 \end{bmatrix} = \begin{bmatrix} 0 \\ 0 \end{bmatrix}$$

or

$$\begin{bmatrix} 2 & 2 \\ 4 & 4 \end{bmatrix}\begin{bmatrix} x_1 \\ y_1 \end{bmatrix} = \begin{bmatrix} 0 \\ 0 \end{bmatrix}$$

or, equivalently,

$$\begin{aligned} 2x_1 + 2y_1 &= 0, \\ 4x_1 + 4y_1 &= 0. \end{aligned}$$

A nontrivial solution to this set of equations is $x_1 = -y_1$, y_1 arbitrary; hence, the eigenvector is

$$\mathbf{x}_1 = \begin{bmatrix} x_1 \\ y_1 \end{bmatrix} = \begin{bmatrix} -y_1 \\ y_1 \end{bmatrix} = y_1\begin{bmatrix} -1 \\ 1 \end{bmatrix}, \qquad y_1 \text{ arbitrary}.$$

By choosing different values of y_1, different eigenvectors for $\lambda_1 = -1$ can be obtained. Note, however, that any two such eigenvectors would be scalar multiples of each other, hence linearly dependent. Thus, there is only one linearly independent eigenvector corresponding to $\lambda_1 = -1$. For convenience we choose $y_1 = 1$, which gives us the eigenvector

$$\mathbf{x}_1 = \begin{bmatrix} -1 \\ 1 \end{bmatrix}.$$

Many times, however, the scalar y_1 is chosen in such a manner that the resulting eigenvector becomes a unit vector. If we wished to achieve this result for the above vector, we would have to choose $y_1 = 1/\sqrt{2}$.

Having found the eigenvector corresponding to $\lambda_1 = -1$, we now proceed to find the eigenvector $\mathbf{x}_2$ corresponding to $\lambda_2 = 5$. Designating the unknown vector $\mathbf{x}_2$ by

$$\begin{bmatrix} x_2 \\ y_2 \end{bmatrix}$$

and substituting it with λ_2 into (6), we obtain

$$\left\{\begin{bmatrix} 1 & 2 \\ 4 & 3 \end{bmatrix} - 5\begin{bmatrix} 1 & 0 \\ 0 & 1 \end{bmatrix}\right\}\begin{bmatrix} x_2 \\ y_2 \end{bmatrix} = \begin{bmatrix} 0 \\ 0 \end{bmatrix},$$

or

$$\begin{bmatrix} -4 & 2 \\ 4 & -2 \end{bmatrix}\begin{bmatrix} x_2 \\ y_2 \end{bmatrix} = \begin{bmatrix} 0 \\ 0 \end{bmatrix},$$

or, equivalently,

$$\begin{aligned} -4x_2 + 2y_2 &= 0, \\ 4x_2 - 2y_2 &= 0. \end{aligned}$$

A nontrivial solution to this set of equations is $x_2 = \frac{1}{2}y_2$, where y_2 is arbitrary; hence

$$\mathbf{x}_2 = \begin{bmatrix} x_2 \\ y_2 \end{bmatrix} = \begin{bmatrix} y_2/2 \\ y_2 \end{bmatrix} = y_2\begin{bmatrix} \frac{1}{2} \\ 1 \end{bmatrix}.$$

For convenience, we choose $y_2 = 2$, thus

$$\mathbf{x}_2 = \begin{bmatrix} 1 \\ 2 \end{bmatrix}.$$

In order to check whether or not $\mathbf{x}_2$ is an eigenvector corresponding to $\lambda_2 = 5$, we need only check if $\mathbf{A}\mathbf{x}_2 = \lambda_2 \mathbf{x}_2$:

$$\mathbf{A}\mathbf{x}_2 = \begin{bmatrix} 1 & 2 \\ 4 & 3 \end{bmatrix}\begin{bmatrix} 1 \\ 2 \end{bmatrix} = \begin{bmatrix} 5 \\ 10 \end{bmatrix} = 5\begin{bmatrix} 1 \\ 2 \end{bmatrix} = \lambda_2 \mathbf{x}_2.$$

Again note that $\mathbf{x}_2$ is *not* unique! Any scalar multiple of $\mathbf{x}_2$ is also an eigenvector corresponding to λ_2. However, in this case, there is just one *linearly independent* eigenvector corresponding to λ_2.

EXAMPLE 2 Find the eigenvectors of

$$\mathbf{A} = \begin{bmatrix} 2 & 0 & 0 \\ 0 & 2 & 5 \\ 0 & -1 & -2 \end{bmatrix}.$$

Solution By using the method of the previous section, we find the eigenvalues to be $\lambda_1 = 2$, $\lambda_2 = i$, $\lambda_3 = -i$. We first calculate the eigenvector corresponding to $\lambda_1 = 2$. Designate $\mathbf{x}_1$ by

$$\begin{bmatrix} x_1 \\ y_1 \\ z_1 \end{bmatrix}.$$

Then (6) becomes

$$\left\{\begin{bmatrix} 2 & 0 & 0 \\ 0 & 2 & 5 \\ 0 & -1 & -2 \end{bmatrix} - 2\begin{bmatrix} 1 & 0 & 0 \\ 0 & 1 & 0 \\ 0 & 0 & 1 \end{bmatrix}\right\}\begin{bmatrix} x_1 \\ y_1 \\ z_1 \end{bmatrix} = \begin{bmatrix} 0 \\ 0 \\ 0 \end{bmatrix},$$

or

$$\begin{bmatrix} 0 & 0 & 0 \\ 0 & 0 & 5 \\ 0 & -1 & -4 \end{bmatrix}\begin{bmatrix} x_1 \\ y_1 \\ z_1 \end{bmatrix} = \begin{bmatrix} 0 \\ 0 \\ 0 \end{bmatrix},$$

or, equivalently,

$$\begin{aligned} 0 &= 0, \\ 5z_1 &= 0, \\ -y_1 - 4z_1 &= 0. \end{aligned}$$

A nontrivial solution to this set of equations is $y_1 = z_1 = 0$, x_1 arbitrary; hence

$$\mathbf{x}_1 = \begin{bmatrix} x_1 \\ y_1 \\ z_1 \end{bmatrix} = \begin{bmatrix} x_1 \\ 0 \\ 0 \end{bmatrix} = x_1\begin{bmatrix} 1 \\ 0 \\ 0 \end{bmatrix}.$$

We now find the eigenvector corresponding to $\lambda_2 = i$. If we designate $\mathbf{x}_2$ by

$$\begin{bmatrix} x_2 \\ y_2 \\ z_2 \end{bmatrix},$$

Eq. (6) becomes

$$\begin{bmatrix} 2-i & 0 & 0 \\ 0 & 2-i & 5 \\ 0 & -1 & -2-i \end{bmatrix}\begin{bmatrix} x_2 \\ y_2 \\ z_2 \end{bmatrix} = \begin{bmatrix} 0 \\ 0 \\ 0 \end{bmatrix}$$

or

$$(2-i)x_2 = 0,$$
$$(2-i)y_2 + 5z_2 = 0,$$
$$-y_2 + (-2-i)z_2 = 0.$$

A nontrivial solution to this set of equations is $x_2 = 0$, $y_2 = (-2-i)z_2$, z_2 arbitrary; hence,

$$\mathbf{x}_2 = \begin{bmatrix} x_2 \\ y_2 \\ z_2 \end{bmatrix} = \begin{bmatrix} 0 \\ (-2-i)z_2 \\ z_2 \end{bmatrix} = z_2 \begin{bmatrix} 0 \\ -2-i \\ 1 \end{bmatrix}.$$

The eigenvector corresponding to $\lambda_3 = -i$ is found in a similar manner to be

$$\mathbf{x}_3 = z_3 \begin{bmatrix} 0 \\ -2+i \\ 1 \end{bmatrix}.$$

It should be noted that even if a mistake is made in finding the eigenvalues of a matrix, the error will become apparent when the eigenvector corresponding to the incorrect eigenvalue is found. For instance, suppose that λ_1 in Example 2 was calculated erroneously to be 3. If we now try to find $\mathbf{x}_1$ we obtain the equations

$$-x_1 = 0,$$
$$-y_1 + 5z_1 = 0,$$
$$-y_1 - 5z_1 = 0.$$

The only solution to this set of equations is $x_1 = y_1 = z_1 = 0$, hence

$$\mathbf{x}_1 = \begin{bmatrix} 0 \\ 0 \\ 0 \end{bmatrix}.$$

However, by definition, an eigenvector can not be the zero vector. Since every eigenvalue must have a corresponding eigenvector, there is a mistake. A quick check shows that all the calculations above are valid, hence the error must lie in the value of the eigenvalue.

PROBLEMS 5.3

Find the eigenvalues and a set of corresponding eigenvectors for **A** if:

1. $\mathbf{A} = \begin{bmatrix} 2 & -1 \\ 3 & -2 \end{bmatrix}$

2. $\mathbf{A} = \begin{bmatrix} 2 & 1 \\ 3 & -2 \end{bmatrix}$

3. $$\mathbf{A} = \begin{bmatrix} 2t & t \\ t & 2t \end{bmatrix}$$

4. $$\mathbf{A} = \begin{bmatrix} 3 & 1 & 2 \\ 0 & 0 & 0 \\ 1 & 1 & 2 \end{bmatrix}$$

5. $$\mathbf{A} = \begin{bmatrix} 3 & 2 & 1 \\ 0 & 4 & 0 \\ 0 & 1 & 5 \end{bmatrix}$$

6. $$\mathbf{A} = \begin{bmatrix} 2 & 4 & 2 & -2 \\ 0 & 1 & 0 & 0 \\ 0 & 3 & 3 & -1 \\ 0 & 2 & 0 & 4 \end{bmatrix}.$$

5.4 PROPERTIES OF EIGENVALUES AND EIGENVECTORS

Definition 1 The *trace* of a matrix $\mathbf{A}$, designated by $\operatorname{tr}(\mathbf{A})$, is the sum of the elements on the main diagonal.

EXAMPLE 1 Find the $\operatorname{tr}(\mathbf{A})$ if

$$\mathbf{A} = \begin{bmatrix} 3 & -1 & 2 \\ 0 & 4 & 1 \\ 1 & -1 & -5 \end{bmatrix}.$$

Solution $\operatorname{tr}(\mathbf{A}) = 3 + 4 + (-5) = 2.$

Property 1 The sum of the eigenvalues of a matrix equals the trace of the matrix.

Although the proof of Property 1 is too involved to be presented here,[10] the property itself is quite simple and provides us with a quick and useful procedure for checking eigenvalues.

EXAMPLE 2 Verify Property 1 for

$$\mathbf{A} = \begin{bmatrix} 11 & 3 \\ -5 & -5 \end{bmatrix}.$$

Solution The eigenvalues of $\mathbf{A}$ are $\lambda_1 = 10$, $\lambda_2 = -4$.

$$\operatorname{tr}(\mathbf{A}) = 11 + (-5) = 6 = \lambda_1 + \lambda_2.$$

Property 2 A matrix is singular if and only if it has a zero eigenvalue.

[10] For a proof of Property 1, see F. R. Gantmacher, "The Theory of Matrices," p. 87. Chelsea, New York, 1960.

Proof A matrix $\mathbf{A}$ has a zero eigenvalue if and only if $\det(\mathbf{A} - 0\mathbf{I}) = 0$, or (since $0\mathbf{I} = \mathbf{0}$) if and only if $\det(\mathbf{A}) = 0$. But $\det(\mathbf{A}) = 0$ if and only if $\mathbf{A}$ is singular, thus, the result is immediate.

Property 3 The eigenvalues of an upper (or lower) triangular matrix are the elements on the main diagonal.

Proof See Problem 1.

EXAMPLE 3 Find the eigenvalues of

$$\mathbf{A} = \begin{bmatrix} 1 & 0 & 0 \\ 2 & 1 & 0 \\ 3 & 4 & -1 \end{bmatrix}.$$

Solution Since $\mathbf{A}$ is lower triangular, the eigenvalues must be $\lambda_1 = \lambda_2 = 1$, $\lambda_3 = -1$.

Property 4 If λ is an eigenvalue of $\mathbf{A}$ and if $\mathbf{A}$ is invertible, then $1/\lambda$ is an eigenvalue of $\mathbf{A}^{-1}$.

Proof Since $\mathbf{A}$ is invertible, Property 2 implies that $\lambda \neq 0$; hence $1/\lambda$ exists. Since λ is an eigenvalue of $\mathbf{A}$ there must exist an eigenvector $\mathbf{x}$ such that $\mathbf{Ax} = \lambda\mathbf{x}$. Premultiplying both sides of this equation by $\mathbf{A}^{-1}$, we obtain

$$\mathbf{x} = \lambda\mathbf{A}^{-1}\mathbf{x}$$

or, equivalently, $\mathbf{A}^{-1}\mathbf{x} = (1/\lambda)\mathbf{x}$. Thus, $1/\lambda$ is an eigenvalue of $\mathbf{A}^{-1}$.

Observation 1 If $\mathbf{x}$ is an eigenvector of $\mathbf{A}$ corresponding to the eigenvalue λ and if $\mathbf{A}$ is invertible, then $\mathbf{x}$ is an eigenvector of $\mathbf{A}^{-1}$ corresponding to the eigenvalue $1/\lambda$.

Property 5 If λ is an eigenvalue of $\mathbf{A}$, then $\alpha\lambda$ is an eigenvalue of $\alpha\mathbf{A}$ where α is any arbitrary scalar.

Proof If λ is an eigenvalue of $\mathbf{A}$, then there must exist an eigenvector $\mathbf{x}$ such that $\mathbf{Ax} = \lambda\mathbf{x}$. Multiplying both sides of this equation by α, we obtain $(\alpha\mathbf{A})\mathbf{x} = (\alpha\lambda)\mathbf{x}$ which implies Property 5.

Obervation 2 If $\mathbf{x}$ is an eigenvector of $\mathbf{A}$ corresponding to the eigenvalue λ, then $\mathbf{x}$ is an eigenvector of $\alpha\mathbf{A}$ corresponding to eigenvalue $\alpha\lambda$.

Property 6 If λ is an eigenvalue of $\mathbf{A}$, then λ is an eigenvalue of $\mathbf{A}'$.

Proof Since λ is an eigenvalue of $\mathbf{A}$, $\det(\mathbf{A} - \lambda\mathbf{I}) = 0$.
Hence

$$\begin{aligned} 0 = |\mathbf{A} - \lambda\mathbf{I}| &= |(\mathbf{A}')' - \lambda\mathbf{I}'| && \{\text{(Property 1, Section 1.6)} \\ &= |(\mathbf{A}' - \lambda\mathbf{I})'| && \{\text{(Property 3, Section 1.6)} \\ &= |\mathbf{A}' - \lambda\mathbf{I}| && \{\text{(Property 7, Section 2.3)} \end{aligned}$$

Thus, $\det(\mathbf{A}' - \lambda\mathbf{I}) = 0$, which implies that λ is an eigenvalue of $\mathbf{A}'$.

Property 7 The product of the eigenvalues (counting multiplicity) of a matrix equals the determinant of the matrix.[11]

EXAMPLE 4 Verify Property 7 for the matrix $\mathbf{A}$ given in Example 2.

Solution For this $\mathbf{A}$, $\lambda_1 = 10$, $\lambda_2 = -4$, $\det(\mathbf{A}) = -55 + 15 = -40 = \lambda_1\lambda_2$.

PROBLEMS 5.4

1. Prove Property 3.
2. Show, by example, that if $\mathbf{x}$ is an eigenvector of $\mathbf{A}$, it need not be an eigenvector of $\mathbf{A}'$.
3. Verify Property 2 for

$$\mathbf{A} = \begin{bmatrix} 12 & 16 \\ -3 & -7 \end{bmatrix}.$$

4. Verify Property 2 for

$$\mathbf{A} = \begin{bmatrix} 1 & 3 & 6 \\ -1 & 2 & -1 \\ 2 & 1 & 7 \end{bmatrix}.$$

5. Verify Properties 4–7 for the matrix given in Problem 3.

5.5 LINEARLY INDEPENDENT EIGENVECTORS

Since every eigenvalue has an infinite number of eigenvectors associated with it (recall that if $\mathbf{x}$ is an eigenvector, then any scalar multiple of $\mathbf{x}$ is also

[11] For a proof of Property 7, see F. R. Gantmacher, "The Theory of Matrices," p. 70. Chelsea, New York, 1960. This proof requires a knowledge of elementary symmetry functions which can be found in B. L. van der Waerden, "Modern Algebra," p. 78. Ungar, New York, 1953.

an eigenvector), it becomes academic to ask how many different eigenvectors can a matrix have? The answer is clearly an infinite number. A more revealing question is how many linearly independent eigenvectors can a matrix have? Theorem 2 of Section 4.8 provides us with a partial answer.

Theorem 1 In an n-dimensional vector space, every set of $n + 1$ vectors is linearly dependent.

Therefore, since all of the eigenvectors of an $n \times n$ matrix must be n-dimensional (why?), it follows from Theorem 1 that an $n \times n$ matrix can have *at most n* linearly independent eigenvectors. The following three examples shed more light on the subject.

EXAMPLE 1 Find the eigenvectors of

$$\mathbf{A} = \begin{bmatrix} 2 & 1 & 0 \\ 0 & 2 & 1 \\ 0 & 0 & 2 \end{bmatrix}.$$

Solution The eigenvalues of $\mathbf{A}$ are $\lambda_1 = \lambda_2 = \lambda_3 = 2$, therefore $\lambda = 2$ is an eigenvalue of multiplicity 3. If we designate the unknown eigenvector $\mathbf{x}$ by

$$\begin{bmatrix} x \\ y \\ z \end{bmatrix},$$

then Eq. (6) gives rise to the three equations

$$y = 0,$$
$$z = 0,$$
$$0 = 0.$$

Thus, $y = z = 0$ and x is arbitrary; hence

$$\mathbf{x} = \begin{bmatrix} x \\ y \\ z \end{bmatrix} = \begin{bmatrix} x \\ 0 \\ 0 \end{bmatrix} = x \begin{bmatrix} 1 \\ 0 \\ 0 \end{bmatrix}.$$

Setting $x = 1$, we see that $\lambda = 2$ generates only one linearly independent eigenvector,

$$\mathbf{x} = \begin{bmatrix} 1 \\ 0 \\ 0 \end{bmatrix}.$$

EXAMPLE 2 Find the eigenvectors of

$$\mathbf{A} = \begin{bmatrix} 2 & 1 & 0 \\ 0 & 2 & 0 \\ 0 & 0 & 2 \end{bmatrix}.$$

Solution Again, the eigenvalues are $\lambda_1 = \lambda_2 = \lambda_3 = 2$, therefore $\lambda = 2$ is an eigenvalue of multiplicity 3. Designate the unknown eigenvector $\mathbf{x}$ by

$$\begin{bmatrix} x \\ y \\ z \end{bmatrix}.$$

Equation (6) now gives rise to the equations

$$\begin{aligned} y &= 0, \\ 0 &= 0, \\ 0 &= 0. \end{aligned}$$

Thus, $y = 0$ and both x and z are arbitrary; hence

$$\mathbf{x} = \begin{bmatrix} x \\ y \\ z \end{bmatrix} = \begin{bmatrix} x \\ 0 \\ z \end{bmatrix} = \begin{bmatrix} x \\ 0 \\ 0 \end{bmatrix} + \begin{bmatrix} 0 \\ 0 \\ z \end{bmatrix} = x\begin{bmatrix} 1 \\ 0 \\ 0 \end{bmatrix} + z\begin{bmatrix} 0 \\ 0 \\ 1 \end{bmatrix}.$$

Since x and z can be chosen arbitrarily, we can first choose $x = 1$ and $z = 0$ to obtain

$$\mathbf{x}_1 = \begin{bmatrix} 1 \\ 0 \\ 0 \end{bmatrix}$$

and then choose $x = 0$ and $z = 1$ to obtain

$$\mathbf{x}_2 = \begin{bmatrix} 0 \\ 0 \\ 1 \end{bmatrix}.$$

$\mathbf{x}_1$ and $\mathbf{x}_2$ can easily be shown to be linearly independent vectors, hence we see that $\lambda = 2$ generates the two linearly independent eigenvectors

$$\begin{bmatrix} 1 \\ 0 \\ 0 \end{bmatrix} \quad \text{and} \quad \begin{bmatrix} 0 \\ 0 \\ 1 \end{bmatrix}.$$

EXAMPLE 3 Find the eigenvectors of

$$\mathbf{A} = \begin{bmatrix} 2 & 0 & 0 \\ 0 & 2 & 0 \\ 0 & 0 & 2 \end{bmatrix}.$$

Solution Again the eigenvalues are $\lambda_1 = \lambda_2 = \lambda_3 = 2$ so again $\lambda = 2$ is an eigenvalue of multiplicity three. Designate the unknown eigenvector $\mathbf{x}$ by

$$\begin{bmatrix} x \\ y \\ z \end{bmatrix}.$$

Equation (6) gives rise to the equations

$$\begin{aligned} 0 &= 0, \\ 0 &= 0, \\ 0 &= 0, \end{aligned}$$

Thus, x, y and z are all arbitrary; hence

$$\mathbf{x} = \begin{bmatrix} x \\ y \\ z \end{bmatrix} = \begin{bmatrix} x \\ 0 \\ 0 \end{bmatrix} + \begin{bmatrix} 0 \\ y \\ 0 \end{bmatrix} + \begin{bmatrix} 0 \\ 0 \\ z \end{bmatrix} = x\begin{bmatrix} 1 \\ 0 \\ 0 \end{bmatrix} + y\begin{bmatrix} 0 \\ 1 \\ 0 \end{bmatrix} + z\begin{bmatrix} 0 \\ 0 \\ 1 \end{bmatrix}.$$

Since x, y, and z can be chosen arbitrarily, we can first choose $x = 1, y = z = 0$, then choose $x = z = 0$, $y = 1$ and finally choose $y = x = 0$, $z = 1$ to generate the three linearly independent eigenvectors

$$\begin{bmatrix} 1 \\ 0 \\ 0 \end{bmatrix}, \quad \begin{bmatrix} 0 \\ 1 \\ 0 \end{bmatrix}, \quad \begin{bmatrix} 0 \\ 0 \\ 1 \end{bmatrix}.$$

In this case we see that three linearly independent eigenvectors are generated by $\lambda = 2$. (Note that, from Theorem 1, this is the maximal number that could be generated.)

The preceding examples are illustrations of

Theorem 2 If λ is an eigenvalue of multiplicity k of an $n \times n$ matrix $\mathbf{A}$, then the number of linearly independent eigenvectors of $\mathbf{A}$ associated with λ is given by $\rho = n - r(\mathbf{A} - \lambda\mathbf{I})$. Furthermore, $1 \leq \rho \leq k$.

Proof Let $\mathbf{x}$ be an n-dimensional vector. If $\mathbf{x}$ is an eigenvector, then it must satisfy the vector equation $\mathbf{Ax} = \lambda\mathbf{x}$ or, equivalently, $(\mathbf{A} - \lambda\mathbf{I})\mathbf{x} = \mathbf{0}$. This system is homogeneous, hence consistent, so by Theorem 2 of Section 4.6, we have that the solution vector $\mathbf{x}$ will be in terms of $n - r(\mathbf{A} - \lambda\mathbf{I})$ arbitrary unknowns. Since these unknowns can be picked independently of each other, it follows that the number of linearly independent eigenvectors of $\mathbf{A}$ associated with λ is also $\rho = n - r(\mathbf{A} - \lambda\mathbf{I})$. We defer a proof that $1 \leq \rho \leq k$ until Chapter 8.

In Example 1, $\mathbf{A}$ is 3×3; hence $n = 3$, and $r(\mathbf{A} - 2\mathbf{I}) = 2$. Thus, there should be $3 - 2 = 1$ linearly independent eigenvector associated with $\lambda = 2$

which is indeed the case. In Example 2, once again $n = 3$ but $r(\mathbf{A} - 2\mathbf{I}) = 1$. Thus, there should be $3 - 1 = 2$ linearly independent eigenvectors associated with $\lambda = 2$ which also is the case.

The next theorem gives the relationship between eigenvectors that correspond to different eigenvalues.

Theorem 3 Eigenvectors corresponding to distinct (that is, different) eigenvalues are linearly independent.

Proof For the sake of clarity, we consider the case of three distinct eigenvectors and leave the more general proof as an exercise (see Problem 4). Therefore, let λ_1, λ_2, λ_3, be distinct eigenvalues of the matrix $\mathbf{A}$ and let $\mathbf{x}_1$, $\mathbf{x}_2$, $\mathbf{x}_3$ be the associated eigenvectors. That is

$$\begin{aligned} \mathbf{A}\mathbf{x}_1 &= \lambda_1 \mathbf{x}_1, \\ \mathbf{A}\mathbf{x}_2 &= \lambda_2 \mathbf{x}_2, \\ \mathbf{A}\mathbf{x}_3 &= \lambda_3 \mathbf{x}_3, \end{aligned} \tag{8}$$

and $\lambda_1 \neq \lambda_2 \neq \lambda_3 \neq \lambda_1$.

Since we want to show that $\mathbf{x}_1$, $\mathbf{x}_2$, $\mathbf{x}_3$ are linearly independent, we must show that the only solution to

$$c_1\mathbf{x}_1 + c_2\mathbf{x}_2 + c_3\mathbf{x}_3 = \mathbf{0} \tag{9}$$

is $c_1 = c_2 = c_3 = 0$. By premultiplying (9) by $\mathbf{A}$, we obtain

$$c_1\mathbf{A}\mathbf{x}_1 + c_2\mathbf{A}\mathbf{x}_2 + c_3\mathbf{A}\mathbf{x}_3 = \mathbf{A} \cdot \mathbf{0} = \mathbf{0}.$$

It follows from (8), therefore, that

$$c_1\lambda_1\mathbf{x}_1 + c_2\lambda_2\mathbf{x}_2 + c_3\lambda_3\mathbf{x}_3 = \mathbf{0}. \tag{10}$$

By premultiplying (10) by $\mathbf{A}$ and again using Eq. (8), we obtain

$$c_1\lambda_1^2\mathbf{x}_1 + c_2\lambda_2^2\mathbf{x}_2 + c_3\lambda_3^2\mathbf{x}_3 = \mathbf{0}. \tag{11}$$

Equations (9)–(11) can be written in the matrix form

$$\begin{bmatrix} 1 & 1 & 1 \\ \lambda_1 & \lambda_2 & \lambda_3 \\ \lambda_1^2 & \lambda_2^2 & \lambda_3^2 \end{bmatrix} \begin{bmatrix} c_1\mathbf{x}_1 \\ c_2\mathbf{x}_2 \\ c_3\mathbf{x}_3 \end{bmatrix} = \begin{bmatrix} \mathbf{0} \\ \mathbf{0} \\ \mathbf{0} \end{bmatrix}.$$

Define

$$\mathbf{B} = \begin{bmatrix} 1 & 1 & 1 \\ \lambda_1 & \lambda_2 & \lambda_3 \\ \lambda_1^2 & \lambda_2^2 & \lambda_3^2 \end{bmatrix}.$$

It can be shown that $\det(\mathbf{B}) = (\lambda_2 - \lambda_1)(\lambda_3 - \lambda_2)(\lambda_3 - \lambda_1)$. Thus, since all the eigenvalues are distinct, $\det(\mathbf{B}) \neq 0$. Therefore,

$$\begin{bmatrix} c_1\mathbf{x}_1 \\ c_2\mathbf{x}_2 \\ c_3\mathbf{x}_3 \end{bmatrix} = \mathbf{B}^{-1}\begin{bmatrix} \mathbf{0} \\ \mathbf{0} \\ \mathbf{0} \end{bmatrix} = \begin{bmatrix} \mathbf{0} \\ \mathbf{0} \\ \mathbf{0} \end{bmatrix}$$

or

$$\begin{aligned} c_1\mathbf{x}_1 &= \mathbf{0} \\ c_2\mathbf{x}_2 &= \mathbf{0} \\ c_3\mathbf{x}_3 &= \mathbf{0} \end{aligned} \tag{12}$$

But since $\mathbf{x}_1$, $\mathbf{x}_2$, $\mathbf{x}_3$ are eigenvectors, they are nonzero, therefore, it follows from (12) that $c_1 = c_2 = c_3 = 0$. This result together with (9) implies Theorem 3.

Theorems 2 and 3 together completely determine the number of linearly independent eigenvectors of a matrix.

EXAMPLE 4 Find a set of linearly independent eigenvectors for

$$\mathbf{A} = \begin{bmatrix} 1 & 0 & 0 \\ 4 & 3 & 2 \\ 4 & 2 & 3 \end{bmatrix}.$$

Solution The eigenvalues of $\mathbf{A}$ are $\lambda_1 = \lambda_2 = 1$, and $\lambda_3 = 5$. For this matrix, $n = 3$ and $r(\mathbf{A} - 1\mathbf{I}) = 1$, hence $n - r(\mathbf{A} - 1\mathbf{I}) = 2$. Thus, from Theorem 2, we know that $\mathbf{A}$ has two linearly independent eigenvectors corresponding to $\lambda = 1$ and one linearly independent eigenvector corresponding to $\lambda = 5$ (why?). Furthermore, Theorem 3 guarantees that the two eigenvectors corresponding to $\lambda = 1$ will be linearly independent of the eigenvector corresponding $\lambda = 5$ and vice versa. It only remains to produce these vectors.

For $\lambda = 1$, the unknown vector

$$\mathbf{x}_1 = \begin{bmatrix} x_1 \\ y_1 \\ z_1 \end{bmatrix}$$

must satisfy the vector equation $(\mathbf{A} - 1\mathbf{I})\mathbf{x}_1 = \mathbf{0}$, or equivalently, the set of equations

$$\begin{aligned} 0 &= 0, \\ 4x_1 + 2y_1 + 2z_1 &= 0, \\ 4x_1 + 2y_1 + 2z_1 &= 0. \end{aligned}$$

A solution to this equation is $z_1 = -2x_1 - y_1$, x_1 and y_1 arbitrary. Thus,

$$\mathbf{x}_1 = \begin{bmatrix} x_1 \\ y_1 \\ z_1 \end{bmatrix} = \begin{bmatrix} x_1 \\ y_1 \\ -2x_1 - y_1 \end{bmatrix} = x_1 \begin{bmatrix} 1 \\ 0 \\ -2 \end{bmatrix} + y_1 \begin{bmatrix} 0 \\ 1 \\ -1 \end{bmatrix}.$$

By first choosing $x_1 = 1$, $y_1 = 0$ and then $x_1 = 0$, $y_1 = 1$, we see that $\lambda = 1$ generates the two linearly independent eigenvectors

$$\begin{bmatrix} 1 \\ 0 \\ -2 \end{bmatrix}, \quad \begin{bmatrix} 0 \\ 1 \\ -1 \end{bmatrix}.$$

An eigenvector corresponding to $\lambda_3 = 5$ is found to be

$$\begin{bmatrix} 0 \\ 1 \\ 1 \end{bmatrix}.$$

Therefore, **A** possesses the three linearly independent eigenvectors,

$$\begin{bmatrix} 1 \\ 0 \\ -2 \end{bmatrix}, \quad \begin{bmatrix} 0 \\ 1 \\ -1 \end{bmatrix}, \quad \begin{bmatrix} 0 \\ 1 \\ 1 \end{bmatrix}.$$

PROBLEMS 5.5

In Problems 1–3 find a set of linearly independent eigenvectors for the given matrix.

1. $\begin{bmatrix} 2 & 1 & 1 \\ 0 & 1 & 0 \\ 1 & 1 & 2 \end{bmatrix}$ **2.** $\begin{bmatrix} 2 & 1 & 1 \\ 0 & 1 & 0 \\ 1 & 2 & 2 \end{bmatrix}$ **3.** $\begin{bmatrix} 3 & 1 & 1 & 2 \\ 0 & 3 & 1 & 1 \\ 0 & 0 & 2 & 0 \\ 0 & 0 & 0 & 2 \end{bmatrix}$

4. The Vandermonde determinant

$$\begin{vmatrix} 1 & 1 & \cdots & 1 \\ x_1 & x_2 & \cdots & x_n \\ x_1^2 & x_2^2 & \cdots & x_n^2 \\ \vdots & \vdots & & \vdots \\ x_1^{n-1} & x_2^{n-1} & \cdots & x_n^{n-1} \end{vmatrix}$$

is known to equal the product

$$(x_2 - x_1)(x_3 - x_2)(x_3 - x_1)(x_4 - x_3)(x_4 - x_2) \cdots (x_n - x_1).$$

Using this result, prove Theorem 3 for n distinct eigenvalues.

Chapter 6 MATRIX CALCULUS

6.1 DEFINITIONS

The student should be aware of the vast importance of polynomials and exponentials to calculus and differential equations. He should not be surprised to find that polynomials and exponentials of matrices play an equally important role in matrix calculus and matrix differential equations. Since we will be interested in using matrices to solve linear differential equations, we shall devote this entire chapter to defining matrix functions, specifically polynomials and exponentials, developing techniques for calculating these functions, and discussing some of their important properties.

Let $p_k(x)$ denote an arbitrary polynomial in x of degree k,

$$p_k(x) = a_k x^k + a_{k-1}x^{k-1} + \cdots + a_1 x + a_0, \tag{1}$$

where the coefficients $a_k, a_{k-1}, \ldots, a_1, a_0$ are real numbers. Once $p_k(x)$ is defined, let

$$p_k(\mathbf{A}) = a_k \mathbf{A}^k + a_{k-1}\mathbf{A}^{k-1} + \cdots + a_1\mathbf{A} + a_0 \mathbf{I}. \tag{2}$$

Recall from Chapter 1, that $\mathbf{A}^2 = \mathbf{A} \cdot \mathbf{A}$, $\mathbf{A}^3 = \mathbf{A}^2 \cdot \mathbf{A} = \mathbf{A} \cdot \mathbf{A} \cdot \mathbf{A}$ and, in general, $\mathbf{A}^k = \mathbf{A}^{k-1} \cdot \mathbf{A}$. Also, $\mathbf{A}^0 = \mathbf{I}$.

Two observations are now immediate. Whereas a_0 in (1) is actually multiplied by $x^0 = 1$, a_0 in (2) is multiplied by $\mathbf{A}^0 = \mathbf{I}$. Also, if $\mathbf{A}$ is an $n \times n$ matrix, then $p_k(\mathbf{A})$ is an $n \times n$ matrix since the right-hand side of (2) may be summed.

EXAMPLE 1 Find $p_2(\mathbf{A})$ for

$$\mathbf{A} = \begin{bmatrix} 0 & 1 & 0 \\ 0 & 0 & 1 \\ 0 & 0 & 0 \end{bmatrix}$$

if $p_2(x) = 2x^2 + 3x + 4$.

Solution In this case, $p_2(\mathbf{A}) = 2\mathbf{A}^2 + 3\mathbf{A} + 4\mathbf{I}$. Thus,

$$\begin{aligned} p_2(\mathbf{A}) &= 2\begin{bmatrix} 0 & 1 & 0 \\ 0 & 0 & 1 \\ 0 & 0 & 0 \end{bmatrix}^2 + 3\begin{bmatrix} 0 & 1 & 0 \\ 0 & 0 & 1 \\ 0 & 0 & 0 \end{bmatrix} + 4\begin{bmatrix} 1 & 0 & 0 \\ 0 & 1 & 0 \\ 0 & 0 & 1 \end{bmatrix} \\ &= 2\begin{bmatrix} 0 & 0 & 1 \\ 0 & 0 & 0 \\ 0 & 0 & 0 \end{bmatrix} + 3\begin{bmatrix} 0 & 1 & 0 \\ 0 & 0 & 1 \\ 0 & 0 & 0 \end{bmatrix} + 4\begin{bmatrix} 1 & 0 & 0 \\ 0 & 1 & 0 \\ 0 & 0 & 1 \end{bmatrix} = \begin{bmatrix} 4 & 3 & 2 \\ 0 & 4 & 3 \\ 0 & 0 & 4 \end{bmatrix}. \end{aligned}$$

Note that had we defined $p_2(\mathbf{A}) = 2\mathbf{A}^2 + 3\mathbf{A} + 4$ (that is, without the $\mathbf{I}$ term), we could not have performed the addition since addition of a matrix and a scalar is undefined.

Since a matrix commutes with itself, many of the properties of polynomials (addition, subtraction, multiplication, and factoring but *not* division) are still valid for polynomials of a matrix. For instance, if $f(x)$, $d(x)$, $q(x)$, and $r(x)$ represent polynomials in x and if

$$f(x) = d(x)q(x) + r(x) \tag{3}$$

then it must be the case that

$$f(\mathbf{A}) = d(\mathbf{A})q(\mathbf{A}) + r(\mathbf{A}). \tag{4}$$

Equation (4) follows from (3) only because **A** commutes with itself; thus, we can multiply together two polynomials in **A** precisely in the same manner that we would multiply together two polynomials in x.

If we recall from calculus that many functions can be written as a Maclaurin series, then we can define functions of matrices quite easily. For instance, the Maclaurin series for e^x is

$$e^x = \sum_{k=0}^{\infty} \frac{x^k}{k!} = 1 + \frac{x}{1!} + \frac{x^2}{2!} + \frac{x^3}{3!} + \cdots. \tag{5}$$

Thus, we define the exponential of a matrix **A** as

$$e^{\mathbf{A}} = \sum_{k=0}^{\infty} \frac{\mathbf{A}^k}{k!} = \mathbf{I} + \frac{\mathbf{A}}{1!} + \frac{\mathbf{A}^2}{2!} + \frac{\mathbf{A}^3}{3!} + \cdots. \tag{6}$$

The question of convergence now arises. For an infinite series of matrices we define convergence as follows:

Definition 1 A sequence $\{\mathbf{B}_k\}$ of matrices, $\mathbf{B}_k = [b_{ij}^k]$, is said to *converge* to a matrix $\mathbf{B} = [b_{ij}]$ if the elements b_{ij}^k converge to b_{ij} for every i and j.

Definition 2 The infinite series $\sum_{n=0}^{\infty} \mathbf{B}_n$, converges to **B** if the sequence $\{\mathbf{S}_k\}$ of partial sums, where $\mathbf{S}_k = \sum_{n=0}^{k} \mathbf{B}_n$, converges to **B**.

It can be shown (see Theorem 1, this section) that the infinite series given in (6) converges for any matrix **A**. Thus $e^{\mathbf{A}}$ is defined for every matrix.

EXAMPLE 2 Find $e^{\mathbf{A}}$ if

$$\mathbf{A} = \begin{bmatrix} 2 & 0 \\ 0 & 0 \end{bmatrix}.$$

Solution

$$e^{\mathbf{A}} = e^{\begin{bmatrix} 2 & 0 \\ 0 & 0 \end{bmatrix}} = \begin{bmatrix} 1 & 0 \\ 0 & 1 \end{bmatrix} + \frac{1}{1!}\begin{bmatrix} 2 & 0 \\ 0 & 0 \end{bmatrix} + \frac{1}{2!}\begin{bmatrix} 2 & 0 \\ 0 & 0 \end{bmatrix}^2 + \frac{1}{3!}\begin{bmatrix} 2 & 0 \\ 0 & 0 \end{bmatrix}^3 + \cdots$$

$$= \begin{bmatrix} 1 & 0 \\ 0 & 1 \end{bmatrix} + \begin{bmatrix} 2/1! & 0 \\ 0 & 0 \end{bmatrix} + \begin{bmatrix} 2^2/2! & 0 \\ 0 & 0 \end{bmatrix} + \begin{bmatrix} 2^3/3! & 0 \\ 0 & 0 \end{bmatrix} + \cdots$$

$$= \begin{bmatrix} \sum_{k=0}^{\infty} 2^k/k! & 0 \\ 0 & 1 \end{bmatrix} = \begin{bmatrix} e^2 & 0 \\ 0 & e^0 \end{bmatrix}.$$

In general, if $\mathbf{A}$ is the diagonal matrix

$$\mathbf{A}=\begin{bmatrix}\lambda_1 & 0 & \cdots & 0\\ 0 & \lambda_2 & \cdots & 0\\ \vdots & \vdots & & \vdots\\ 0 & 0 & \cdots & \lambda_n\end{bmatrix},$$

then we can show (see Problem 4) that

$$e^{\mathbf{A}}=\begin{bmatrix}e^{\lambda_1} & 0 & \cdots & 0\\ 0 & e^{\lambda_2} & \cdots & 0\\ \vdots & \vdots & & \vdots\\ 0 & 0 & \cdots & e^{\lambda n}\end{bmatrix}. \tag{7}$$

If $\mathbf{A}$ is not a diagonal matrix, then it is very difficult to find $e^{\mathbf{A}}$ directly from the definition given in (6). For an arbitrary $\mathbf{A}$, $e^{\mathbf{A}}$ does not have the form exhibited in (7). For example, if

$$\mathbf{A}=\begin{bmatrix}1 & 2\\ 4 & 3\end{bmatrix},$$

it can be shown (however, not yet by us) that

$$e^{\mathbf{A}}=\frac{1}{6}\begin{bmatrix}2e^5+4e^{-1} & 2e^5-2e^{-1}\\ 4e^5-4e^{-1} & 4e^5+2e^{-1}\end{bmatrix}$$

For the purposes of this book, the exponential is the only function that is needed. However, it may be of some value to know how other functions of matrices, sines, cosines, etc., are defined. The following theorem provides this information.

Theorem 1 Let z represent the complex variable $x+iy$. If $f(z)$ has the Taylor series $\sum_{k=0}^{\infty} a_k z^k$, which converges for $|z|<R$, and if the eigenvalues $\lambda_1, \lambda_2, \ldots, \lambda_n$ of an $n\times n$ matrix $\mathbf{A}$ have the property that $|\lambda_i|<R$ $(i=1, 2, \ldots, n)$, then $\sum_{k=0}^{\infty} a_k \mathbf{A}^k$ will converge to an $n\times n$ matrix which is defined to be $f(\mathbf{A})$.[12]

EXAMPLE 3 Define sin $\mathbf{A}$.

Solution A Taylor series for sin z is

$$\sin z=\sum_{k=0}^{\infty}\frac{(-1)^k z^{2k+1}}{(2k+1)!}$$

$$=z-\frac{z^3}{3!}+\frac{z^5}{5!}-\frac{z^7}{7!}+\cdots.$$

[12] For a proof of Theorem 1, see D. T. Finkbeiner II, "Introduction to Matrices and Linear Transformations," p. 196. Freeman, San Francisco, 1960.

This series can be shown to converge for all z (that is, $R = \infty$). Hence, since any eigenvalue λ of $\mathbf{A}$ must have the property $|\lambda| < \infty$ (that is, λ is finite) sin $\mathbf{A}$ can be defined for every $\mathbf{A}$ as

$$\sin \mathbf{A} = \sum_{k=0}^{\infty} \frac{(-1)^k \mathbf{A}^{2k+1}}{(2k+1)!} = \mathbf{A} - \frac{\mathbf{A}^3}{3!} + \frac{\mathbf{A}^5}{5!} - \frac{\mathbf{A}^7}{7!} + \cdots. \tag{8}$$

PROBLEMS 6.1

1. Let $q(x) = x - 1$. Find $p_k(\mathbf{A})$ and $q(\mathbf{A})p_k(\mathbf{A})$ if

(a) $$\mathbf{A} = \begin{bmatrix} 1 & 2 & 3 \\ 0 & -1 & 4 \\ 0 & 0 & 1 \end{bmatrix}, \quad k = 2, \quad \text{and} \quad p_2(x) = x^2 - 2x + 1,$$

(b) $$\mathbf{A} = \begin{bmatrix} 1 & 2 \\ 3 & 4 \end{bmatrix}, \quad k = 3, \quad \text{and} \quad p_3(x) = 2x^3 - 3x^2 + 4.$$

2. If $p_k(x)$ is defined by (1), find $p_k(\mathbf{A})$ for the diagonal matrix

$$\mathbf{A} = \begin{bmatrix} \lambda_1 & 0 & 0 \\ 0 & \lambda_2 & 0 \\ 0 & 0 & \lambda_3 \end{bmatrix}.$$ Can you generalize?

3. By actually computing both sides of the following equation separately, verify that $(\mathbf{A} - 3\mathbf{I})(\mathbf{A} + 2\mathbf{I}) = \mathbf{A}^2 - \mathbf{A} - 6\mathbf{I}$ for

(a) $$\mathbf{A} = \begin{bmatrix} 1 & 2 \\ 3 & 4 \end{bmatrix},$$ (b) $$\mathbf{A} = \begin{bmatrix} 1 & 0 & -2 \\ 3 & 1 & 1 \\ -2 & -2 & 3 \end{bmatrix}.$$

The above equation is an example of matrix factoring.

4. Derive Eq. (7).

5. Using Theorem 1, give a definition for cos $\mathbf{A}$ and use this definition to find

$$\cos \begin{bmatrix} 1 & 0 \\ 0 & 2 \end{bmatrix}.$$

6.2 CAYLEY–HAMILTON THEOREM

We now state one of the most powerful theorems of matrix theory, the proof of which can be found in the appendix to this chapter.

Cayley–Hamilton Theorem A matrix satisfies its own characteristic equation. That is, if the characteristic equation of an $n \times n$ matrix $\mathbf{A}$ is

$\lambda^n + a_{n-1}\lambda^{n-1} + \cdots + a_1\lambda + a_0 = 0$, then $\mathbf{A}^n + a_{n-1}\mathbf{A}^{n-1} + \cdots + a_1\mathbf{A} + a_0\mathbf{I} = \mathbf{0}$.

Note once again that when we change a scalar equation to a matrix equation, the unity element 1 is replaced by the identity matrix **I**.

EXAMPLE 1 Verify the Cayley–Hamilton theorem for

$$\mathbf{A} = \begin{bmatrix} 1 & 2 \\ 4 & 3 \end{bmatrix}.$$

Solution The characteristic equation for **A** is $\lambda^2 - 4\lambda - 5 = 0$.

$$\begin{aligned}\mathbf{A}^2 - 4\mathbf{A} - 5\mathbf{I} &= \begin{bmatrix} 1 & 2 \\ 4 & 3 \end{bmatrix}\begin{bmatrix} 1 & 2 \\ 4 & 3 \end{bmatrix} - 4\begin{bmatrix} 1 & 2 \\ 4 & 3 \end{bmatrix} - 5\begin{bmatrix} 1 & 0 \\ 0 & 1 \end{bmatrix} \\ &= \begin{bmatrix} 9 & 8 \\ 16 & 17 \end{bmatrix} - \begin{bmatrix} 4 & 8 \\ 16 & 12 \end{bmatrix} - \begin{bmatrix} 5 & 0 \\ 0 & 5 \end{bmatrix} \\ &= \begin{bmatrix} 9-4-5 & 8-8-0 \\ 16-16-0 & 17-12-5 \end{bmatrix} = \begin{bmatrix} 0 & 0 \\ 0 & 0 \end{bmatrix} = \mathbf{0}.\end{aligned}$$

EXAMPLE 2 Verify the Cayley–Hamilton theorem for

$$\mathbf{A} = \begin{bmatrix} 3 & 0 & -1 \\ 2 & 0 & 1 \\ 0 & 0 & 4 \end{bmatrix}.$$

Solution The characteristic equation of **A** is $(3 - \lambda)(-\lambda)(4 - \lambda) = 0$.

$(3\mathbf{I} - \mathbf{A})(-\mathbf{A})(4\mathbf{I} - \mathbf{A})$

$$\begin{aligned} &= \left(\begin{bmatrix} 3 & 0 & 0 \\ 0 & 3 & 0 \\ 0 & 0 & 3 \end{bmatrix} - \begin{bmatrix} 3 & 0 & -1 \\ 2 & 0 & 1 \\ 0 & 0 & 4 \end{bmatrix}\right)\left(-\begin{bmatrix} 3 & 0 & -1 \\ 2 & 0 & 1 \\ 0 & 0 & 4 \end{bmatrix}\right) \\ &\quad \left(\begin{bmatrix} 4 & 0 & 0 \\ 0 & 4 & 0 \\ 0 & 0 & 4 \end{bmatrix} - \begin{bmatrix} 3 & 0 & -1 \\ 2 & 0 & 1 \\ 0 & 0 & 4 \end{bmatrix}\right) \\ &= \begin{bmatrix} 0 & 0 & 1 \\ -2 & 3 & -1 \\ 0 & 0 & -1 \end{bmatrix}\begin{bmatrix} -3 & 0 & 1 \\ -2 & 0 & -1 \\ 0 & 0 & -4 \end{bmatrix}\begin{bmatrix} 1 & 0 & 1 \\ -2 & 4 & -1 \\ 0 & 0 & 0 \end{bmatrix} \\ &= \begin{bmatrix} 0 & 0 & 1 \\ -2 & 3 & -1 \\ 0 & 0 & -1 \end{bmatrix}\begin{bmatrix} -3 & 0 & -3 \\ -2 & 0 & -2 \\ 0 & 0 & 0 \end{bmatrix} = \begin{bmatrix} 0 & 0 & 0 \\ 0 & 0 & 0 \\ 0 & 0 & 0 \end{bmatrix} = \mathbf{0}.\end{aligned}$$

One immediate consequence of the Cayley–Hamilton theorem is a new method for finding the inverse of a nonsingular matrix.

If $\lambda^n + a_{n-1}\lambda^{n-1} + \cdots + a_1\lambda + a_0 = 0$ is the characteristic equation of a matrix **A**,[13] it can be shown that the product of the eigenvalues is $(-1)^n a_0$[14]. Hence by Property 7 of Section 5.4 [the product of the eigenvalues of **A** equals det(**A**)], we have that $\det(\mathbf{A}) = (-1)^n a_0$. Thus, **A** is invertible if and only if $a_0 \neq 0$.

Now assume that $a_0 \neq 0$. By the Cayley–Hamilton theorem, we have

$$\mathbf{A}^n + a_{n-1}\mathbf{A}^{n-1} + \cdots + a_1\mathbf{A} + a_0\mathbf{I} = \mathbf{0},$$

$$\mathbf{A}[\mathbf{A}^{n-1} + a_{n-1}\mathbf{A}^{n-2} + \cdots + a_1\mathbf{I}] = -a_0\mathbf{I},$$

or

$$\mathbf{A}\left[-\frac{1}{a_0}(\mathbf{A}^{n-1} + a_{n-1}\mathbf{A}^{n-2} + \cdots + a_1\mathbf{I})\right] = \mathbf{I}.$$

Thus, $(-1/a_0)(\mathbf{A}^{n-1} + a_{n-1}\mathbf{A}^{n-2} + \cdots + a_1\mathbf{I})$ is an inverse of **A**. But since the inverse is unique (see Theorem 1 of Section 3.3), we have that

$$\mathbf{A}^{-1} = \frac{-1}{a_0}(\mathbf{A}^{n-1} + a_{n-1}\mathbf{A}^{n-2} + \cdots + a_1\mathbf{I}). \tag{9}$$

EXAMPLE 3 Using the Cayley–Hamilton theorem, find $\mathbf{A}^{-1}$ for

$$\mathbf{A} = \begin{bmatrix} 1 & -2 & 4 \\ 0 & -1 & 2 \\ 2 & 0 & 3 \end{bmatrix}.$$

Solution The characteristic equation for **A** is $\lambda^3 - 3\lambda^2 - 9\lambda + 3 = 0$. Thus, by the Cayley–Hamilton theorem,

$$\mathbf{A}^3 - 3\mathbf{A}^2 - 9\mathbf{A} + 3\mathbf{I} = \mathbf{0}.$$

Hence

$$\mathbf{A}^3 - 3\mathbf{A}^2 - 9\mathbf{A} = -3\mathbf{I},$$

$$\mathbf{A}(\mathbf{A}^2 - 3\mathbf{A} - 9\mathbf{I}) = -3\mathbf{I},$$

or,

$$\mathbf{A}(\tfrac{1}{3})(-\mathbf{A}^2 + 3\mathbf{A} + 9\mathbf{I})) = \mathbf{I}.$$

[13] See footnote 9 on page 89.
[14] B. L. van der Waerden, "Modern Algebra," p. 78. Ungar, New York, 1953.

Thus,

$$\mathbf{A}^{-1} = (\tfrac{1}{3})(-\mathbf{A}^2 + 3\mathbf{A} + 9\mathbf{I})$$

$$= \frac{1}{3}\left(\begin{bmatrix} -9 & 0 & -12 \\ -4 & -1 & -4 \\ -8 & 4 & -17 \end{bmatrix} + \begin{bmatrix} 3 & -6 & 12 \\ 0 & -3 & 6 \\ 6 & 0 & 9 \end{bmatrix} + \begin{bmatrix} 9 & 0 & 0 \\ 0 & 9 & 0 \\ 0 & 0 & 9 \end{bmatrix}\right)$$

$$= \frac{1}{3}\begin{bmatrix} 3 & -6 & 0 \\ -4 & 5 & 2 \\ -2 & 4 & 1 \end{bmatrix}.$$

PROBLEMS 6.2

Verify the Cayley–Hamilton theorem and use it to find $\mathbf{A}^{-1}$, where possible, for:

1. $\mathbf{A} = \begin{bmatrix} 1 & 2 \\ 3 & 4 \end{bmatrix}$

2. $\mathbf{A} = \begin{bmatrix} 1 & 2 \\ 2 & 4 \end{bmatrix}$

3. $\mathbf{A} = \begin{bmatrix} 2 & 0 & 1 \\ 4 & 0 & 2 \\ 0 & 0 & -1 \end{bmatrix}$

4. $\mathbf{A} = \begin{bmatrix} 1 & -1 & 2 \\ 0 & 3 & 2 \\ 2 & 1 & 2 \end{bmatrix}$

5. $\mathbf{A} = \begin{bmatrix} 1 & 0 & 0 & 0 \\ 0 & -1 & 0 & 0 \\ 0 & 0 & -1 & 0 \\ 0 & 0 & 0 & 1 \end{bmatrix}$

6.3 POLYNOMIALS OF MATRICES—DISTINCT EIGENVALUES

In general, it is very difficult to compute functions of matrices from their definition as infinite series (one exception is the diagonal matrix). The Cayley–Hamilton theorem, however, provides a starting point for the development of an alternate, straightforward method for calculating these functions. In this section, we shall develop the method for polynomials of matrices having distinct eigenvalues. In the ensuing sections, we shall extend the method to functions of matrices having arbitrary eigenvalues.

Let $\mathbf{A}$ represent an $n \times n$ matrix. Define $d(\lambda) = \det(\mathbf{A} - \lambda\mathbf{I})$. Thus, $d(\lambda)$ is an nth degree polynomial in λ and the characteristic equation of $\mathbf{A}$ is $d(\lambda) = 0$. From Chapter 5, we know that if λ_i is an eigenvalue of $\mathbf{A}$, then λ_i is a root of the characteristic equation, hence

$$d(\lambda_i) = 0. \tag{10}$$

From the Cayley–Hamilton theorem, we know that a matrix must satisfy its own characteristic equation, hence

$$d(\mathbf{A}) = \mathbf{0}. \tag{11}$$

Let $f(\mathbf{A})$ be any matrix polynomial of arbitrary degree that we wish to compute. $f(\lambda)$ represents the corresponding polynomial of λ. A theorem of algebra[15] states that there exist polynomials $q(\lambda)$ and $r(\lambda)$ such that

$$f(\lambda) = d(\lambda)q(\lambda) + r(\lambda), \tag{12}$$

where $r(\lambda)$ is called the remainder and is of degree $n - 1$. The degree of $r(\lambda)$ is less than that of $d(\lambda)$, which is n, and must be less than or equal to the degree of $f(\lambda)$ (why?).

EXAMPLE 1 Find $q(\lambda)$ and $r(\lambda)$ if $f(\lambda) = \lambda^4 + 2\lambda^3 - 1$ and $d(\lambda) = \lambda^2 - 1$.

Solution For $\lambda \neq \pm 1$, $d(\lambda) \neq 0$. Dividing $f(\lambda)$ by $d(\lambda)$, we obtain

$$\frac{f(\lambda)}{d(\lambda)} = \frac{\lambda^4 + 2\lambda^3 - 1}{\lambda^2 - 1} = (\lambda^2 + 2\lambda + 1) + \frac{2\lambda}{\lambda^2 - 1},$$

$$\frac{f(\lambda)}{d(\lambda)} = (\lambda^2 + 2\lambda + 1) + \frac{2\lambda}{d(\lambda)},$$

or

$$f(\lambda) = d(\lambda)(\lambda^2 + 2\lambda + 1) + (2\lambda). \tag{13}$$

Hence, if we define $q(\lambda) = \lambda^2 + 2\lambda + 1$ and $r(\lambda) = 2\lambda$, Eq. (13) has the exact form of (12) for all λ except possibly $\lambda = \pm 1$. However, by direct substitution, we find that (13) is also valid for $\lambda = \pm 1$; hence (13) is an identity for all λ.

From (12), (3), and (4), we have

$$f(\mathbf{A}) = d(\mathbf{A})q(\mathbf{A}) + r(\mathbf{A}). \tag{14}$$

Using (11), we obtain

$$f(\mathbf{A}) = r(\mathbf{A}). \tag{15}$$

Therefore, it follows that any polynomial in $\mathbf{A}$ may be written as a polynomial of degree $n - 1$. For example, if $\mathbf{A}$ is a 4×4 matrix and if we wish to compute $f(\mathbf{A}) = \mathbf{A}^{957} - 3\mathbf{A}^{59} + 2\mathbf{A}^3 - 4\mathbf{I}$, then (15) implies that $f(\mathbf{A})$ can be written as a third degree polynomial in $\mathbf{A}$, that is,

$$\mathbf{A}^{957} - 3\mathbf{A}^{59} + 2\mathbf{A}^3 - 4\mathbf{I} = \alpha_3\mathbf{A}^3 + \alpha_2\mathbf{A}^2 + \alpha_1\mathbf{A} + \alpha_0\mathbf{I} \tag{16}$$

[15] B. L. van der Waerden, "Modern Algebra," p. 48. Ungar, New York, 1953.

where α_3, α_2, α_1, α_0 are scalars that still must be determined. Once α_3, α_2, α_1, α_0 are computed, the student should observe that it is much easier to calculate the right-hand side rather than the left-hand side of (16).

If $\mathbf{A}$ is an $n \times n$ matrix, then $r(\lambda)$ will be a polynomial having the form

$$r(\lambda) = \alpha_{n-1}\lambda^{n-1} + \alpha_{n-2}\lambda^{n-2} + \cdots + \alpha_1\lambda + \alpha_0. \tag{17}$$

If λ_i is an eigenvalue of $\mathbf{A}$, then we have, after substituting (10) into (12), that

$$f(\lambda_i) = r(\lambda_i). \tag{18}$$

Thus, using (17), Eq. (18) may be rewritten as

$$f(\lambda_i) = \alpha_{n-1}(\lambda_i)^{n-1} + \alpha_{n-2}(\lambda_i)^{n-2} + \cdots + \alpha_1(\lambda_i) + \alpha_0 \tag{19}$$

if λ_i is an eigenvalue.

If we now assume that $\mathbf{A}$ has distinct eigenvalues, $\lambda_1, \lambda_2, \ldots, \lambda_n$ (note that if the eigenvalues are distinct, there must be n of them), then (19) may be used to generate n simultaneous linear equations for the n unknowns $\alpha_{n-1}, \alpha_{n-2}, \ldots, \alpha_1, \alpha_0$:

$$\begin{aligned}
f(\lambda_1) = r(\lambda_1) &= \alpha_{n-1}(\lambda_1)^{n-1} + \alpha_{n-2}(\lambda_1)^{n-2} + \cdots + \alpha_1(\lambda_1) + \alpha_0, \\
f(\lambda_2) = r(\lambda_2) &= \alpha_{n-1}(\lambda_2)^{n-1} + \alpha_{n-2}(\lambda_2)^{n-2} + \cdots + \alpha_1(\lambda_2) + \alpha_0, \\
&\;\vdots \\
f(\lambda_n) = r(\lambda_n) &= \alpha_{n-1}(\lambda_n)^{n-1} + \alpha_{n-2}(\lambda_n)^{n-2} + \cdots + \alpha_1(\lambda_n) + \alpha_0.
\end{aligned} \tag{20}$$

Note that $f(\lambda)$ and the eigenvalues $\lambda_1, \lambda_2, \ldots, \lambda_n$ are assumed known; hence $f(\lambda_1), f(\lambda_2), \ldots, f(\lambda_n)$ are known, and the only unknowns in (20) are α_{n-1}, $\alpha_{n-2}, \ldots, \alpha_1, \alpha_0$.

EXAMPLE 2 Find $\mathbf{A}^{593}$ if

$$\mathbf{A} = \begin{bmatrix} -3 & -4 \\ 2 & 3 \end{bmatrix}.$$

Solution The eigenvalues of $\mathbf{A}$ are $\lambda_1 = 1$, $\lambda_2 = -1$. For this example, $f(\mathbf{A}) = \mathbf{A}^{593}$, thus, $f(\lambda) = \lambda^{593}$. Since $\mathbf{A}$ is a 2×2 matrix, $r(\mathbf{A})$ will be a polynomial of degree $(2 - 1) = 1$, hence $r(\mathbf{A}) = \alpha_1\mathbf{A} + \alpha_0\mathbf{I}$ and $r(\lambda) = \alpha_1\lambda + \alpha_0$. From (15), we have that $f(\mathbf{A}) = r(\mathbf{A})$, thus, for this example,

$$\mathbf{A}^{593} = \alpha_1\mathbf{A} + \alpha_0\mathbf{I}. \tag{21}$$

From (18), we have that $f(\lambda_i) = r(\lambda_i)$ if λ_i is an eigenvalue of $\mathbf{A}$; thus, for this example, $(\lambda_i)^{593} = \alpha_1\lambda_i + \alpha_0$. Substituting the eigenvalues of $\mathbf{A}$ into this equation, we obtain the following system for α_1 and α_0.

$$\begin{aligned}
(1)^{593} &= \alpha_1(1) + \alpha_0, \\
(-1)^{593} &= \alpha_1(-1) + \alpha_0,
\end{aligned}$$

or

$$
\begin{aligned}
1 &= \alpha_1 + \alpha_0, \\
-1 &= -\alpha_1 + \alpha_0.
\end{aligned}
\tag{22}
$$

Solving (22), we obtain $\alpha_0 = 0, \alpha_1 = 1$. Substituting these values into (21), we obtain $\mathbf{A}^{593} = 1 \cdot \mathbf{A} + 0 \cdot \mathbf{I}$
or

$$
\begin{bmatrix} -3 & -4 \\ 2 & 3 \end{bmatrix}^{593} = \begin{bmatrix} -3 & -4 \\ 2 & 3 \end{bmatrix}.
$$

EXAMPLE 3 Find $\mathbf{A}^{39}$ if

$$
\mathbf{A} = \begin{bmatrix} 4 & 1 \\ 2 & 3 \end{bmatrix}.
$$

Solution The eigenvalues of $\mathbf{A}$ are $\lambda_1 = 5, \lambda_2 = 2$. For this example, $f(\mathbf{A}) = \mathbf{A}^{39}$, thus $f(\lambda) = \lambda^{39}$. Since $\mathbf{A}$ is a 2×2 matrix, $r(\mathbf{A})$ will be a polynomial of degree 1, hence $r(\mathbf{A}) = \alpha_1 \mathbf{A} + \alpha_0 \mathbf{I}$ and $r(\lambda) = \alpha_1 \lambda + \alpha_0$. From (15), we have that $f(\mathbf{A}) = r(\mathbf{A})$, thus, for this example,

$$
\mathbf{A}^{39} = \alpha_1 \mathbf{A} + \alpha_0 \mathbf{I}. \tag{23}
$$

From (18) we have that $f(\lambda_i) = r(\lambda_i)$ if λ_i is an eigenvalue of $\mathbf{A}$, thus for this example, $(\lambda_i)^{39} = \alpha_1 \lambda_i + \alpha_0$. Substituting the eigenvalues of $\mathbf{A}$ into this equation, we obtain the following system for α_1 and α_0:

$$
\begin{aligned}
(5)^{39} &= 5\alpha_1 + \alpha_0, \\
(2)^{39} &= 2\alpha_1 + \alpha_0.
\end{aligned}
\tag{24}
$$

Solving (24), we obtain

$$
\alpha_1 = \frac{(5)^{39} - 2^{39}}{3}, \qquad \alpha_0 = \frac{-2(5)^{39} + 5(2)^{39}}{3},
$$

Substituting these values into (23), we obtain

$$
\begin{aligned}
\mathbf{A}^{39} &= \frac{(5)^{39} - (2)^{39}}{3} \begin{bmatrix} 4 & 1 \\ 2 & 3 \end{bmatrix} + \frac{-2(5)^{39} + 5(2)^{39}}{3} \begin{bmatrix} 1 & 0 \\ 0 & 1 \end{bmatrix}. \\
&= \frac{1}{3} \begin{bmatrix} 2(5)^{39} + (2)^{39} & (5)^{39} - (2)^{39} \\ 2(5)^{39} - 2(2)^{39} & (5)^{39} + 2(2)^{39} \end{bmatrix}.
\end{aligned}
\tag{25}
$$

Here, the numbers $(5)^{39}$ and $(2)^{39}$ must be computed. This can be accomplished in any number of ways including logarithms. For our purposes, however, the form of (25) is sufficient and no further simplification is needed.

EXAMPLE 4 Find $\mathbf{A}^{602} - 3\mathbf{A}^3$ if

$$\mathbf{A} = \begin{bmatrix} 1 & 4 & -2 \\ 0 & 0 & 0 \\ 0 & -3 & 3 \end{bmatrix}.$$

Solution The eigenvalues of $\mathbf{A}$ are $\lambda_1 = 0$, $\lambda_2 = 1$, $\lambda_3 = 3$.

$$f(\mathbf{A}) = \mathbf{A}^{602} - 3\mathbf{A}^3, \qquad r(\mathbf{A}) = \alpha_2 \mathbf{A}^2 + \alpha_1 \mathbf{A} + \alpha_0 \mathbf{I},$$
$$f(\lambda) = \lambda^{602} - 3\lambda^3, \qquad r(\lambda) = \alpha_2 \lambda^2 + \alpha_1 \lambda + \alpha_0 .$$

Note that since $\mathbf{A}$ is a 3×3 matrix, $r(\mathbf{A})$ must be a second degree polynomial.

$$f(\mathbf{A}) = r(\mathbf{A});$$

thus,

$$\mathbf{A}^{602} - 3\mathbf{A}^3 = \alpha_2 \mathbf{A}^2 + \alpha_1 \mathbf{A} + \alpha_0 \mathbf{I}. \tag{26}$$

If λ_i is an eigenvalue of $\mathbf{A}$, then $f(\lambda_i) = r(\lambda_i)$. Thus,

$$(\lambda_i)^{602} - 3(\lambda_i)^3 = \alpha_2(\lambda_i)^2 + \alpha_1\lambda_i + \alpha_0;$$

hence,

$$(0)^{602} - 3(0)^3 = \alpha_2(0)^2 + \alpha_1(0) + \alpha_0,$$
$$(1)^{602} - 3(1)^3 = \alpha_2(1)^2 + \alpha_1(1) + \alpha_0,$$
$$(3)^{602} - 3(3)^3 = \alpha_2(3)^2 + \alpha_1(3) + \alpha_0,$$

or

$$0 = \alpha_0,$$
$$-2 = \alpha_2 + \alpha_1 + \alpha_0,$$
$$(3)^{602} - 81 = 9\alpha_2 + 3\alpha_1 + \alpha_0.$$

Thus,

$$\alpha_2 = \frac{(3)^{602} - 75}{6}, \qquad \alpha_1 = \frac{-(3)^{602} + 63}{6}, \qquad \alpha_0 = 0. \tag{27}$$

Substituting (27) into (26), we obtain

$$\mathbf{A}^{602} - 3\mathbf{A}^3 = \frac{(3)^{602} - 75}{6}\begin{bmatrix} 1 & 10 & -8 \\ 0 & 0 & 0 \\ 0 & -9 & 9 \end{bmatrix} + \frac{-(3)^{602} + 63}{6}\begin{bmatrix} 1 & 4 & -2 \\ 0 & 0 & 0 \\ 0 & -3 & 3 \end{bmatrix}$$
$$= \frac{1}{6}\begin{bmatrix} -12 & 6(3)^{602} - 498 & -6(3)^{602} + 474 \\ 0 & 0 & 0 \\ 0 & -6(3)^{602} + 486 & 6(3)^{602} - 486 \end{bmatrix}.$$

Finally, the student should note that if the polynomial to be calculated is already of degree less than or equal to $n-1$, then this method affords no simplification and the polynomial must still be computed directly.

PROBLEMS 6.3

1. Find $\mathbf{A}^7$ for

$$\mathbf{A} = \begin{bmatrix} -2 & 3 \\ -1 & 2 \end{bmatrix}.$$

check your results by direct calculations.

2. Find $\mathbf{A}^{202} - 3\mathbf{A}^{147} + 2\mathbf{I}$ for the $\mathbf{A}$ of Problem 1.
3. Find $\mathbf{A}^{735}$ for

$$\mathbf{A} = \begin{bmatrix} 0 & 1 \\ 0 & -1 \end{bmatrix}.$$

4. Find $\mathbf{A}^{1025} - 4\mathbf{A}^5$ for the $\mathbf{A}$ of Problem 1.
5. Find $\mathbf{A}^{78}$ for

$$\mathbf{A} = \begin{bmatrix} 2 & -1 \\ 2 & 5 \end{bmatrix}.$$

6. Find $\mathbf{A}^{593} - 2\mathbf{A}^{15}$ for

$$\mathbf{A} = \begin{bmatrix} -2 & 4 & 3 \\ 0 & 0 & 0 \\ -1 & 5 & 2 \end{bmatrix}.$$

7. Find $\mathbf{A}^{222}$ for

$$\mathbf{A} = \begin{bmatrix} 1 & -1 & 2 \\ 0 & -1 & 2 \\ 0 & 0 & 2 \end{bmatrix}.$$

6.4 POLYNOMIALS OF MATRICES—GENERAL CASE

The only restriction in the previous section was that the eigenvalues of $\mathbf{A}$ had to be distinct. The following theorem suggests how to obtain n equations for the unknown α's even if some of the eigenvalues are identical.

Theorem 1 Let $f(\lambda)$ and $r(\lambda)$ be defined as in Eq. (12). If λ_i is an eigenvalue of multiplicity k, then

$$
\begin{aligned}
f(\lambda_i) &= r(\lambda_i), \\
\frac{df(\lambda_i)}{d\lambda} &= \frac{dr(\lambda_i)}{d\lambda}, \\
\frac{d^2f(\lambda_i)}{d\lambda^2} &= \frac{d^2r(\lambda_i)}{d\lambda^2}, \\
&\vdots \\
\frac{d^{k-1}f(\lambda_i)}{d\lambda^{k-1}} &= \frac{d^{k-1}r(\lambda_i)}{d\lambda^{k-1}},
\end{aligned}
\tag{28}
$$

where the notation $d^nf(\lambda_i)/d\lambda^n$ denotes the nth derivative of $f(\lambda)$ with respect to λ evaluated at $\lambda = \lambda_i$.[16]

Thus, for example, if λ_i is an eigenvalue of multiplicity 3, Theorem 1 implies that $f(\lambda)$ and its first two derivatives evaluated at $\lambda = \lambda_i$ are equal to $r(\lambda)$ and its first two derivatives also evaluated at $\lambda = \lambda_i$. If λ_i is an eigenvalue of multiplicity 5, then $f(\lambda)$ and the first four derivatives of $f(\lambda)$ evaluated at $\lambda = \lambda_i$ are equal respectively to $r(\lambda)$ and the first four derivatives of $r(\lambda)$ evaluated at $\lambda = \lambda_i$. Note, furthermore, that if λ_i is an eigenvalue of multiplicity 1, then Theorem 1 implies that $f(\lambda_i) = r(\lambda_i)$, which is Eq. (18).

EXAMPLE 1 Find $\mathbf{A}^{24} - 3\mathbf{A}^{15}$ if

$$
\mathbf{A} = \begin{bmatrix} 3 & 2 & 4 \\ 0 & 1 & 0 \\ -1 & -3 & -1 \end{bmatrix}.
$$

Solution The eigenvalues of $\mathbf{A}$ are $\lambda_1 = \lambda_2 = \lambda_3 = 1$; hence, $\lambda = 1$ is an eigenvalue of multiplicity three.

$$
\begin{aligned}
f(\mathbf{A}) &= \mathbf{A}^{24} - 3\mathbf{A}^{15} & r(\mathbf{A}) &= \alpha_2\mathbf{A}^2 + \alpha_1\mathbf{A} + \alpha_0\mathbf{I} \\
f(\lambda) &= \lambda^{24} - 3\lambda^{15} & r(\lambda) &= \alpha_2\lambda^2 + \alpha_1\lambda + \alpha_0 \\
f'(\lambda) &= 24\lambda^{23} - 45\lambda^{14} & r'(\lambda) &= 2\alpha_2\lambda + \alpha_1 \\
f''(\lambda) &= 552\lambda^{22} - 630\lambda^{13} & r''(\lambda) &= 2\alpha_2.
\end{aligned}
$$

Now $f(\mathbf{A}) = r(\mathbf{A})$, hence

$$
\mathbf{A}^{24} - 3\mathbf{A}^{15} = \alpha_2\mathbf{A}^2 + \alpha_1\mathbf{A} + \alpha_0\mathbf{I}. \tag{29}
$$

[16] Theorem 1 is proved by differentiating Eq. (12) $k-1$ times and noting that if λ_i is an eigenvalue of multiplicity k, then

$$
d(\lambda_i) = \frac{d[d(\lambda_i)]}{d\lambda} = \cdots = \frac{d^{(k-1)}d(\lambda_i)}{d\lambda^{k-1}} = 0.
$$

See B. L. van der Waerden, "Modern Algebra," p. 65. Ungar, New York, 1953.

Also, since $\lambda = 1$ is an eigenvalue of multiplicity 3, it follows from Theorem 1 that

$$f(1) = r(1),$$
$$f'(1) = r'(1),$$
$$f''(1) = r''(1).$$

Hence,

$$(1)^{24} - 3(1)^{15} = \alpha_2(1)^2 + \alpha_1(1) + \alpha_0,$$
$$24(1)^{23} - 45(1)^{14} = 2\alpha_2(1) + \alpha_1,$$
$$552(1)^{22} - 630(1)^{13} = 2\alpha_2,$$

or

$$-2 = \alpha_2 + \alpha_1 + \alpha_0,$$
$$-21 = 2\alpha_2 + \alpha_1,$$
$$-78 = 2\alpha_2.$$

Thus, $\alpha_2 = -39$, $\alpha_1 = 57$, $\alpha_0 = -20$, and from Eq. (29)

$$\mathbf{A}^{24} - 3\mathbf{A}^{15} = -39\mathbf{A}^2 + 57\mathbf{A} - 20\mathbf{I} = \begin{bmatrix} -44 & 270 & -84 \\ 0 & -2 & 0 \\ 21 & -93 & 40 \end{bmatrix}.$$

EXAMPLE 2 Set up the necessary equation to find $\mathbf{A}^{15} - 6\mathbf{A}^2$ if

$$\mathbf{A} = \begin{bmatrix} 1 & 4 & 3 & 2 & 1 & -7 \\ 0 & 0 & 2 & 11 & 1 & 0 \\ 0 & 0 & 1 & -1 & 0 & 1 \\ 0 & 0 & 0 & -1 & 2 & 1 \\ 0 & 0 & 0 & 0 & -1 & 17 \\ 0 & 0 & 0 & 0 & 0 & 1 \end{bmatrix}.$$

Solution The eigenvalues of $\mathbf{A}$ are $\lambda_1 = \lambda_2 = \lambda_3 = 1$, $\lambda_4 = \lambda_5 = -1$, $\lambda_6 = 0$.

$$f(\mathbf{A}) = \mathbf{A}^{15} - 6\mathbf{A}^2 \qquad r(\mathbf{A}) = \alpha_5\mathbf{A}^5 + \alpha_4\mathbf{A}^4 + \alpha_3\mathbf{A}^3 + \alpha_2\mathbf{A}^2 + \alpha_1\mathbf{A} + \alpha_0\mathbf{I}$$
$$f(\lambda) = \lambda^{15} - 6\lambda^2 \qquad r(\lambda) = \alpha_5\lambda^5 + \alpha_4\lambda^4 + \alpha_3\lambda^3 + \alpha_2\lambda^2 + \alpha_1\lambda^1 + \alpha_0$$
$$f'(\lambda) = 15\lambda^{14} - 12\lambda \qquad r'(\lambda) = 5\alpha_5\lambda^4 + 4\alpha_4\lambda^3 + 3\alpha_3\lambda^2 + 2\alpha_2\lambda + \alpha_1$$
$$f''(\lambda) = 210\lambda^{13} - 12 \qquad r''(\lambda) = 20\alpha_5\lambda^3 + 12\alpha_4\lambda^2 + 6\alpha_3\lambda + 2\alpha_2.$$

Since $f(\mathbf{A}) = r(\mathbf{A})$,

$$\mathbf{A}^{15} - 6\mathbf{A}^2 = \alpha_5\mathbf{A}^5 + \alpha_4\mathbf{A}^4 + \alpha_3\mathbf{A}^3 + \alpha_2\mathbf{A}^2 + \alpha_1\mathbf{A} + \alpha_0\mathbf{I}. \qquad (30)$$

Since $\lambda = 1$ is an eigenvalue of multiplicity 3, $\lambda = -1$ is an eigenvalue of multiplicity 2 and $\lambda = 0$ is an eigenvalue of multiplicity 1, it follows from Theorem 1 that

$$\begin{aligned} f(1) &= r(1), \\ f'(1) &= r'(1), \\ f''(1) &= r''(1), \\ f(-1) &= r(-1), \\ f'(-1) &= r'(-1), \\ f(0) &= r(0). \end{aligned} \tag{31}$$

Hence,

$$\begin{aligned} (1)^{15} - 6(1)^2 &= \alpha_5(1)^5 + \alpha_4(1)^4 + \alpha_3(1)^3 + \alpha_2(1)^2 + \alpha_1(1) + \alpha_0 \\ 15(1)^{14} - 12(1) &= 5\alpha_5(1)^4 + 4\alpha_4(1)^3 + 3\alpha_3(1)^2 + 2\alpha_2(1) + \alpha_1 \\ 210(1)^{13} - 12 &= 20\alpha_5(1)^3 + 12\alpha_4(1)^2 + 6\alpha_3(1) + 2\alpha_2 \\ (-1)^{15} - 6(-1)^2 &= \alpha_5(-1)^5 + \alpha_4(-1)^4 + \alpha_3(-1)^3 + \alpha_2(-1)^2 + \alpha_1(-1) \\ &\quad + \alpha_0 \\ 15(-1)^{14} - 12(-1) &= 5\alpha_5(-1)^4 + 4\alpha_4(-1)^3 + 3\alpha_3(-1)^2 + 2\alpha_2(-1) + \alpha_1 \\ (0)^{15} - 12(0)^2 &= \alpha_5(0)^5 + \alpha_4(0)^4 + \alpha_3(0)^3 + \alpha_2(0)^2 + \alpha_1(0) + \alpha_0 \end{aligned}$$

or

$$\begin{aligned} -5 &= \alpha_5 + \alpha_4 + \alpha_3 + \alpha_2 + \alpha_1 + \alpha_0 \\ 3 &= 5\alpha_5 + 4\alpha_4 + 3\alpha_3 + 2\alpha_2 + \alpha_1 \\ 198 &= 20\alpha_5 + 12\alpha_4 + 6\alpha_3 + 2\alpha_2 \\ -7 &= -\alpha_5 + \alpha_4 - \alpha_3 + \alpha_2 - \alpha_1 + \alpha_0 \\ 27 &= 5\alpha_5 - 4\alpha_4 + 3\alpha_3 - 2\alpha_2 + \alpha_1 \\ 0 &= \alpha_0 . \end{aligned} \tag{32}$$

System (32) can now be solved uniquely for $\alpha_5, \alpha_4, \ldots, \alpha_0$; the results are then substituted into (30) to obtain $f(\mathbf{A})$.

PROBLEMS 6.4

1. Find $\mathbf{A}^6$ in two different ways if

$$\mathbf{A} = \begin{bmatrix} 5 & 8 \\ -2 & -5 \end{bmatrix}.$$

(First find $\mathbf{A}^6$ using Theorem 1, and then by direct multiplication.)

2. Find $\mathbf{A}^{521}$ if

$$\mathbf{A} = \begin{bmatrix} 4 & 1 & -3 \\ 0 & -1 & 0 \\ 5 & 1 & -4 \end{bmatrix}.$$

3. Find $\mathbf{A}^{14} - 3\mathbf{A}^{13}$ if

$$\mathbf{A} = \begin{bmatrix} 4 & 1 & 2 \\ 0 & 0 & 0 \\ -8 & 1 & -4 \end{bmatrix}.$$

4. Set up (but do not solve) the necessary equations to find $\mathbf{A}^{10} - 3\mathbf{A}^{5}$ if

$$\mathbf{A} = \begin{bmatrix} 5 & -2 & 1 & 1 & 5 & -7 \\ 0 & 5 & 2 & 1 & -1 & 1 \\ 0 & 0 & 5 & 0 & 1 & -3 \\ 0 & 0 & 0 & 2 & 1 & 2 \\ 0 & 0 & 0 & 0 & 2 & 0 \\ 0 & 0 & 0 & 0 & 0 & 5 \end{bmatrix}.$$

6.5 FUNCTIONS OF A MATRIX

Once the student understands how to compute polynomials of a matrix, computing exponentials, sines, and other functions of a matrix will be easy since the methods developed in the previous two sections remain valid for these more general functions.

Let $f(\lambda)$ represent a *function* of λ and suppose we wish to compute $f(\mathbf{A})$. It can be shown, for a large class of problems,[17] that there exists a function $q(\lambda)$ and an $n - 1$ degree polynomial $r(\lambda)$ (we assume $\mathbf{A}$ is of order $n \times n$) such that

$$f(\lambda) = q(\lambda)d(\lambda) + r(\lambda), \tag{33}$$

where $d(\lambda) = \det(\mathbf{A} - \lambda\mathbf{I})$. Hence, it follows that

$$f(\mathbf{A}) = q(\mathbf{A})d(\mathbf{A}) + r(\mathbf{A}). \tag{34}$$

Since (33) and (34) are exactly Eqs. (12) and (14), where $f(\lambda)$ is now understood to be a general function and not restricted to polynomials, the analysis of Section 6.3 and 6.4 can again be applied. It then follows that

(a) $f(\mathbf{A}) = r(\mathbf{A})$, and (b) Theorem 1 of Section 6.4 remains valid.

Thus, the methods used to compute a polynomial of a matrix can be generalized and used to compute arbitrary functions of a matrix.

EXAMPLE 1 Find $e^{\mathbf{A}}$ if

$$\mathbf{A} = \begin{bmatrix} 1 & 2 \\ 4 & 3 \end{bmatrix}.$$

[17] See B. Friedman, "Principles and Techniques of Applied Mathematics," p. 120. Wiley, New York, 1956.

Solution The eigenvalues of **A** are $\lambda_1 = 5$, $\lambda_2 = -1$; thus,

$$f(\mathbf{A}) = e^{\mathbf{A}} \qquad r(\mathbf{A}) = \alpha_1 \mathbf{A} + \alpha_0 \mathbf{I}$$
$$f(\lambda) = e^{\lambda} \qquad r(\lambda) = \alpha_1 \lambda + \alpha_0 .$$

Now $f(\mathbf{A}) = r(\mathbf{A})$; hence

$$e^{\mathbf{A}} = \alpha_1 \mathbf{A} + \alpha_0 \mathbf{I}. \tag{35}$$

Also, since Theorem 1 of Section 6.4 is still valid,

$$f(5) = r(5),$$

and

$$f(-1) = r(-1);$$

hence,

$$e^5 = 5\alpha_1 + \alpha_0 ,$$
$$e^{-1} = -\alpha_1 + \alpha_0 .$$

Thus,

$$\alpha_1 = \frac{e^5 - e^{-1}}{6} \quad \text{and} \quad \alpha_0 = \frac{e^5 + 5e^{-1}}{6}.$$

Substituting these values into (35), we obtain

$$e^{\mathbf{A}} = \frac{1}{6}\begin{bmatrix} 2e^5 + 4e^{-1} & 2e^5 - 2e^{-1} \\ 4e^5 - 4e^{-1} & 4e^5 + 2e^{-1} \end{bmatrix}.$$

EXAMPLE 2 Find $e^{\mathbf{A}}$ if

$$\mathbf{A} = \begin{bmatrix} 2 & 1 & 0 \\ 0 & 2 & 1 \\ 0 & 0 & 2 \end{bmatrix}.$$

Solution The eigenvalues of **A** are $\lambda_1 = \lambda_2 = \lambda_3 = 2$, thus,

$$f(\mathbf{A}) = e^{\mathbf{A}} \qquad r(\mathbf{A}) = \alpha_2 \mathbf{A}^2 + \alpha_1 \mathbf{A} + \alpha_0 \mathbf{I}$$
$$f(\lambda) = e^{\lambda} \qquad r(\lambda) = \alpha_2 \lambda^2 + \alpha_1 \lambda + \alpha_0$$
$$f'(\lambda) = e^{\lambda} \qquad r'(\lambda) = 2\alpha_2 \lambda + \alpha_1$$
$$f''(\lambda) = e^{\lambda} \qquad r''(\lambda) = 2\alpha_2 .$$

Since $f(\mathbf{A}) = r(\mathbf{A})$,

$$e^{\mathbf{A}} = \alpha_2 \mathbf{A}^2 + \alpha_1 \mathbf{A} + \alpha_0 \mathbf{I}. \tag{36}$$

Since $\lambda = 2$ is an eigenvalue of multiplicity three,

$$f(2) = r(2),$$
$$f'(2) = r'(2),$$
$$f''(2) = r''(2);$$

hence,

$$e^2 = 4\alpha_2 + 2\alpha_1 + \alpha_0,$$
$$e^2 = 4\alpha_2 + \alpha_1,$$
$$e^2 = 2\alpha_2,$$

or

$$\alpha_2 = \frac{e^2}{2}, \qquad \alpha_1 = -e^2, \qquad \alpha_0 = e^2.$$

Substituting these values into (36), we obtain

$$e^{\mathbf{A}} = \frac{e^2}{2}\begin{bmatrix} 4 & 4 & 1 \\ 0 & 4 & 4 \\ 0 & 0 & 4 \end{bmatrix} - e^2\begin{bmatrix} 2 & 1 & 0 \\ 0 & 2 & 1 \\ 0 & 0 & 2 \end{bmatrix} + e^2\begin{bmatrix} 1 & 0 & 0 \\ 0 & 1 & 0 \\ 0 & 0 & 1 \end{bmatrix} = \begin{bmatrix} e^2 & e^2 & e^2/2 \\ 0 & e^2 & e^2 \\ 0 & 0 & e^2 \end{bmatrix}.$$

EXAMPLE 3 Find sin **A** if

$$\mathbf{A} = \begin{bmatrix} \pi & 1 & 0 \\ 0 & \pi & 0 \\ 4 & 1 & \pi/2 \end{bmatrix}.$$

Solution The eigenvalues of **A** are $\lambda_1 = \pi/2$, $\lambda_2 = \lambda_3 = \pi$; thus

$$f(\mathbf{A}) = \sin \mathbf{A} \qquad r(\mathbf{A}) = \alpha_2 \mathbf{A}^2 + \alpha_1 \mathbf{A} + \alpha_0 \mathbf{I}$$
$$f(\lambda) = \sin \lambda \qquad r(\lambda) = \alpha_2 \lambda^2 + \alpha_1 \lambda + \alpha_0$$
$$f'(\lambda) = \cos \lambda \qquad r'(\lambda) = 2\alpha_2 \lambda + \alpha_1.$$

But $f(\mathbf{A}) = r(\mathbf{A})$, hence

$$\sin \mathbf{A} = \alpha_2 \mathbf{A}^2 + \alpha_1 \mathbf{A} + \alpha_0 \mathbf{I}. \tag{37}$$

Since $\lambda = \pi/2$ is an eigenvalue of multiplicity 1 and $\lambda = \pi$ is an eigenvalue of multiplicity 2, it follows that

$$f(\pi/2) = r(\pi/2),$$
$$f(\pi) = r(\pi),$$
$$'(\pi) = r'(\pi);$$

hence,

$$\sin \pi/2 = \alpha_2(\pi/2)^2 + \alpha_1(\pi/2) + \alpha_0,$$
$$\sin \pi = \alpha_2(\pi)^2 + \alpha_1(\pi) + \alpha_0,$$
$$\cos \pi = 2\alpha_2 \pi + \alpha_1,$$

or simplifying

$$4 = \alpha_2 \pi^2 + 2\alpha_1 \pi + 4\alpha_0,$$
$$0 = \alpha_2 \pi^2 + \alpha_1 \pi + \alpha_0,$$
$$-1 = 2\alpha_2 \pi + \alpha_1.$$

Thus, $\alpha_2 = (1/\pi^2)(4 - 2\pi)$, $\alpha_1 = (1/\pi^2)(-8\pi + 3\pi^2)$, $\alpha_0 = (1/\pi^2)(4\pi^2 - \pi^3)$. Substituting these values into (37), we obtain

$$\sin \mathbf{A} = 1/\pi^2 \begin{bmatrix} 0 & -\pi^2 & 0 \\ 0 & 0 & 0 \\ -8\pi & 16 - 10\pi & \pi^2 \end{bmatrix}.$$

In closing, we point out that although exponentials of any square matrix can always be computed by the above methods, not all functions of all matrices can; $f(\mathbf{A})$ must first be "well defined" where by "well defined" (see Theorem 1 of Section 6.1) we mean that $f(z)$ has a Taylor series which converges for $|z| < R$ and all the eigenvalues of $\mathbf{A}$ have the property that their absolute values are also less than R.

PROBLEMS 6.5

In Problems 1–4, determine $e^{\mathbf{A}}$:

1. $\mathbf{A} = \begin{bmatrix} 1 & 3 \\ 4 & 2 \end{bmatrix}$

2. $\mathbf{A} = \begin{bmatrix} 4 & -1 \\ 1 & 2 \end{bmatrix}$

3. $\mathbf{A} = \begin{bmatrix} 1 & 1 & 2 \\ -1 & 3 & 4 \\ 0 & 0 & 2 \end{bmatrix}$

4. $\mathbf{A} = \begin{bmatrix} 1 & 1 & 2 \\ 3 & -1 & 4 \\ 0 & 0 & 2 \end{bmatrix}.$

5. Find $\cos \mathbf{A}$ if

$$\mathbf{A} = \begin{bmatrix} \pi & 3\pi \\ 2\pi & 2\pi \end{bmatrix}.$$

6. The function $f(z) = \log(1 + z)$ has the Taylor series

$$\sum_{k=1}^{\infty} \frac{(-1)^{k-1} z^k}{k}$$

which converges for $|z| < 1$. For the following matrices, $\mathbf{A}$, determine whether or not $\log(\mathbf{A} + \mathbf{I})$ is defined and, if so, find it.

(a) $\begin{bmatrix} \frac{1}{2} & 1 \\ 0 & -\frac{1}{2} \end{bmatrix}$ (b) $\begin{bmatrix} -6 & 9 \\ -2 & 3 \end{bmatrix}$ (c) $\begin{bmatrix} 3 & 5 \\ -1 & -3 \end{bmatrix}$ (d) $\begin{bmatrix} 0 & 0 \\ 0 & 0 \end{bmatrix}.$

6.6 THE FUNCTION $e^{\mathbf{A}t}$

A very important function in the matrix calculus is $e^{\mathbf{A}t}$, where $\mathbf{A}$ is a square constant matrix (that is, all of its entries are constants) and t is a variable. This function may be calculated easily by defining a new matrix $\mathbf{B} = \mathbf{A}t$ and then computing $e^{\mathbf{B}}$ by the methods of the previous section.

EXAMPLE 1 Find $e^{\mathbf{A}t}$ if

$$\mathbf{A} = \begin{bmatrix} 1 & 2 \\ 4 & 3 \end{bmatrix}.$$

Solution Define

$$\mathbf{B} = \mathbf{A}t = \begin{bmatrix} t & 2t \\ 4t & 3t \end{bmatrix}.$$

The problem then reduces to finding $e^{\mathbf{B}}$. The eigenvalues of $\mathbf{B}$ are $\lambda_1 = 5t$, $\lambda_2 = -t$. Note that the eigenvalues now depend on t.

$$f(\mathbf{B}) = e^{\mathbf{B}} \qquad r(\mathbf{B}) = \alpha_1 \mathbf{B} + \alpha_0 \mathbf{I}$$
$$f(\lambda) = e^{\lambda} \qquad r(\lambda) = \alpha_1 \lambda + \alpha_0 .$$

Since $f(\mathbf{B}) = r(\mathbf{B})$,

$$e^{\mathbf{B}} = \alpha_1 \mathbf{B} + \alpha_0 \mathbf{I}. \tag{38}$$

Also, $f(\lambda_i) = r(\lambda_i)$; hence

$$e^{5t} = \alpha_1(5t) + \alpha_0 ,$$
$$e^{-t} = \alpha_1(-t) + \alpha_0 .$$

Thus, $\alpha_1 = (1/6t)(e^{5t} - e^{-t})$ and $\alpha_0 = (1/6)(e^{5t} + 5e^{-t})$. Substituting these values into (38), we obtain

$$e^{\mathbf{A}t} = e^{\mathbf{B}} = \left(\frac{1}{6t}\right)(e^{5t} - e^{-t})\begin{bmatrix} t & 2t \\ 4t & 3t \end{bmatrix} + \left(\frac{1}{6}\right)(e^{5t} + 5e^{-t})\begin{bmatrix} 1 & 0 \\ 0 & 1 \end{bmatrix}$$
$$= \frac{1}{6}\begin{bmatrix} 2e^{5t} + 4e^{-t} & 2e^{5t} - 2e^{-t} \\ 4e^{5t} - 4e^{-t} & 4e^{5t} + 2e^{-t} \end{bmatrix}.$$

EXAMPLE 2 Find $e^{\mathbf{A}t}$ if

$$\mathbf{A} = \begin{bmatrix} 3 & 1 & 0 \\ 0 & 3 & 1 \\ 0 & 0 & 3 \end{bmatrix}.$$

Solution Define

$$\mathbf{B} = \mathbf{A}t = \begin{bmatrix} 3t & t & 0 \\ 0 & 3t & t \\ 0 & 0 & 3t \end{bmatrix}.$$

The problem again reduces to finding $e^{\mathbf{B}}$. The eigenvalues of $\mathbf{B}$ are

$$\lambda_1 = \lambda_2 = \lambda_3 = 3t$$

thus,

$$f(\mathbf{B}) = e^{\mathbf{B}} \qquad r(\mathbf{B}) = \alpha_2 \mathbf{B}^2 + \alpha_1 \mathbf{B} + \alpha_0 \mathbf{I}$$
$$f(\lambda) = e^{\lambda} \qquad r(\lambda) = \alpha_2 \lambda^2 + \alpha_1 \lambda + \alpha_0 \tag{39}$$

$$f'(\lambda) = e^{\lambda} \qquad r'(\lambda) = 2\alpha_2 \lambda + \alpha_1 \tag{40}$$

$$f''(\lambda) = e^{\lambda} \qquad r''(\lambda) = 2\alpha_2. \tag{41}$$

Since $f(\mathbf{B}) = r(\mathbf{B})$,

$$e^{\mathbf{B}} = \alpha_2 \mathbf{B}^2 + \alpha_1 \mathbf{B} + \alpha_0 \mathbf{I}. \tag{42}$$

Since $\lambda = 3t$ is an eigenvalue of multiplicity 3,

$$f(3t) = r(3t), \tag{43}$$

$$f'(3t) = r'(3t), \tag{44}$$

$$f''(3t) = r''(3t). \tag{45}$$

Thus, using (39)–(41), we obtain

$$e^{3t} = (3t)^2\alpha_2 + (3t)\alpha_1 + \alpha_0,$$
$$e^{3t} = 2(3t)\alpha_2 + \alpha_1,$$
$$e^{3t} = 2\alpha_2$$

or

$$e^{3t} = 9t^2\alpha_2 + 3t\alpha_1 + \alpha_0, \tag{46}$$

$$e^{3t} = 6t\alpha_2 + \alpha_1, \tag{47}$$

$$e^{3t} = 2\alpha_2. \tag{48}$$

Solving (46)–(48) simultaneously, we obtain

$$\alpha_2 = \tfrac{1}{2}e^{3t}, \qquad \alpha_1 = (1 - 3t)e^{3t}, \qquad \alpha_0 = (1 - 3t + \tfrac{9}{2}t^2)e^{3t}.$$

From (42), it follows that

$$e^{\mathbf{A}t} = e^{\mathbf{B}} = \tfrac{1}{2}e^{3t}\begin{bmatrix} 9t^2 & 6t^2 & t^2 \\ 0 & 9t^2 & 6t^2 \\ 0 & 0 & 9t^2 \end{bmatrix} + (1 - 3t)e^{3t}\begin{bmatrix} 3t & t & 0 \\ 0 & 3t & t \\ 0 & 0 & 3t \end{bmatrix}$$

$$+ (1 - 3t + \tfrac{9}{2}t^2)e^{3t}\begin{bmatrix} 1 & 0 & 0 \\ 0 & 1 & 0 \\ 0 & 0 & 1 \end{bmatrix}$$

$$= e^{3t}\begin{bmatrix} 1 & t & t^2/2 \\ 0 & 1 & t \\ 0 & 0 & 1 \end{bmatrix}.$$

PROBLEMS 6.6

Find $e^{\mathbf{A}t}$ if $\mathbf{A}$ is given by:

1. $\begin{bmatrix} 4 & 4 \\ 3 & 5 \end{bmatrix}$

2. $\begin{bmatrix} 2 & 1 \\ -1 & -2 \end{bmatrix}$

3. $\begin{bmatrix} 4 & 1 \\ -1 & 2 \end{bmatrix}$

4. $\begin{bmatrix} 0 & 1 & 0 \\ 0 & 0 & 1 \\ 0 & 0 & 0 \end{bmatrix}$

5. $\begin{bmatrix} 1 & 0 & 0 \\ 4 & 1 & 2 \\ -1 & 4 & -1 \end{bmatrix}.$

6.7 COMPLEX EIGENVALUES

When computing $e^{\mathbf{A}t}$, it is often the case that the eigenvalues of $\mathbf{B} = \mathbf{A}t$ turn out to be complex. If this occurs the complex eigenvalues will appear in conjugate pairs, assuming the elements of $\mathbf{A}$ to be real, and these can be combined to produce real functions.

Let z represent a complex variable. Define e^z by

$$e^z = \sum_{k=0}^{\infty} \frac{z^k}{k!} = 1 + z + \frac{z^2}{2!} + \frac{z^3}{3!} + \frac{z^4}{4!} + \frac{z^5}{5!} + \cdots \tag{49}$$

(see Eq. (5)). Setting $z = i\theta$, θ real, we obtain

$$\begin{aligned} e^{i\theta} &= 1 + i\theta + \frac{(i\theta)^2}{2!} + \frac{(i\theta)^3}{3!} + \frac{(i\theta)^4}{4!} + \frac{(i\theta)^5}{5!} + \cdots \\ &= 1 + i\theta - \frac{\theta^2}{2!} - \frac{i\theta^3}{3!} + \frac{\theta^4}{4!} + \frac{i\theta^5}{5!} - \cdots \end{aligned}$$

Combining real and imaginary terms, we obtain

$$e^{i\theta} = \left(1 - \frac{\theta^2}{2!} + \frac{\theta^4}{4!} - \cdots\right) + i\left(\theta - \frac{\theta^3}{3!} + \frac{\theta^5}{5!} - \cdots\right). \tag{50}$$

But the Maclaurin series expansions for $\sin\theta$ and $\cos\theta$ are

$$\sin\theta = \frac{\theta}{1!} - \frac{\theta^3}{3!} + \frac{\theta^5}{5!} + \cdots$$

$$\cos\theta = 1 - \frac{\theta^2}{2!} + \frac{\theta^4}{4!} - \frac{\theta^6}{6!} + \cdots;$$

hence, Eq. (50) may be rewritten as

$$e^{i\theta} = \cos\theta + i\sin\theta. \tag{51}$$

Equation (51) is referred to as DeMoivre's formula. If the same analysis is applied to $z = -i\theta$, it follows that

$$e^{-i\theta} = \cos\theta - i\sin\theta. \tag{52}$$

Adding (51) and (52), we obtain

$$\cos\theta = \frac{e^{i\theta} + e^{-i\theta}}{2}, \tag{53}$$

while subtracting (52) from (51), we obtain

$$\sin\theta = \frac{e^{i\theta} - e^{-i\theta}}{2i}. \tag{54}$$

Equations (53) and (54) are Euler's relations and can be used to reduce complex exponentials to expressions involving real numbers.

EXAMPLE 1 Find $e^{\mathbf{A}t}$ if

$$\mathbf{A} = \begin{bmatrix} -1 & 5 \\ -2 & 1 \end{bmatrix}.$$

Solution

$$\mathbf{B} = \mathbf{A}t = \begin{bmatrix} -t & 5t \\ -2t & t \end{bmatrix}.$$

Hence the eigenvalues of $\mathbf{B}$ are $\lambda_1 = 3ti$ and $\lambda_2 = -3ti$; thus

$$f(\mathbf{B}) = e^{\mathbf{B}} \qquad r(\mathbf{B}) = \alpha_1\mathbf{B} + \alpha_0\mathbf{I}$$
$$f(\lambda) = e^{\lambda} \qquad r(\lambda) = \alpha_1\lambda + \alpha_0$$

Since $f(\mathbf{B}) = r(\mathbf{B})$,

$$e^{\mathbf{B}} = \alpha_1\mathbf{B} + \alpha_0\mathbf{I}, \tag{55}$$

and since $f(\lambda_i) = r(\lambda_i)$,

$$e^{3ti} = \alpha_1(3ti) + \alpha_0,$$
$$e^{-3ti} = \alpha_1(-3ti) + \alpha_0.$$

Thus,

$$\alpha_0 = \frac{e^{3ti} + e^{-3ti}}{2} \qquad \text{and} \qquad \alpha_1 = \frac{1}{3t}\left(\frac{e^{3ti} - e^{-3ti}}{2i}\right).$$

If we now use (53) and (54), where in this case $\theta = 3t$, it follows that

$$\alpha_0 = \cos 3t \qquad \text{and} \qquad \alpha_1 = (1/3t)\sin 3t.$$

Substituting these values into (55), we obtain

$$e^{\mathbf{A}t} = e^{\mathbf{B}} = \begin{bmatrix} -\frac{1}{3}\sin 3t + \cos 3t & \frac{5}{3}\sin 3t \\ -\frac{2}{3}\sin 3t & \frac{1}{3}\sin 3t + \cos 3t \end{bmatrix}.$$

In Example 1, the eigenvalues of $\mathbf{B}$ are pure imaginary permitting the application of (53) and (54) in a straightforward manner. In the general case, where the eigenvalues are complex numbers, we can still use Euler's relations providing we note the following:

$$\frac{e^{\beta+i\theta}+e^{\beta-i\theta}}{2}=\frac{e^{\beta}e^{i\theta}+e^{\beta}e^{-i\theta}}{2}=\frac{e^{\beta}(e^{i\theta}+e^{-i\theta})}{2}=e^{\beta}\cos\theta, \tag{56}$$

and

$$\frac{e^{\beta+i\theta}-e^{\beta-i\theta}}{2i}=\frac{e^{\beta}e^{i\theta}-e^{\beta}e^{-i\theta}}{2i}=\frac{e^{\beta}(e^{i\theta}-e^{-i\theta})}{2i}=e^{\beta}\sin\theta. \tag{57}$$

EXAMPLE 2 Find $e^{\mathbf{A}t}$ if

$$\mathbf{A}=\begin{bmatrix}2 & -1\\ 4 & 1\end{bmatrix}.$$

Solution

$$\mathbf{B}=\mathbf{A}t=\begin{bmatrix}2t & -t\\ 4t & t\end{bmatrix};$$

hence, the eigenvalues of **B** are

$$\lambda_1=\left(\frac{3}{2}+i\frac{\sqrt{15}}{2}\right)t, \qquad \lambda_2=\left(\frac{3}{2}-i\frac{\sqrt{15}}{2}\right)t.$$

Thus,

$$f(\mathbf{B})=e^{\mathbf{B}} \qquad r(\mathbf{B})=\alpha_1\mathbf{B}+\alpha_0\mathbf{I}$$
$$f(\lambda)=e^{\lambda} \qquad r(\lambda)=\alpha_1\lambda+\alpha_0.$$

Since $f(\mathbf{B})=r(\mathbf{B})$,

$$e^{\mathbf{B}}=\alpha_1\mathbf{B}+\alpha_0\mathbf{I}, \tag{58}$$

and since $f(\lambda_i)=r(\lambda_i)$,

$$e^{[3/2+i(\sqrt{15}/2)]t}=\alpha_1[\tfrac{3}{2}+i(\sqrt{15}/2)]t+\alpha_0,$$
$$e^{[3/2-i(\sqrt{15}/2)]t}=\alpha_1[\tfrac{3}{2}-i(\sqrt{15}/2)]t+\alpha_0.$$

Putting this system into matrix form, and solving for α_1 and α_0 by inversion, we obtain

$$\alpha_1=\frac{2}{\sqrt{15}t}\left[\frac{e^{[(3/2)t+\sqrt{15}/2)ti]}-e^{[(3/2)t-(\sqrt{15}/2)ti]}}{2i}\right]$$

$$\alpha_0=\frac{-3}{\sqrt{15}}\left(\frac{e^{[(3/2)t+(\sqrt{15}/2)ti]}-e^{[(3/2)t-(\sqrt{15}/2)ti]}}{2i}\right)$$
$$+\left(\frac{e^{[(3/2)t+(\sqrt{15}/2)ti]}+e^{[(3/2)t-(\sqrt{15}/2)ti]}}{2}\right).$$

Using (56) and (57) where, $\beta = \frac{3}{2}t$ and $\theta = (\sqrt{15}/2)t$, we obtain

$$\alpha_1 = \frac{2}{\sqrt{15t}} e^{3t/2} \sin \frac{\sqrt{15t}}{2}$$

$$\alpha_0 = -\frac{3}{\sqrt{15}} e^{3t/2} \sin \frac{\sqrt{15}}{2} t + e^{3t/2} \cos \frac{\sqrt{15}}{2} t.$$

Substituting these values into (58), we obtain

$$e^{\mathbf{A}t} = e^{3t/2} \begin{bmatrix} \frac{1}{\sqrt{15}} \sin \frac{\sqrt{15}}{2} t + \cos \frac{\sqrt{15}}{2} t & \frac{-2}{\sqrt{15}} \sin \frac{\sqrt{15}}{2} t \\ \frac{8}{\sqrt{15}} \sin \frac{\sqrt{15}}{2} t & \frac{-1}{\sqrt{15}} \sin \frac{\sqrt{15}}{2} t + \cos \frac{\sqrt{15}}{2} t \end{bmatrix}.$$

PROBLEMS 6.7

Find $e^{\mathbf{A}t}$ if $\mathbf{A}$ is given by:

1. $\begin{bmatrix} 1 & -1 \\ 5 & -1 \end{bmatrix}$ **2.** $\begin{bmatrix} 2 & -2 \\ 3 & -2 \end{bmatrix}$ **3.** $\begin{bmatrix} 3 & 1 \\ -2 & 5 \end{bmatrix}$.

6.8 PROPERTIES OF $e^{\mathbf{A}}$

Since the scalar function e^x and the matrix function $e^{\mathbf{A}}$ are defined similarly (see Eqs. (5) and (6)), it should not be surprising to find that they possess some similar properties. What might be surprising, however, is that *not* all properties of e^x are common to $e^{\mathbf{A}}$. For example, while it is always true that $e^x e^y = e^{x+y} = e^y e^x$, the same cannot be said for matrices $e^{\mathbf{A}}$ and $e^{\mathbf{B}}$ *unless* **A** *and* **B** *commute.*

EXAMPLE 1 Find $e^{\mathbf{A}}e^{\mathbf{B}}$, $e^{\mathbf{A}+\mathbf{B}}$, and $e^{\mathbf{B}}e^{\mathbf{A}}$ if

$$\mathbf{A} = \begin{bmatrix} 1 & 1 \\ 0 & 0 \end{bmatrix} \quad \text{and} \quad \mathbf{B} = \begin{bmatrix} 0 & 0 \\ 0 & 1 \end{bmatrix}.$$

Solution Using the methods developed in Section 6.5, we find

$$e^{\mathbf{A}} = \begin{bmatrix} e & e-1 \\ 0 & 1 \end{bmatrix}, \quad e^{\mathbf{B}} = \begin{bmatrix} 1 & 0 \\ 0 & e \end{bmatrix}, \quad e^{\mathbf{A}+\mathbf{B}} = \begin{bmatrix} e & e \\ 0 & e \end{bmatrix}.$$

Therefore,

$$e^{\mathbf{A}}e^{\mathbf{B}} = \begin{bmatrix} e & e-1 \\ 0 & 1 \end{bmatrix} \begin{bmatrix} 1 & 0 \\ 0 & e \end{bmatrix} = \begin{bmatrix} e & e^2 - e \\ 0 & e \end{bmatrix}$$

and

$$e^{\mathbf{B}}e^{\mathbf{A}} = \begin{bmatrix} 1 & 0 \\ 0 & e \end{bmatrix} \begin{bmatrix} e & e-1 \\ 0 & 1 \end{bmatrix} = \begin{bmatrix} e & e-1 \\ 0 & e \end{bmatrix};$$

hence

$$e^{\mathbf{A}}e^{\mathbf{B}} \neq e^{\mathbf{A}+\mathbf{B}} \neq e^{\mathbf{B}}e^{\mathbf{A}}.$$

Two properties that both e^x and $e^{\mathbf{A}}$ do have in common are given by the following:

Property 1 $e^{\mathbf{0}} = \mathbf{I}$, where $\mathbf{0}$ represents the zero matrix.

Proof From (6) we have that

$$e^{\mathbf{A}} = \sum_{k=0}^{\infty} \left(\frac{\mathbf{A}^k}{k!}\right) = \mathbf{I} + \sum_{k=1}^{\infty} \left(\frac{\mathbf{A}^k}{k!}\right).$$

Hence,

$$e^{\mathbf{0}} = \mathbf{I} + \sum_{k=1}^{\infty} \frac{\mathbf{0}^k}{k!} = \mathbf{I}.$$

Property 2 $(e^{\mathbf{A}})^{-1} = e^{-\mathbf{A}}$.

Proof

$$(e^{\mathbf{A}})(e^{-\mathbf{A}})$$

$$= \left[\sum_{k=0}^{\infty} \left(\frac{\mathbf{A}^k}{k!}\right)\right] \left[\sum_{k=0}^{\infty} \frac{(-\mathbf{A})^k}{k!}\right]$$

$$= \left[\mathbf{I} + \mathbf{A} + \frac{\mathbf{A}^2}{2!} + \frac{\mathbf{A}^3}{3!} + \cdots\right]\left[\mathbf{I} - \mathbf{A} + \frac{\mathbf{A}^2}{2!} - \frac{\mathbf{A}^3}{3!} + \cdots\right]$$

$$= \mathbf{II} + \mathbf{A}[1 - 1] + \mathbf{A}^2[\tfrac{1}{2}! - 1 + \tfrac{1}{2}!] + \mathbf{A}^3[-\tfrac{1}{3}! + \tfrac{1}{2}! - \tfrac{1}{2}! + \tfrac{1}{3}!] + \cdots$$

$$= \mathbf{I}.$$

Thus, $e^{-\mathbf{A}}$ is an inverse of $e^{\mathbf{A}}$. However, by definition, an inverse of $e^{\mathbf{A}}$ is $(e^{\mathbf{A}})^{-1}$; hence, from the uniqueness of the inverse (Theorem 1 of Section 3.3), we have that $e^{-\mathbf{A}} = (e^{\mathbf{A}})^{-1}$.

EXAMPLE 2 Verify Property 2 for

$$\mathbf{A} = \begin{bmatrix} 0 & 1 \\ 0 & 0 \end{bmatrix}.$$

Solution

$$-\mathbf{A} = \begin{bmatrix} 0 & -1 \\ 0 & 0 \end{bmatrix},$$

$$e^{\mathbf{A}} = \begin{bmatrix} 1 & 1 \\ 0 & 1 \end{bmatrix}, \qquad \text{and} \qquad e^{-\mathbf{A}} = \begin{bmatrix} 1 & -1 \\ 0 & 1 \end{bmatrix}.$$

Thus,

$$(e^{\mathbf{A}})^{-1} = \begin{bmatrix} 1 & 1 \\ 0 & 1 \end{bmatrix}^{-1} = \begin{bmatrix} 1 & -1 \\ 0 & 1 \end{bmatrix} = e^{-\mathbf{A}}.$$

Note that Property 2 implies that $e^{\mathbf{A}}$ *is always invertible* even if $\mathbf{A}$ itself is not.

Property 3 $(e^{\mathbf{A}})' = e^{\mathbf{A}'}$.

Proof The proof of this property is left as an exercise for the reader (see Problem 3).

EXAMPLE 3 Verify Property 3 for

$$\mathbf{A} = \begin{bmatrix} 1 & 2 \\ 4 & 3 \end{bmatrix}.$$

Solution

$$\mathbf{A}' = \begin{bmatrix} 1 & 4 \\ 2 & 3 \end{bmatrix},$$

$$e^{\mathbf{A}'} = \frac{1}{6}\begin{bmatrix} 2e^5 + 4e^{-1} & 4e^5 - 4e^{-1} \\ 2e^5 - 2e^{-1} & 4e^5 + 2e^{-1} \end{bmatrix},$$

and

$$e^{\mathbf{A}} = \frac{1}{6}\begin{bmatrix} 2e^5 + 4e^{-1} & 2e^5 - 2e^{-1} \\ 4e^5 - 4e^{-1} & 4e^5 + 2e^{-1} \end{bmatrix};$$

Hence, $(e^{\mathbf{A}})' = e^{\mathbf{A}'}$.

PROBLEMS 6.8

1. Verify Property 2 for

$$\mathbf{A} = \begin{bmatrix} 1 & 3 \\ 0 & 1 \end{bmatrix}.$$

2. Verify Property 3 for

$$\mathbf{A} = \begin{bmatrix} 2 & 1 & 0 \\ 0 & 2 & 0 \\ 1 & -1 & 1 \end{bmatrix}.$$

3. Prove Property (3). (Hint: Using the fact that $\det(\mathbf{B}) = \det(\mathbf{B}')$, first show that the eigenvalues of $\mathbf{A}$ are identical to eigenvalues of $\mathbf{A}'$. Next show that if $e^{\mathbf{A}} = \alpha_{n-1}\mathbf{A}^{n-1} + \cdots + \alpha_1\mathbf{A} + \alpha_0\mathbf{I}$, and if

$$e^{\mathbf{A}'} = \beta_{n-1}(\mathbf{A}')^{n-1} + \cdots + \beta_1\mathbf{A}' + \beta_0\mathbf{I},$$

then $\alpha_j = \beta_j$ for $j = 0, 1, \ldots, n-1$.)

4. Find $e^{\mathbf{A}}e^{\mathbf{B}}$, $e^{\mathbf{B}}e^{\mathbf{A}}$, and $e^{\mathbf{A}+\mathbf{B}}$ if

$$\mathbf{A} = \begin{bmatrix} 1 & 1 \\ 0 & 0 \end{bmatrix} \quad \text{and} \quad \mathbf{B} = \begin{bmatrix} 0 & 1 \\ 0 & 1 \end{bmatrix}$$

and show that $e^{\mathbf{A}}e^{\mathbf{B}} \neq e^{\mathbf{A}+\mathbf{B}} \neq e^{\mathbf{B}}e^{\mathbf{A}}$.

5. Find two matrices $\mathbf{A}$ and $\mathbf{B}$ such that

$$e^{\mathbf{A}}e^{\mathbf{B}} = e^{\mathbf{A}+\mathbf{B}}.$$

6. By using the definition of $e^{\mathbf{A}}$, prove that if $\mathbf{A}$ and $\mathbf{B}$ commute, then $e^{\mathbf{A}}e^{\mathbf{B}} = e^{\mathbf{A}+\mathbf{B}}$.

6.9 DERIVATIVES OF A MATRIX

Definition 1 An $n \times n$ matrix $\mathbf{A}(t) = [a_{ij}(t)]$ is *continuous* at $t = t_0$ if each of its elements $a_{ij}(t)$ $(i, j = 1, 2, \ldots, n)$ is continuous at $t = t_0$.

For example, the matrix given in (59) is continuous everywhere since each of its elements is continuous everywhere while the matrix given in (60) is not continuous at $t = 0$ since the (1, 2) element, $\sin(1/t)$, is not continuous at $t = 0$.

$$\begin{bmatrix} e^t & t^2 - 1 \\ 2 & \sin^2 t \end{bmatrix} \tag{59}$$

$$\begin{bmatrix} t^3 - 3t & \sin(1/t) \\ 2t & 45 \end{bmatrix} \tag{60}$$

We shall use the notation $\mathbf{A}(t)$ to emphasize that the matrix $\mathbf{A}$ may depend on the variable t.

Definition 2 An $n \times n$ matrix $\mathbf{A}(t) = [a_{ij}(t)]$ is *differentiable* at $t = t_0$ if each of the elements $a_{ij}(t)$ $(i, j = 1, 2, \ldots, n)$ is differentiable at $t = t_0$ and

$$\frac{d\mathbf{A}(t)}{dt} = \left[\frac{da_{ij}(t)}{dt}\right]. \tag{61}$$

Generally we will use the notation $\dot{\mathbf{A}}(t)$ to represent $d\mathbf{A}(t)/dt$.

EXAMPLE 1 Find $\dot{\mathbf{A}}(t)$ if

$$\mathbf{A}(t) = \begin{bmatrix} t^2 & \sin t \\ \ln t & e^{t^2} \end{bmatrix}.$$

Solution

$$\dot{\mathbf{A}}(t) = \frac{d\mathbf{A}(t)}{dt} = \begin{bmatrix} \dfrac{d(t^2)}{dt} & \dfrac{d(\sin t)}{dt} \\ \dfrac{d(\ln t)}{dt} & \dfrac{d(e^{t^2})}{dt} \end{bmatrix} = \begin{bmatrix} 2t & \cos t \\ \dfrac{1}{t} & 2te^{t^2} \end{bmatrix}.$$

EXAMPLE 2 Find $\dot{\mathbf{A}}(t)$ if

$$\mathbf{A}(t) = \begin{bmatrix} 3t \\ 45 \\ t^2 \end{bmatrix}.$$

Solution

$$\dot{\mathbf{A}}(t) = \begin{bmatrix} \dfrac{d(3t)}{dt} \\ \dfrac{d(45)}{dt} \\ \dfrac{d(t^2)}{dt} \end{bmatrix} = \begin{bmatrix} 3 \\ 0 \\ 2t \end{bmatrix}.$$

EXAMPLE 3 Find $\dot{\mathbf{x}}(t)$ if

$$\mathbf{x}(t) = \begin{bmatrix} x_1(t) \\ x_2(t) \\ \vdots \\ x_n(t) \end{bmatrix}.$$

Solution

$$\dot{\mathbf{x}}(t) = \begin{bmatrix} \dot{x}_1(t) \\ \dot{x}_2(t) \\ \vdots \\ \dot{x}_n(t) \end{bmatrix}.$$

The following properties of the derivative can be verified:

(P1) $\dfrac{d(\mathbf{A}(t)+\mathbf{B}(t))}{dt} = \dfrac{d\mathbf{A}(t)}{dt} + \dfrac{d\mathbf{B}(t)}{dt}$.

(P2) $\dfrac{d[\alpha\mathbf{A}(t)]}{dt} = \alpha\dfrac{d\mathbf{A}(t)}{dt}$, where α is a constant.

(P3) $\dfrac{d[\beta(t)\mathbf{A}(t)]}{dt} = \left(\dfrac{d\beta(t)}{dt}\right)\mathbf{A}(t) + \beta(t)\left(\dfrac{d\mathbf{A}(t)}{dt}\right)$, when $\beta(t)$ is a scalar function of t.

(P4) $\dfrac{d[\mathbf{A}(t)\mathbf{B}(t)]}{dt} = \left(\dfrac{d\mathbf{A}(t)}{dt}\right)\mathbf{B}(t) + \mathbf{A}(t)\left(\dfrac{d\mathbf{B}(t)}{dt}\right)$.

We warn the student to be very careful about the order of the matrices in (P4). Any commutation of the matrices on the right-hand side will generally yield a wrong answer. For instance, it generally is not true that

$$\frac{d}{dt}[\mathbf{A}(t)\mathbf{B}(t)] = \left(\frac{d\mathbf{A}(t)}{dt}\right)\mathbf{B}(t) + \left(\frac{d\mathbf{B}(t)}{dt}\right)\mathbf{A}(t).$$

EXAMPLE 4 Verify Property (P4) for

$$\mathbf{A}(t) = \begin{bmatrix} 2t & 3t^2 \\ 1 & t \end{bmatrix} \quad \text{and } \mathbf{B}(t) = \begin{bmatrix} 1 & 2t \\ 3t & 2 \end{bmatrix}.$$

Solution

$$\frac{d}{dt}[\mathbf{A}(t)\mathbf{B}(t)] = \frac{d}{dt}\left(\begin{bmatrix} 2t & 3t^2 \\ 1 & t \end{bmatrix}\begin{bmatrix} 1 & 2t \\ 3t & 2 \end{bmatrix}\right)$$

$$= \frac{d}{dt}\begin{bmatrix} 2t+9t^3 & 10t^2 \\ 1+3t^2 & 4t \end{bmatrix} = \begin{bmatrix} 2+27t^2 & 20t \\ 6t & 4 \end{bmatrix},$$

and

$$\left[\frac{d\mathbf{A}(t)}{dt}\right]\mathbf{B}(t) + \mathbf{A}(t)\left[\frac{d\mathbf{B}(t)}{dt}\right] = \begin{bmatrix} 2 & 6t \\ 0 & 1 \end{bmatrix}\begin{bmatrix} 1 & 2t \\ 3t & 2 \end{bmatrix} + \begin{bmatrix} 2t & 3t^2 \\ 1 & t \end{bmatrix}\begin{bmatrix} 0 & 2 \\ 3 & 0 \end{bmatrix}$$

$$= \begin{bmatrix} 2+27t^2 & 20t \\ 6t & 4 \end{bmatrix} = \frac{d[\mathbf{A}(t)\mathbf{B}(t)]}{dt}.$$

We are now in a position to establish one of the more important properties of $e^{\mathbf{A}t}$. It is this property that makes the exponential so useful in differential equations (as we shall see in Chapter 7) and hence so fundamental in analysis.

Theorem 1 If $\mathbf{A}$ is a constant matrix then

$$\frac{de^{\mathbf{A}t}}{dt} = \mathbf{A}e^{\mathbf{A}t} = e^{\mathbf{A}t}\mathbf{A}.$$

Proof From (6) we have that

$$e^{\mathbf{A}t} = \sum_{k=0}^{\infty} \frac{(\mathbf{A}t)^k}{k!}$$

or

$$e^{\mathbf{A}t} = \mathbf{I} + t\mathbf{A} + \frac{t^2\mathbf{A}^2}{2!} + \frac{t^3\mathbf{A}^3}{3!} + \cdots + \frac{t^{n-1}\mathbf{A}^{n-1}}{(n-1)!} + \frac{t^n\mathbf{A}^n}{n!} + \frac{t^{n+1}\mathbf{A}^{n+1}}{(n+1)!} + \cdots.$$

Therefore,

$$\begin{aligned}\frac{de^{\mathbf{A}t}}{dt} &= \mathbf{0} + \frac{\mathbf{A}}{1!} + \frac{2t\mathbf{A}^2}{2!} + \frac{3t^2\mathbf{A}^3}{3!} + \cdots + \frac{nt^{n-1}\mathbf{A}^n}{n!} + \frac{(n+1)t^n\mathbf{A}^{n+1}}{(n+1)!} + \cdots \\ &= \mathbf{A} + \frac{t\mathbf{A}^2}{1!} + \frac{t^2\mathbf{A}^3}{2!} + \cdots + \frac{t^{n-1}\mathbf{A}^n}{(n-1)!} + \frac{t^n\mathbf{A}^{n+1}}{n!} + \cdots \\ &= \left[\mathbf{I} + \frac{t\mathbf{A}}{1!} + \frac{t^2\mathbf{A}^2}{2!} + \cdots + \frac{t^{n-1}\mathbf{A}^{n-1}}{(n-1)!} + \frac{t^n\mathbf{A}^n}{n!} + \cdots\right]\mathbf{A} \\ &= e^{\mathbf{A}t}\mathbf{A}.\end{aligned}$$

If we had factored $\mathbf{A}$ on the left, instead of on the right, we would have obtained the other identity,

$$\frac{de^{\mathbf{A}t}}{dt} = \mathbf{A}e^{\mathbf{A}t}.$$

Corollary 1 If $\mathbf{A}$ is a constant matrix, then

$$\frac{de^{-\mathbf{A}t}}{dt} = -\mathbf{A}e^{-\mathbf{A}t} = -e^{-\mathbf{A}t}\mathbf{A}.$$

Proof Define $\mathbf{C} = -\mathbf{A}$. Hence, $e^{-\mathbf{A}t} = e^{\mathbf{C}t}$. Since $\mathbf{C}$ is a constant matrix, using Theorem 1, we have

$$\frac{de^{\mathbf{C}t}}{dt} = \mathbf{C}e^{\mathbf{C}t} = e^{\mathbf{C}t}\mathbf{C}.$$

If we now substitute for $\mathbf{C}$ its value, $-\mathbf{A}$, Corollary 1 is immediate.

Definition 3 An $n \times n$ matrix $\mathbf{A}(t) = [a_{ij}(t)]$ is *integrable* if each of its elements $a_{ij}(t)$ $(i, 1, 2, \ldots, n)$ is integrable, and if this is the case,

$$\int \mathbf{A}(t)\, dt = \left[\int a_{ij}(t)\, dt\right].$$

EXAMPLE 5 Find $\int \mathbf{A}(t)\,dt$ if

$$\mathbf{A}(t) = \begin{bmatrix} 3t & 2 \\ t^2 & e^t \end{bmatrix}.$$

Solution

$$\int \mathbf{A}(t)\,dt = \begin{bmatrix} \int 3t\,dt & \int 2\,dt \\ \int t^2\,dt & \int e^t\,dt \end{bmatrix} = \begin{bmatrix} (\frac{3}{2})t^2 + c_1 & 2t + c_2 \\ (\frac{1}{3})t^3 + c_3 & e^t + c_4 \end{bmatrix}.$$

EXAMPLE 6 Find $\int_0^1 \mathbf{A}(t)\,dt$ if

$$\mathbf{A}(t) = \begin{bmatrix} 2t & 1 & 2 \\ e^t & 6t^2 & -1 \\ \sin \pi t & 0 & 1 \end{bmatrix}.$$

Solution

$$\int_0^1 A(t)\,dt = \begin{bmatrix} \int_0^1 2t\,dt & \int_0^1 1\,dt & \int_0^1 2\,dt \\ \int_0^1 e^t\,dt & \int_0^1 6t^2\,dt & \int_0^1 -1\,dt \\ \int_0^1 \sin \pi t\,dt & \int_0^1 0\,dt & \int_0^1 1\,dt \end{bmatrix}$$

$$= \begin{bmatrix} 1 & 1 & 2 \\ e - 1 & 2 & -1 \\ 2/\pi & 0 & 1 \end{bmatrix}.$$

The following property of the integral can be verified:

(P5) $\int [\alpha\mathbf{A}(t) + \beta\mathbf{B}(t)]\,dt = \alpha\int \mathbf{A}(t)\,dt + \beta\int \mathbf{B}(t)\,dt,$

where α and β are constants.

PROBLEMS 6.9

1. Find $\dot{\mathbf{A}}(t)$ if

(a) $\mathbf{A}(t) = \begin{bmatrix} \cos t & t^2 - 1 \\ 2t & e^{(t-1)} \end{bmatrix}.$

(b) $\mathbf{A}(t) = \begin{bmatrix} 2e^{t^3} & t(t-1) & 17 \\ t^2 + 3t - 1 & \sin 2t & t \\ \cos^3(3t^2) & 4 & \ln t \end{bmatrix}.$

2. Verify Properties (P1)–(P4) for

$$\alpha = 7, \quad \beta(t) = t^2, \quad \mathbf{A}(t) = \begin{bmatrix} t^3 & 3t^2 \\ 1 & 2t \end{bmatrix}, \quad \text{and} \quad \mathbf{B}(t) = \begin{bmatrix} t & -2t \\ t^3 & t^5 \end{bmatrix}.$$

3. Prove that if $d\mathbf{A}(t)/dt = \mathbf{0}$, then $\mathbf{A}(t)$ is a constant matrix. (That is, a matrix independent of t).

4. Find $\int \mathbf{A}(t)\, dt$ for the $\mathbf{A}(t)$ given in Problem 1(a).

5. Verify Property (P5) for

$$\alpha = 2, \quad \beta = 10, \quad \mathbf{A}(t) = \begin{bmatrix} 6t & t^2 \\ 2t & 1 \end{bmatrix}, \quad \text{and} \quad \mathbf{B}(t) = \begin{bmatrix} t & 4t^2 \\ 1 & 2t \end{bmatrix}.$$

6. Using Property (P4), derive a formula for differentiating $\mathbf{A}^2(t)$. Use this formula to find $d\mathbf{A}^2(t)/dt$, where

$$\mathbf{A}(t) = \begin{bmatrix} t & 2t^2 \\ 4t^3 & e^t \end{bmatrix};$$

hence, show that $d\mathbf{A}^2(t)/dt \neq 2\mathbf{A}(t)\, d\mathbf{A}(t)/dt$.

The power rule of differentiation *does not hold* for matrices unless a matrix commutes with its derivative.

APPENDIX TO CHAPTER 6

We begin a proof of the Cayley–Hamilton theorem by noting that if $\mathbf{B}$ is an $n \times n$ matrix having elements which are polynomials in λ with constant coefficients, then $\mathbf{B}$ can be expressed as a matrix polynomial in λ whose coefficients are $n \times n$ constant matrices. As an example, consider the following decomposition:

$$\begin{bmatrix} \lambda^3 + 2\lambda^2 + 3\lambda + 4 & 2\lambda^3 + 3\lambda^2 + 4\lambda + 5 \\ 3\lambda^3 + 4\lambda^2 + 5\lambda & 2\lambda + 3 \end{bmatrix}$$
$$= \begin{bmatrix} 1 & 2 \\ 3 & 0 \end{bmatrix}\lambda^3 + \begin{bmatrix} 2 & 3 \\ 4 & 0 \end{bmatrix}\lambda^2 + \begin{bmatrix} 3 & 4 \\ 5 & 2 \end{bmatrix}\lambda + \begin{bmatrix} 4 & 5 \\ 0 & 3 \end{bmatrix}.$$

In general, if the elements of $\mathbf{B}$ are polynomials of degree k or less, then

$$\mathbf{B} = \mathbf{B}_k \lambda^k + \mathbf{B}_{k-1}\lambda^{k-1} + \cdots + \mathbf{B}_1\lambda + \mathbf{B}_0,$$

where $\mathbf{B}_j$ $(j = 0, 1, \ldots, k)$ is an $n \times n$ constant matrix.

Now let $\mathbf{A}$ be any arbitrary $n \times n$ matrix. Define

$$\mathbf{C} = (\mathbf{A} - \lambda\mathbf{I}) \tag{62}$$

and let

$$d(\lambda) = \lambda_n + a_{n-1}\lambda^{n-1} + \cdots + a_1\lambda + a_0 \tag{63}$$

represent the characteristic polynomial of **A**. Thus,

$$d(\lambda) = \det(\mathbf{A} - \lambda\mathbf{I}) = \det \mathbf{C}. \tag{64}$$

Since **C** is an $n \times n$ matrix, it follows that the elements of $\mathbf{C}^a$ (see Definition 3 of Section 3.1) will be polynomials in λ of either degree $n - 1$ or $n - 2$. (Elements on the diagonal of $\mathbf{C}^a$ will be polynomials of degree $n - 1$ while all other elements will be polynomials of degree $n - 2$.) Thus, $\mathbf{C}^a$ can be written as

$$\mathbf{C}^a = \mathbf{C}_{n-1}\lambda^{n-1} + \mathbf{C}_{n-2}\,\lambda^{n-2} + \cdots + \mathbf{C}_1\,\lambda + \mathbf{C}_0, \tag{65}$$

where $\mathbf{C}_j$ ($j = 0, 1, \ldots, n-1$) is an $n \times n$ constant matrix.

From Theorem 1 of Section 3.1 and Eq. (64), we have that

$$\mathbf{C}^a\mathbf{C} = [\det \mathbf{C}]\mathbf{I} = d(\lambda)\,\mathbf{I}. \tag{66}$$

From (62), we have that

$$\mathbf{C}^a\mathbf{C} = \mathbf{C}^a(\mathbf{A} - \lambda\mathbf{I}) = \mathbf{C}^a\mathbf{A} - \lambda\mathbf{C}^a. \tag{67}$$

Equating (66) and (67), we obtain

$$d(\lambda)\,\mathbf{I} = \mathbf{C}^a\mathbf{A} - \lambda\mathbf{C}^a. \tag{68}$$

Substituting (63) and (65) into (68), we find that

$$\begin{aligned}
&\mathbf{I}\lambda_n + a_{n-1}\mathbf{I}\lambda^{n-1} + \cdots + a_1\mathbf{I}\lambda + a_0\,\mathbf{I} \\
&\quad = \mathbf{C}_{n-1}\mathbf{A}\lambda^{n-1} + \mathbf{C}_{n-2}\mathbf{A}\lambda^{n-2} + \cdots + \mathbf{C}_1\mathbf{A}\lambda + \mathbf{C}_0\mathbf{A} \\
&\qquad - \mathbf{C}_{n-1}\lambda^n - \mathbf{C}_{n-2}\,\lambda^{n-1} - \cdots - \mathbf{C}_1\lambda^2 - \mathbf{C}_0\,\lambda.
\end{aligned}$$

Both sides of this equation are matrix polynomials in λ of degree n. Since two polynomials are equal if and only if their corresponding coefficients are equal we have

$$\begin{aligned}
\mathbf{I} &= -\mathbf{C}_{n-1} \\
a_{n-1}\mathbf{I} &= -\mathbf{C}_{n-2} + \mathbf{C}_{n-1}\mathbf{A} \\
&\;\;\vdots \\
a_1\mathbf{I} &= -\mathbf{C}_0 + \mathbf{C}_1\mathbf{A} \\
a_0\,\mathbf{I} &= \mathbf{C}_0\,\mathbf{A}.
\end{aligned} \tag{69}$$

Multiplying the first equation in (69) by $\mathbf{A}^n$, the second equation by $\mathbf{A}^{n-1}, \ldots,$ and the last equation by $\mathbf{A}^0 = \mathbf{I}$ and adding, we obtain (note that the terms on the right-hand side cancel out)

$$\mathbf{A}^n + a_{n-1}\mathbf{A}^{n-1} + \cdots + a_1\mathbf{A} + \mathbf{A}_0\,\mathbf{I} = \mathbf{0}. \tag{70}$$

Equation (70) is the Cayley–Hamilton theorem.

Chapter 7 LINEAR DIFFERENTIAL EQUATIONS

7.1 FUNDAMENTAL FORM

We are now ready to solve linear differential equations. The method that we shall use involves reducing a given system of differential equations by introducing new variables $x_1(t), x_2(t), \ldots, x_n(t)$, to the system

$$\begin{aligned}
\frac{dx_1(t)}{dt} &= a_{11}(t)x_1(t) + a_{12}(t)x_2(t) + \cdots + a_{1n}(t)x_n(t) + f_1(t) \\
\frac{dx_2(t)}{dt} &= a_{21}(t)x_1(t) + a_{22}(t)x_2(t) + \cdots + a_{2n}(t)x_n(t) + f_2(t) \\
&\vdots \\
\frac{dx_n(t)}{dt} &= a_{n1}(t)x_1(t) + a_{n2}(t)x_2(t) + \cdots + a_{nn}(t)x_n(t) + f_n(t).
\end{aligned} \tag{1}$$

If we define

$$\mathbf{x}(t) = \begin{bmatrix} x_1(t) \\ x_2(t) \\ \vdots \\ x_n(t) \end{bmatrix},$$

$$\mathbf{A}(t) = \begin{bmatrix} a_{11}(t) & a_{12}(t) & \cdots & a_{1n}(t) \\ a_{21}(t) & a_{22}(t) & \cdots & a_{2n}(t) \\ \vdots & \vdots & & \vdots \\ a_{n1}(t) & a_{n2}(t) & \cdots & a_{nn}(t) \end{bmatrix}, \quad \text{and} \quad \mathbf{f}(t) = \begin{bmatrix} f_1(t) \\ f_2(t) \\ \vdots \\ f_n(t) \end{bmatrix}, \tag{2}$$

then (1) can be rewritten in the matrix form

$$\frac{d\mathbf{x}(t)}{dt} = \mathbf{A}(t)\mathbf{x}(t) + \mathbf{f}(t). \tag{3}$$

EXAMPLE 1 Put the following system into matrix form:

$$\begin{aligned} \dot{y}(t) &= t^2 y(t) + 3z(t) + \sin t, \\ \dot{z}(t) &= -e^t y(t) + tz(t) - t^2 + 1. \end{aligned}$$

Note that we are using the standard notation $\dot{y}(t)$ and $\dot{z}(t)$ to represent

$$\frac{dy(t)}{dt} \quad \text{and} \quad \frac{dz(t)}{dt}.$$

Solution Define $x_1(t) = y(t)$ and $x_2(t) = z(t)$. This system is then equivalent to the matrix equation

$$\begin{bmatrix} \dot{x}_1(t) \\ \dot{x}_2(t) \end{bmatrix} = \begin{bmatrix} t^2 & 3 \\ -e^t & t \end{bmatrix} \begin{bmatrix} x_1(t) \\ x_2(t) \end{bmatrix} + \begin{bmatrix} \sin t \\ -t^2 + 1 \end{bmatrix}. \tag{4}$$

If we define

$$\mathbf{x}(t) = \begin{bmatrix} x_1(t) \\ x_2(t) \end{bmatrix}, \quad \mathbf{A}(t) = \begin{bmatrix} t^2 & 3 \\ -e^t & t \end{bmatrix}, \quad \text{and} \quad \mathbf{f}(t) = \begin{bmatrix} \sin t \\ -t^2 + 1 \end{bmatrix}$$

then (4) is in the required form, $\dot{\mathbf{x}}(t) = \mathbf{A}(t)\mathbf{x}(t) + \mathbf{f}(t)$.

In practice, we are usually interested in solving an initial value problem; that is, we seek functions $x_1(t), x_2(t), \ldots, x_n(t)$ that satisfy not only the differential equations given by (1) but also a set of initial conditions of the form

$$x_1(t_0) = c_1, \qquad x_2(t_0) = c_2, \ldots, x_n(t_0) = c_n, \tag{5}$$

where $c_1, c_2, \ldots, c_n$, and t_0 are known constants. Upon defining

$$\mathbf{c} = \begin{bmatrix} c_1 \\ c_2 \\ \vdots \\ c_n \end{bmatrix},$$

it follows from the definition of $\mathbf{x}(t)$ (see Eqs. (2) and (5)) that

$$\mathbf{x}(t_0)=\begin{bmatrix} x_1(t_0)\\ x_2(t_0)\\ \vdots\\ x_n(t_0)\end{bmatrix}=\begin{bmatrix} c_1\\ c_2\\ \vdots\\ c_n\end{bmatrix}=\mathbf{c}.$$

Thus, the initial conditions can be put into the matrix form

$$\mathbf{x}(t_0)=\mathbf{c}. \tag{6}$$

Definition 1 A system of differential equations is in *fundamental form* if it is given by the matrix equations

$$\begin{aligned}\dot{\mathbf{x}}(t)&=\mathbf{A}(t)\mathbf{x}(t)+\mathbf{f}(t)\\ \mathbf{x}(t_0)&=\mathbf{c}.\end{aligned} \tag{7}$$

EXAMPLE 2 Put the following system into fundamental form:

$$\begin{aligned}\dot{x}(t)&=2x(t)-ty(t)\\ \dot{y}(t)&=t^2x(t)+e^t\\ x(2)&=3,\qquad y(2)=1.\end{aligned}$$

Solution Define $x_1(t)=x(t)$ and $x_2(t)=y(t)$. This system is then equivalent to the matrix equations

$$\begin{aligned}\begin{bmatrix}\dot{x}_1(t)\\ \dot{x}_2(t)\end{bmatrix}&=\begin{bmatrix}2 & -t\\ t^2 & 0\end{bmatrix}\begin{bmatrix}x_1(t)\\ x_2(t)\end{bmatrix}+\begin{bmatrix}0\\ e^t\end{bmatrix}\\ \begin{bmatrix}x_1(2)\\ x_2(2)\end{bmatrix}&=\begin{bmatrix}3\\ 1\end{bmatrix}.\end{aligned} \tag{8}$$

Consequently, if we define

$$\mathbf{x}(t)=\begin{bmatrix}x_1(t)\\ x_2(t)\end{bmatrix},\qquad \mathbf{A}(t)=\begin{bmatrix}2 & -t\\ t^2 & 0\end{bmatrix},$$

$$\mathbf{f}(t)=\begin{bmatrix}0\\ e^t\end{bmatrix},\qquad \mathbf{c}=\begin{bmatrix}3\\ 1\end{bmatrix},\qquad \text{and}\qquad t_0=2,$$

then (8) is in fundamental form.

EXAMPLE 3 Put the following system into fundamental form:

$$\begin{aligned}\dot{l}(t)&=2l(t)+3m(t)-n(t)\\ \dot{m}(t)&=l(t)-m(t)\\ \dot{n}(t)&=m(t)-n(t)\end{aligned}$$

$$l(15)=0,\qquad m(15)=-170,\qquad n(15)=1007.$$

Solution Define $x_1(t) = l(t)$, $x_2(t) = m(t)$, $x_3(t) = n(t)$. This system is then equivalent to the matrix equations

$$\begin{bmatrix} \dot{x}_1(t) \\ \dot{x}_2(t) \\ \dot{x}_3(t) \end{bmatrix} = \begin{bmatrix} 2 & 3 & -1 \\ 1 & -1 & 0 \\ 0 & 1 & -1 \end{bmatrix} \begin{bmatrix} x_1(t) \\ x_2(t) \\ x_3(t) \end{bmatrix},$$

$$\begin{bmatrix} x_1(15) \\ x_2(15) \\ x_3(15) \end{bmatrix} = \begin{bmatrix} 0 \\ -170 \\ 1007 \end{bmatrix}. \tag{9}$$

Thus, if we define

$$\mathbf{x}(t) = \begin{bmatrix} x_1(t) \\ x_2(t) \\ x_3(t) \end{bmatrix}, \qquad \mathbf{A}(t) = \begin{bmatrix} 2 & 3 & -1 \\ 1 & -1 & 0 \\ 0 & 1 & -1 \end{bmatrix}, \qquad \mathbf{f}(t) = \begin{bmatrix} 0 \\ 0 \\ 0 \end{bmatrix},$$

$$\mathbf{c} = \begin{bmatrix} 0 \\ -170 \\ 1007 \end{bmatrix} \qquad \text{and} \qquad t_0 = 15,$$

then (9) is in fundamental form.

Definition 2 A system in fundamental form is *homogeneous* if $\mathbf{f}(t) = \mathbf{0}$ (that is, if $f_1(t) = f_2(t) = \cdots = f_n(t) = 0$) and *nonhomogeneous* if $\mathbf{f}(t) \neq \mathbf{0}$ (that is, if at least one component of $\mathbf{f}(t)$ differs from zero).

The systems given in Examples 2 and 3 are nonhomogeneous and homogeneous respectively.

Since we will be attempting to solve differential equations, it is important to know exactly what is meant by a solution.

Definition 3 $\mathbf{x}(t)$ is a *solution* of (7) if

(a) both $\mathbf{x}(t)$ and $\dot{\mathbf{x}}(t)$ are continuous in some neighborhood J of the initial time $t = t_0$,

(b) the substitution of $\mathbf{x}(t)$ into the differential equation $\dot{\mathbf{x}}(t) = \mathbf{A}(t)\mathbf{x}(t) + \mathbf{f}(t)$ makes the equation an identity in t on the interval J; that is, the equation is valid for each t in J, and

(c) $\mathbf{x}(t_0) = \mathbf{c}$.

It would also seem advantageous, before trying to find the solutions, to know whether or not a given system has any solutions at all, and if it does, how many. The following theorem from differential equations answers both of these questions.

Theorem 1 Consider a system given by (7). If $\mathbf{A}(t)$ and $\mathbf{f}(t)$ are continuous in some interval containing $t = t_0$, then this system possesses a unique continuous solution on that interval.[18]

Hence, to insure the applicability of this theorem, we assume for the remainder of the chapter that $\mathbf{A}(t)$ and $\mathbf{f}(t)$ are both continuous on some common interval containing $t = t_0$.

PROBLEMS 7.1

Put the following systems into fundamental form:

1. $\dot{y}(t) = 3y(t) + 2z(t), \qquad y(0) = 1,\ z(0) = 1.$
$\dot{z}(t) = 4y(t) + z(t),$

2. $\dot{r}(t) = t^2 r(t) - 3s(t) - (\sin t)u(t) + \sin t,$
$\dot{s}(t) = r(t) - s(t) + t^2 - 1,$
$\dot{u}(t) = 2r(t) + e^t s(t) + (t^2 - 1)u(t) + \cos t,$
$r(1) = 4, \quad s(1) = -2, \quad u(1) = 5.$

3. Determine which of the following are solutions to the system

$$\begin{bmatrix} \dot{x}_1 \\ \dot{x}_2 \end{bmatrix} = \begin{bmatrix} 0 & 1 \\ -1 & 0 \end{bmatrix} \begin{bmatrix} x_1 \\ x_2 \end{bmatrix}, \qquad \begin{bmatrix} x_1(0) \\ x_2(0) \end{bmatrix} = \begin{bmatrix} 1 \\ 0 \end{bmatrix}:$$

(a) $\begin{bmatrix} \sin t \\ \cos t \end{bmatrix}$ (b) $\begin{bmatrix} e^t \\ 0 \end{bmatrix}$ (c) $\begin{bmatrix} \cos t \\ -\sin t \end{bmatrix}$.

7.2 REDUCTION OF AN *n*th ORDER EQUATION

Before seeking solutions to linear differential equations, we will first develop techniques for reducing these equations to fundamental form. In this section, we consider the initial-value problems given by

$$a_n(t)\frac{d^n x(t)}{dt^n} + a_{n-1}(t)\frac{d^{n-1}x(t)}{dt^{n-1}} + \cdots + a_1(t)\frac{dx(t)}{dt} + a_0(t)x(t) = f(t)$$
$$x(t_0) = c_1, \quad \frac{dx(t_0)}{dt} = c_2, \ldots, \quad \frac{d^{n-1}x(t_0)}{dt^{n-1}} = c_n. \tag{10}$$

Equation (10) is an nth order differential equation for $x(t)$ where $a_0(t)$, $a_1(t), \ldots, a_n(t)$ and $f(t)$ are assumed known and continuous on some interval containing t_0. Furthermore, we assume that $a_n(t) \neq 0$ on this interval.

[18] For a proof of Theorem 1, see E. A. Coddington and N. Levinson, p. 20. "Theory of Ordinary Differential Equations," McGraw-Hill, New York, 1955.

A method of reduction is as follows:[19]

STEP 1 Rewrite (10) so that the nth derivative of $x(t)$ appears by itself;

$$\frac{d^n x(t)}{dt^n} = -\frac{a_{n-1}(t)}{a_n(t)}\frac{d^{n-1}x(t)}{dt^{n-1}} - \cdots - \frac{a_1(t)}{a_n(t)}\frac{dx(t)}{dt} - \frac{a_0(t)}{a_n(t)}x(t) + \frac{f(t)}{a_n(t)}. \tag{11}$$

STEP 2 Define n new variables (the same number as the order of the differential equation), $x_1(t)$, $x_2(t)$, ..., $x_n(t)$ by the equations[20]

$$\begin{aligned} x_1 &= x(t),\\ x_2 &= \frac{dx_1}{dt},\\ x_3 &= \frac{dx_2}{dt},\\ &\vdots\\ x_{n-1} &= \frac{dx_{n-2}}{dt},\\ x_n &= \frac{dx_{n-1}}{dt}. \end{aligned} \tag{12}$$

Note that from the manner in which the new variables are defined in (12), we also have the following relationships between $x_1, x_2, \ldots, x_n$ and the unknown $x(t)$:

$$\begin{aligned} x_1 &= x,\\ x_2 &= \frac{dx}{dt},\\ x_3 &= \frac{d^2x}{dt^2},\\ &\vdots\\ x_{n-1} &= \frac{d^{n-2}x}{dt^{n-2}},\\ x_n &= \frac{d^{n-1}x}{dt^{n-1}}. \end{aligned} \tag{13}$$

[19] For certain types of differential equations not considered in this book, for example, nonlinear or those with singular points, other reductions may be more useful. See, for example, E. A. Coddington and N. Levinson, "Theory of Ordinary Differential Equations," pp. 182–183. McGraw-Hill, New York, 1955.

[20] Generally, we will write $x_j(t)$ $(j = 1, 2, \ldots, n)$ simply as x_j when the dependence on the variable t is obvious from context.

Hence, by differentiating the last equation of (13), we have

$$\frac{dx_n}{dt} = \frac{d^n x}{dt^n}. \tag{14}$$

STEP 3 Rewrite dx_n/dt in terms of the new variables $x_1, x_2, \ldots, x_n$. Substituting (11) into (14), we have

$$\frac{dx_n}{dt} = -\frac{a_{n-1}(t)}{a_n(t)}\frac{d^{n-1}x}{dt^{n-1}} - \cdots - \frac{a_1(t)}{a_n(t)}\frac{dx}{dt} - \frac{a_0(t)}{a_n(t)}x + \frac{f(t)}{a_n(t)}.$$

Substituting (13) into this equation, we obtain

$$\frac{dx_n}{dt} = -\frac{a_{n-1}(t)}{a_n(t)}x_n - \cdots - \frac{a_1(t)}{a_n(t)}x_2 - \frac{a_0(t)}{a_n(t)}x_1 + \frac{f(t)}{a_n(t)}. \tag{15}$$

STEP 4 Form a system of n first-order differential equations for $x_1, x_2, \ldots, x_n$.

Using (12) and (15), we obtain the system

$$\begin{aligned}
\frac{dx_1}{dt} &= x_2,\\
\frac{dx_2}{dt} &= x_3,\\
&\vdots\\
\frac{dx_{n-2}}{dt} &= x_{n-1},\\
\frac{dx_{n-1}}{dt} &= x_n,\\
\frac{dx_n}{dt} &= -\frac{a_0(t)}{a_n(t)}x_1 - \frac{a_1(t)}{a_n(t)}x_2 - \cdots - \frac{a_{n-1}(t)}{a_n(t)}x_n + \frac{f(t)}{a_n(t)}.
\end{aligned} \tag{16}$$

Note that in the last equation of (16) we have rearranged the order of (15) so that the x_1 term appears first, the x_2 term appears second, etc. This was done in order to simplify the next step.

STEP 5 Put (16) into matrix form.

Define

$$\mathbf{x}(t) = \begin{bmatrix} x_1(t) \\ x_2(t) \\ \vdots \\ x_{n-1}(t) \\ x_n(t) \end{bmatrix},$$

$$\mathbf{A}(t) = \begin{bmatrix} 0 & 1 & 0 & 0 & \cdots & 0 \\ 0 & 0 & 1 & 0 & \cdots & 0 \\ 0 & 0 & 0 & 1 & \cdots & 0 \\ \vdots & \vdots & \vdots & \vdots & & \vdots \\ 0 & 0 & 0 & 0 & \cdots & 1 \\ -\dfrac{a_0(t)}{a_n(t)} & -\dfrac{a_1(t)}{a_n(t} & -\dfrac{a_2(t)}{a_n(t)} & -\dfrac{a_3(t)}{a_n(t)} & \cdots & -\dfrac{a_{n-1}(t)}{a_n(t)} \end{bmatrix},$$

and (17)

$$\mathbf{f}(t) = \begin{bmatrix} 0 \\ 0 \\ \vdots \\ 0 \\ \dfrac{f(t)}{a_n(t)} \end{bmatrix}.$$

Then (16) can be written as

$$\dot{\mathbf{x}}(t) = \mathbf{A}(t)\mathbf{x}(t) + \mathbf{f}(t). \tag{18}$$

STEP 6 Rewrite the initial conditions in matrix form.

From (17), (13), and (10), we have that

$$\mathbf{x}(t_0) = \begin{bmatrix} x_1(t_0) \\ x_2(t_0) \\ \vdots \\ x_n(t_0) \end{bmatrix} = \begin{bmatrix} x(t_0) \\ \dfrac{dx(t_0)}{dt} \\ \vdots \\ \dfrac{d^{n-1}x(t_0)}{dt^{n-1}} \end{bmatrix} = \begin{bmatrix} c_1 \\ c_2 \\ \vdots \\ c_n \end{bmatrix}.$$

Thus, if we define

$$\mathbf{c} = \begin{bmatrix} c_1 \\ c_2 \\ \vdots \\ c_n \end{bmatrix},$$

the initial conditions can be put into matrix form

$$\mathbf{x}(t_0) = \mathbf{c}. \tag{19}$$

Equations (18) and (19) together represent the fundamental form for (10).

Since $\mathbf{A}(t)$ and $\mathbf{f}(t)$ are continuous (why?), Theorem 1 of the previous section guarantees that a unique solution exists to (18) and (19). Once this solution is obtained, $\mathbf{x}(t)$ will be known; hence, the components of $\mathbf{x}(t)$, $x_1(t), \ldots, x_n(t)$ will be known and, consequently, so will $x(t)$, the variable originally sought (from (12), $x_1(t) = x(t)$).

EXAMPLE 1 Put the following initial-value problem into fundamental form:

$$2\ddddot{x} - 4\dddot{x} + 16t\ddot{x} - \dot{x} + 2t^2x = \sin t,$$
$$x(0) = 1, \quad \dot{x}(0) = 2, \quad \ddot{x}(0) = -1, \quad \dddot{x}(0) = 0.$$

Solution The differential equation may be rewritten as

$$\ddddot{x} = 2\dddot{x} - 8t\ddot{x} + \tfrac{1}{2}\dot{x} - t^2x + (\tfrac{1}{2})\sin t.$$

Define

$$\begin{aligned} x_1 &= x \\ x_2 &= \dot{x}_1 = \dot{x}, \\ x_3 &= \dot{x}_2 = \ddot{x}, \\ x_4 &= \dot{x}_3 = \dddot{x} \end{aligned}$$

hence, $\dot{x}_4 = \ddddot{x}$. Thus,

$$\begin{aligned} \dot{x}_1 &= x_2 \\ \dot{x}_2 &= x_3 \\ \dot{x}_3 &= x_4 \\ \dot{x}_4 &= \ddddot{x} = 2\dddot{x} - 8t\ddot{x} + \tfrac{1}{2}\dot{x} - t^2x + \tfrac{1}{2}\sin t \\ &= 2x_4 - 8tx_3 + \tfrac{1}{2}x_2 - t^2x_1 + \tfrac{1}{2}\sin t, \end{aligned}$$

or

$$\begin{aligned} \dot{x}_1 &= \phantom{-t^2x_1 + \tfrac{1}{2}}x_2 \\ \dot{x}_2 &= \phantom{-t^2x_1 + \tfrac{1}{2}x_2 - 8t}x_3 \\ \dot{x}_3 &= \phantom{-t^2x_1 + \tfrac{1}{2}x_2 - 8tx_3 + 2}x_4 \\ \dot{x}_4 &= -t^2x_1 + \tfrac{1}{2}x_2 - 8tx_3 + 2x_4 + \tfrac{1}{2}\sin t. \end{aligned}$$

Define

$$\mathbf{x}(t) = \begin{bmatrix} x_1 \\ x_2 \\ x_3 \\ x_4 \end{bmatrix}, \qquad \mathbf{A}(t) = \begin{bmatrix} 0 & 1 & 0 & 0 \\ 0 & 0 & 1 & 0 \\ 0 & 0 & 0 & 1 \\ -t^2 & \tfrac{1}{2} & -8t & 2 \end{bmatrix},$$

$$\mathbf{f}(t) = \begin{bmatrix} 0 \\ 0 \\ 0 \\ \tfrac{1}{2}\sin t \end{bmatrix}, \qquad \mathbf{c} = \begin{bmatrix} 1 \\ 2 \\ -1 \\ 0 \end{bmatrix}, \qquad \text{and} \qquad t_0 = 0.$$

Thus, the initial value problem may be rewritten in the fundamental form

$$\dot{\mathbf{x}}(t) = \mathbf{A}(t)\mathbf{x}(t) + \mathbf{f}(t),$$
$$\mathbf{x}(t_0) = \mathbf{c}.$$

EXAMPLE 2 Put the following initial value problem into fundamental form:

$$e^t \frac{d^5x}{dt^5} - 2e^{2t}\frac{d^4x}{dt^4} + tx = 4e^t,$$

$$x(2) = 1, \quad \frac{dx(2)}{dt} = -1, \quad \frac{d^2x(2)}{dt^2} = -1, \quad \frac{d^3x(2)}{dt^3} = 2, \quad \frac{d^4x(2)}{dt^4} = 3.$$

The differential equation may be rewritten

$$\frac{d^5x}{dt^5} = 2e^t\frac{d^4x}{dt^4} - te^{-t}x + 4.$$

Define

$$\begin{aligned} x_1 &= x \\ x_2 &= \dot{x}_1 = \dot{x} \\ x_3 &= \dot{x}_2 = \ddot{x} \\ x_4 &= \dot{x}_3 = \dddot{x} \\ x_5 &= \dot{x}_4 = \frac{d^4x}{dt^4}; \end{aligned}$$

hence,

$$\dot{x}_5 = \frac{d^5x}{dt^5}.$$

Thus,

$$\begin{aligned} \dot{x}_1 &= x_2 \\ \dot{x}_2 &= x_3 \\ \dot{x}_3 &= x_4 \\ \dot{x}_4 &= x_5 \\ \dot{x}_5 &= \frac{d^5x}{dt^5} = 2e^t\frac{d^4x}{dt^4} - te^{-t}x + 4 \\ &= 2e^tx_5 - te^{-t}x_1 + 4, \end{aligned}$$

or

$$
\begin{aligned}
\dot{x}_1 &= x_2\\
\dot{x}_2 &= x_3\\
\dot{x}_3 &= x_4\\
\dot{x}_4 &= x_5\\
\dot{x}_5 &= -te^{-t}x_1 + 2e^t x_5 + 4.
\end{aligned}
$$

Define

$$
\mathbf{x}(t) = \begin{bmatrix} x_1 \\ x_2 \\ x_3 \\ x_4 \\ x_5 \end{bmatrix}, \qquad \mathbf{A}(t) = \begin{bmatrix} 0 & 1 & 0 & 0 & 0 \\ 0 & 0 & 1 & 0 & 0 \\ 0 & 0 & 0 & 1 & 0 \\ 0 & 0 & 0 & 0 & 1 \\ -te^{-t} & 0 & 0 & 0 & 2e^t \end{bmatrix},
$$

$$
\mathbf{f}(t) = \begin{bmatrix} 0 \\ 0 \\ 0 \\ 0 \\ 4 \end{bmatrix}, \qquad \mathbf{c} = \begin{bmatrix} 1 \\ -1 \\ -1 \\ 2 \\ 3 \end{bmatrix}, \qquad \text{and} \qquad t_0 = 2.
$$

Thus, the initial value problem may be rewritten in the fundamental form

$$
\begin{aligned}
\dot{\mathbf{x}}(t) &= \mathbf{A}(t)\mathbf{x}(t) + \mathbf{f}(t),\\
\mathbf{x}(t_0) &= \mathbf{c}.
\end{aligned}
$$

PROBLEMS 7.2

Put the following initial-value problems into fundamental form:

1. $\ddot{x} - 3\dot{x} + 2x = e^{-t}, \qquad x(1) = \dot{x}(1) = 2.$

2. $4\dddot{x} + t\ddot{x} - x = 0, \qquad x(-1) = 2, \quad \dot{x}(-1) = 1, \quad \ddot{x}(-1) = -205.$

3. $e^t \dfrac{d^4x}{dt^4} + t\dfrac{d^2x}{dt^2} = 1 + \dfrac{dx}{dt},$

$x(0) = 1, \quad \dfrac{dx(0)}{dt} = 2, \quad \dfrac{d^2x(0)}{dt^2} = \pi, \quad \dfrac{d^3x(0)}{dt^3} = e^3.$

4. $\dfrac{d^6x}{dt^6} + 4\dfrac{d^4x}{dt^4} = t^2 - t,$

$x(\pi) = 2, \quad \dot{x}(\pi) = 1, \quad \ddot{x}(\pi) = 0, \quad \dddot{x}(\pi) = 2,$

$\dfrac{d^4x(\pi)}{dt^4} = 1, \qquad \dfrac{d^5x(\pi)}{dt^5} = 0.$

7.3 REDUCTION OF A SYSTEM

Based on our work in the preceding section, we are now able to reduce systems of higher order linear differential equations to fundamental form. The method, which is a straightforward extension of that used to reduce the nth order differential equation to fundamental form, is best demonstrated by examples.

EXAMPLE 1 Put the following system into fundamental form:

$$\dddot{x} = 5\ddot{x} + \dot{y} - 7y + e^t,$$
$$\ddot{y} = \dot{x} - 2\dot{y} + 3y + \sin t, \tag{20}$$
$$x(1) = 2, \quad \dot{x}(1) = 3, \quad \ddot{x}(1) = -1, \quad y(1) = 0, \quad \dot{y}(1) = -2.$$

STEP 1 Rewrite the differential equations so that the highest derivative of each unknown function appears by itself.

For the above system, this has already been done.

STEP 2 Define new variables $x_1(t)$, $x_2(t)$, $x_3(t)$, $y_1(t)$, and $y_2(t)$. (Since the highest derivative of $x(t)$ is of order 3, and the highest derivative of $y(t)$ is of order 2, we need 3 new variables for $x(t)$ and 2 new variables for $y(t)$. In general, for each unknown function we define a set of k new variables, where k is the order of the highest derivative of the original function appearing in the system under consideration.) The new variables are defined in a manner analogous to that used in the previous section:

$$\begin{aligned} x_1 &= x, \\ x_2 &= \dot{x}_1, \\ x_3 &= \dot{x}_2, \\ y_1 &= y, \\ y_2 &= \dot{y}_1. \end{aligned} \tag{21}$$

From (21), the new variables are related to the functions $x(t)$ and $y(t)$ by the following:

$$\begin{aligned} x_1 &= x, \\ x_2 &= \dot{x}, \\ x_3 &= \ddot{x}, \\ y_1 &= y, \\ y_2 &= \dot{y}. \end{aligned} \tag{22}$$

It follows from (22), by differentiating x_3 and y_2, that

$$\begin{aligned}\dot{x}_3 &= \dddot{x},\\ \dot{y}_2 &= \ddot{y}.\end{aligned} \tag{23}$$

STEP 3 Rewrite $\dot{x}_3$ and $\dot{y}_2$ in terms of the new variables defined in (21).
Substituting (20) into (23), we have

$$\begin{aligned}\dot{x}_3 &= 5\ddot{x} + \dot{y} - 7y + e^t,\\ \dot{y}_2 &= \dot{x} - 2\dot{y} + 3y + \sin t.\end{aligned}$$

Substituting (22) into these equations, we obtain

$$\begin{aligned}\dot{x}_3 &= 5x_3 + y_2 - 7y_1 + e^t,\\ \dot{y}_2 &= x_2 - 2y_2 + 3y_1 + \sin t.\end{aligned} \tag{24}$$

STEP 4 Set up a system of first-order differential equations for x_1, x_2, x_3, y_1, and y_2.
Using (21) and (24), we obtain the system

$$\begin{aligned}\dot{x}_1 &= x_2,\\ \dot{x}_2 &= x_3,\\ \dot{x}_3 &= 5x_3 - 7y_1 + y_2 + e^t,\\ \dot{y}_1 &= y_2,\\ \dot{y}_2 &= x_2 + 3y_1 - 2y_2 + \sin t.\end{aligned} \tag{25}$$

Note, that for convenience we have rearranged terms in some of the equations in order to present them in their natural order.

STEP 5 Write (25) in matrix form.
Define

$$\mathbf{x}(t) = \begin{bmatrix} x_1(t) \\ x_2(t) \\ x_3(t) \\ y_1(t) \\ y_2(t) \end{bmatrix}, \quad \mathbf{A}(t) = \begin{bmatrix} 0 & 1 & 0 & 0 & 0 \\ 0 & 0 & 1 & 0 & 0 \\ 0 & 0 & 5 & -7 & 1 \\ 0 & 0 & 0 & 0 & 1 \\ 0 & 1 & 0 & 3 & -2 \end{bmatrix}, \quad \text{and} \quad \mathbf{f}(t) = \begin{bmatrix} 0 \\ 0 \\ e^t \\ 0 \\ \sin t \end{bmatrix}. \tag{26}$$

Thus, Eq. (25) can be rewritten in the matrix form

$$\dot{\mathbf{x}}(t) = \mathbf{A}(t)\mathbf{x}(t) + \mathbf{f}(t). \tag{27}$$

STEP 6 Rewrite the initial conditions in matrix form.

From Eqs. (26), (22), and (20) we have

$$\mathbf{x}(1) = \begin{bmatrix} x_1(1) \\ x_2(1) \\ x_3(1) \\ y_1(1) \\ y_2(1) \end{bmatrix} = \begin{bmatrix} x(1) \\ \dot{x}(1) \\ \ddot{x}(1) \\ y(1) \\ \dot{y}(1) \end{bmatrix} = \begin{bmatrix} 2 \\ 3 \\ -1 \\ 0 \\ -2 \end{bmatrix}.$$

Thus, if we define

$$c = \begin{bmatrix} 2 \\ 3 \\ -1 \\ 0 \\ -2 \end{bmatrix}$$

and $t_0 = 1$, then the initial conditions can be rewritten as

$$\mathbf{x}(1) = \mathbf{c}. \tag{28}$$

Since $\mathbf{A}(t)$ and $\mathbf{f}(t)$ are continuous, (27) and (28) possess a unique solution. Once $\mathbf{x}(t)$ is known, we immediately have the components of $\mathbf{x}(t)$, namely $x_1(t)$, $x_2(t)$, $x_3(t)$, $y_1(t)$ and $y_2(t)$. Thus, we have the functions $x(t)$ and $y(t)$ (from (21), $x_1(t) = x(t)$ and $y_1(t) = y(t)$).

All similar systems containing higher order derivatives may be put into fundamental form in exactly the same manner as that used here.

EXAMPLE 2 Put the following system into fundamental form:

$$\begin{aligned} \dddot{x} &= 2\dot{x} + t\dot{y} - 3z + t^2\dot{z} + t, \\ \dddot{y} &= \dot{z} + (\sin t)y + x - t, \\ \ddot{z} &= \ddot{x} - \ddot{y} + t^2 + 1; \end{aligned}$$

$$x(\pi) = 15, \quad \dot{x}(\pi) = 59, \quad \ddot{x}(\pi) = -117, \quad y(\pi) = 2, \quad \dot{y}(\pi) = -19,$$
$$\ddot{y}(\pi) = 3, \quad z(\pi) = 36, \quad \dot{z}(\pi) = -212.$$

Solution Define

$$\begin{aligned} & x_1 = x \\ & x_2 = \dot{x}_1 = \dot{x} \\ & x_3 = \dot{x}_2 = \ddot{x}; \text{ hence, } \dot{x}_3 = \dddot{x}. \\ & y_1 = y \\ & y_2 = \dot{y}_1 = \dot{y} \\ & y_3 = \dot{y}_2 = \ddot{y}; \text{ hence, } \dot{y}_3 = \dddot{y}. \\ & z_1 = z \\ & z_2 = \dot{z}_1 = \dot{z}; \text{ hence, } \dot{z}_2 = \ddot{z}. \end{aligned}$$

Thus,

$$\begin{aligned}
\dot{x}_1 &= x_2\\
\dot{x}_2 &= x_3\\
\dot{x}_3 &= \dddot{x} = 2\dot{x} + t\dot{y} - 3z + t^2\dot{z} + t\\
&= 2x_2 + ty_2 - 3z_1 + t^2 z_2 + t;\\
\dot{y}_1 &= y_2\\
\dot{y}_2 &= y_3\\
\dot{y}_3 &= \dddot{y} = \dot{z} + (\sin t)y + x - t\\
&= z_2 + (\sin t)y_1 + x_1 - t;\\
\dot{z}_1 &= z_2\\
\dot{z}_2 &= \ddot{z} = \ddot{x} - \ddot{y} + t^2 + 1\\
&= x_3 - y_3 + t^2 + 1;
\end{aligned}$$

or

$$\begin{aligned}
\dot{x}_1 &= \quad x_2\\
\dot{x}_2 &= \qquad x_3\\
\dot{x}_3 &= \quad 2x_2 \qquad\qquad + ty_2 \qquad -3z_1 + t^2 z_2 + t\\
\dot{y}_1 &= \qquad\qquad\qquad y_2\\
\dot{y}_2 &= \qquad\qquad\qquad\qquad y_3\\
\dot{y}_3 &= x_1 \qquad + (\sin t)y_1 \qquad\qquad\qquad z_2 - t\\
\dot{z}_1 &= \qquad\qquad\qquad\qquad\qquad\qquad z_2\\
\dot{z}_2 &= \qquad x_3 \qquad\qquad\qquad -y_3 \qquad\qquad + t^2 + 1.
\end{aligned}$$

Define

$$\mathbf{x} = \begin{bmatrix} x_1\\ x_2\\ x_3\\ y_1\\ y_2\\ y_3\\ z_1\\ z_2 \end{bmatrix}, \qquad \mathbf{A}(t) = \begin{bmatrix} 0 & 1 & 0 & 0 & 0 & 0 & 0 & 0\\ 0 & 0 & 1 & 0 & 0 & 0 & 0 & 0\\ 0 & 2 & 0 & 0 & t & 0 & -3 & t^2\\ 0 & 0 & 0 & 0 & 1 & 0 & 0 & 0\\ 0 & 0 & 0 & 0 & 0 & 1 & 0 & 0\\ 1 & 0 & 0 & \sin t & 0 & 0 & 0 & 1\\ 0 & 0 & 0 & 0 & 0 & 0 & 0 & 1\\ 0 & 0 & 1 & 0 & 0 & -1 & 0 & 0 \end{bmatrix},$$

$$\mathbf{f}(t) = \begin{bmatrix} 0\\ 0\\ t\\ 0\\ 0\\ -t\\ 0\\ t^2+1 \end{bmatrix}, \qquad \mathbf{c} = \begin{bmatrix} 15\\ 59\\ -117\\ 2\\ -19\\ 3\\ 36\\ -212 \end{bmatrix}, \qquad \text{and} \qquad t_0 = \pi.$$

Thus, the system can now be rewritten in the fundamental form

$$\dot{\mathbf{x}}(t) = \mathbf{A}(t)\mathbf{x}(t) + \mathbf{f}(t),$$
$$\mathbf{x}(t_0) = \mathbf{c}.$$

PROBLEMS 7.3

Put the following initial-value problems into fundamental form:

1. $\ddot{x} = 2\dot{x} + \dddot{y} - t,$
$\ddddot{y} = tx - ty + \ddot{y} - e^t;$
$x(-1) = 2, \quad \dot{x}(-1) = 0, \quad y(-1) = 0, \quad \dot{y}(-1) = 3, \quad \ddot{y}(-1) = 9,$
$\dddot{y}(-1) = 4.$

2. $\dddot{x} = x - y + \dot{y},$
$\ddot{y} = \ddot{x} - x + 2\dot{y};$
$x(0) = 21, \quad \dot{x}(0) = 4, \quad \ddot{x}(0) = -5, \quad y(0) = 5, \quad \dot{y}(0) = 7.$

3. $\dot{x} = y - 2,$
$\ddot{y} = z - 2,$
$\dot{z} = x + y;$
$x(\pi) = 1, \quad y(\pi) = 2, \quad \dot{y}(\pi) = 17, \quad z(\pi) = 0.$

4. $\ddot{x} = y + z + 2,$
$\ddot{y} = x + y - 1,$
$\ddot{z} = x - z + 1;$
$x(20) = 4 \quad \dot{x}(20) = -4, \quad y(20) = 5, \quad \dot{y}(20) = -5, \quad z(20) = 9,$
$\dot{z}(20) = -9.$

7.4 SOLUTIONS OF SYSTEMS WITH CONSTANT COEFFICIENTS

In general, when one reduces a system of differential equations to fundamental form, the matrix $\mathbf{A}(t)$ will depend explicitly on the variable t. For some systems however, $\mathbf{A}(t)$ will not vary with t (that is, every element of $\mathbf{A}(t)$ is a constant). If this is the case, the system is said to have *constant coefficients*. For instance, in Section 7.1, Example 3 illustrates a system having constant coefficients, while Example 2 illustrates a system that does not have constant coefficients.

In this section, we only consider systems having constant coefficients; hence, we shall designate the matrix $\mathbf{A}(t)$ as $\mathbf{A}$ in order to emphasize its independence of t. We seek the solution to the initial-value problem in the fundamental form

$$\begin{aligned}\dot{\mathbf{x}}(t) &= \mathbf{A}\mathbf{x}(t) + \mathbf{f}(t),\\ \mathbf{x}(t_0) &= \mathbf{c}.\end{aligned} \tag{29}$$

The differential equation in (29) can be rewritten as

$$\dot{\mathbf{x}}(t) - \mathbf{A}\mathbf{x}(t) = \mathbf{f}(t). \tag{30}$$

If we premultiply each side of (30) by $e^{-\mathbf{A}t}$, we obtain

$$e^{-\mathbf{A}t}[\dot{\mathbf{x}}(t) - \mathbf{A}\mathbf{x}(t)] = e^{-\mathbf{A}t}\mathbf{f}(t). \tag{31}$$

Using matrix differentiation and Corollary 1 of Section 6.9, we find that

$$\begin{aligned}\frac{d}{dt}[e^{-\mathbf{A}t}\mathbf{x}(t)] &= e^{-\mathbf{A}t}(-\mathbf{A})\mathbf{x}(t) + e^{-\mathbf{A}t}\dot{\mathbf{x}}(t)\\ &= e^{-\mathbf{A}t}[\dot{\mathbf{x}}(t) - \mathbf{A}\mathbf{x}(t)].\end{aligned} \tag{32}$$

Substituting (32) into (31), we obtain

$$\frac{d}{dt}[e^{-\mathbf{A}t}\mathbf{x}(t)] = e^{-\mathbf{A}t}\mathbf{f}(t). \tag{33}$$

Integrating (33) between the limits $t = t_0$ and $t = t$, we have

$$\begin{aligned}\int_{t_0}^{t} \frac{d}{dt}[e^{-\mathbf{A}t}\mathbf{x}(t)]\, dt &= \int_{t_0}^{t} e^{-\mathbf{A}t}\mathbf{f}(t)\, dt\\ &= \int_{t_0}^{t} e^{-\mathbf{A}s}\mathbf{f}(s)\, ds.\end{aligned} \tag{34}$$

Note that we have replaced the dummy variable t by the dummy variable s in the right-hand integral of (34), which *in no way* alters the definite integral (see Problem 1).

Upon evaluating the left-hand integral, it follows from (34) that

$$e^{-\mathbf{A}t}\mathbf{x}(t)\Big|_{t_0}^{t} = \int_{t_0}^{t} e^{-\mathbf{A}s}\mathbf{f}(s)\, ds$$

or that

$$e^{-\mathbf{A}t}\mathbf{x}(t) = e^{-\mathbf{A}t_0}\mathbf{x}(t_0) + \int_{t_0}^{t} e^{-\mathbf{A}s}\mathbf{f}(s)\, ds. \tag{35}$$

But $\mathbf{x}(t_0) = \mathbf{c}$, hence

$$e^{-\mathbf{A}t}\mathbf{x}(t) = e^{-\mathbf{A}t_0}\mathbf{c} + \int_{t_0}^{t} e^{-\mathbf{A}s}\mathbf{f}(s)\, ds, \tag{36}$$

Premultiplying both sides of (36) by $(e^{-\mathbf{A}t})^{-1}$, we obtain

$$\mathbf{x}(t) = (e^{-\mathbf{A}t})^{-1}e^{-\mathbf{A}t_0}\mathbf{c} + (e^{-\mathbf{A}t})^{-1}\int_{t_0}^{t} e^{-\mathbf{A}s}\mathbf{f}(s)\, ds. \tag{37}$$

Using Property 2 of Section 6.8, we have

$$(e^{-\mathbf{A}t})^{-1} = e^{\mathbf{A}t},$$

whereupon we can rewrite (37) as

$$\mathbf{x}(t) = e^{\mathbf{A}t}e^{-\mathbf{A}t_0}\mathbf{c} + e^{\mathbf{A}t}\int_{t_0}^{t} e^{-\mathbf{A}s}\mathbf{f}(s)\,ds. \tag{38}$$

Since $\mathbf{A}t$ and $-\mathbf{A}t_0$ commute (why?), we have from Problem 6 of Section 6.8,

$$e^{\mathbf{A}t}e^{-\mathbf{A}t_0} = e^{\mathbf{A}(t-t_0)}. \tag{39}$$

Finally using (39), we can rewrite (38) as

$$\boxed{\mathbf{x}(t) = e^{\mathbf{A}(t-t_0)}\mathbf{c} + e^{\mathbf{A}t}\int_{t_0}^{t} e^{-\mathbf{A}s}\mathbf{f}(s)\,ds.} \tag{40}$$

Equation (40) is the unique solution to the initial-value problem given by (29).

A simple method for calculating the quantities $e^{\mathbf{A}(t-t_0)}$, and $e^{-\mathbf{A}s}$ is to first compute $e^{\mathbf{A}t}$ (see Section 6.6) and then replace the variable t wherever it appears by the variables $t - t_0$ and $(-s)$.

EXAMPLE 1 Find $e^{\mathbf{A}(t-t_0)}$ and $e^{-\mathbf{A}s}$ for

$$\mathbf{A} = \begin{bmatrix} -1 & 1 \\ 0 & -1 \end{bmatrix}.$$

Solution Using the method of Section 6.6, we calculate $e^{\mathbf{A}t}$ as

$$e^{\mathbf{A}t} = \begin{bmatrix} e^{-t} & te^{-t} \\ 0 & e^{-t} \end{bmatrix}.$$

Hence,

$$e^{\mathbf{A}(t-t_0)} = \begin{bmatrix} e^{-(t-t_0)} & (t-t_0)e^{-(t-t_0)} \\ 0 & e^{-(t-t_0)} \end{bmatrix}$$

and

$$e^{-\mathbf{A}s} = \begin{bmatrix} e^{s} & -se^{s} \\ 0 & e^{s} \end{bmatrix}.$$

Note that when t is replaced by $(t - t_0)$ in e^{-t}, the result is $e^{-(t-t_0)} = e^{-t+t_0}$ and not e^{-t-t_0}. That is, we replaced the *quantity* t by the *quantity* $(t - t_0)$; we did not simply add $-t_0$ to the variable t wherever it appeared. Also note that the same result could have been obtained for $e^{-\mathbf{A}s}$ by first computing $e^{\mathbf{A}s}$ and then inverting by the method of cofactors (recall that $e^{-\mathbf{A}s}$ is the inverse of $e^{\mathbf{A}s}$) or by computing $e^{-\mathbf{A}s}$ directly (define $\mathbf{B} = -\mathbf{A}s$ and calculate $e^{\mathbf{B}}$). However, if $e^{\mathbf{A}t}$ is already known, the above method is by far the most expedient one for obtaining $e^{-\mathbf{A}s}$.

We can now derive an alternate representation for the solution vector $\mathbf{x}(t)$ if we note that if $e^{\mathbf{A}t}$ depends only on t and the integration is with respect to s. Hence, $e^{\mathbf{A}t}$ can be brought inside the integral, and (40) can be rewritten as

$$\mathbf{x}(t) = e^{\mathbf{A}(t-t_0)}\mathbf{c} + \int_{t_0}^{t} e^{\mathbf{A}t}e^{-\mathbf{A}s}\mathbf{f}(s)\,ds.$$

Since $\mathbf{A}t$ and $-\mathbf{A}s$ commute, we have that

$$e^{\mathbf{A}t}e^{-\mathbf{A}s} = e^{\mathbf{A}(t-s)}$$

Thus, the solution to (29) can be written as

$$\boxed{\mathbf{x}(t) = e^{\mathbf{A}(t-t_0)}\mathbf{c} + \int_{t_0}^{t} e^{\mathbf{A}(t-s)}\mathbf{f}(s)\,ds.} \tag{41}$$

Again the quantity $e^{\mathbf{A}(t-s)}$ can be obtained by replacing the variable t in $e^{\mathbf{A}t}$ by the variable $(t - s)$.

In general, the solution $\mathbf{x}(t)$ may be obtained quicker by using (41) than by using (40), since there is one less multiplication involved. (Note that in (40) one must premultiply the integral by $e^{\mathbf{A}t}$ while in (41) this step is eliminated.) However, since the integration in (41) is more difficult than that in (40), the reader who is not confident of his integrating abilities will probably be more comfortable using (40).

If one has a homogeneous initial-value problem with constant coefficients, that is, a system defined by

$$\begin{aligned} \dot{\mathbf{x}}(t) &= \mathbf{A}\mathbf{x}(t), \\ \mathbf{x}(t_0) &= \mathbf{c}, \end{aligned} \tag{42}$$

a great simplification of (40) is effected. In this case, $\mathbf{f}(t) \equiv \mathbf{0}$. The integral in (40), therefore, becomes the zero vector, and the solution to the system given by (42) is

$$\boxed{\mathbf{x}(t) = e^{\mathbf{A}(t-t_0)}\mathbf{c}.} \tag{43}$$

Occasionally, we are interested in just solving a differential equation and not an entire initial-value problem. In this case, the general solution can be shown to be (see Problem 2)

$$\boxed{\mathbf{x}(t) = e^{\mathbf{A}t}\mathbf{k} + e^{\mathbf{A}t}\int e^{-\mathbf{A}t}\mathbf{f}(t)\,dt,} \tag{44}$$

where $\mathbf{k}$ is an arbitrary n-dimensional constant vector. The general solution to the homogeneous differential equation by itself is given by

$$\boxed{\mathbf{x}(t) = e^{\mathbf{A}t}\mathbf{k}.} \tag{45}$$

PROBLEMS 7.4

1. Show by direct integration that

$$\int_{t_0}^{t} t^2\, dt = \int_{t_0}^{t} s^2\, ds = \int_{t_0}^{t} p^2\, dp.$$

In general, show that if $f(t)$ is integrable on $[a, b]$, then

$$\int_a^b f(t)\, dt = \int_a^b f(s)\, ds.$$

(Hint: Assume $\int f(t)\, dt = F(t) + c$. Hence, $\int f(s)\, ds = F(s) + c$. Use the fundamental theorem of integral calculus to obtain result.)

2. Derive Eq. (44). (Hint: Follow steps (30)–(33). For step (34) use indefinite integration and note that

$$\int \frac{d}{dt}\left[e^{-\mathbf{A}t}\mathbf{x}(t)\right] dt = e^{-\mathbf{A}t}\mathbf{x}(t) + \mathbf{k},$$

where **k** is an arbitrary constant vector of integration.)

3. Find (a) $e^{-\mathbf{A}t}$ (b) $e^{\mathbf{A}(t-2)}$ (c) $e^{\mathbf{A}(t-s)}$ (d) $e^{-\mathbf{A}(t-2)}$, if

$$e^{\mathbf{A}t} = e^{3t}\begin{bmatrix} 1 & t & t^2/2 \\ 0 & 1 & t \\ 0 & 0 & 1 \end{bmatrix}.$$

4. For $e^{\mathbf{A}t}$ as given in Problem 3, invert by the method of cofactors to obtain $e^{-\mathbf{A}t}$ and hence verify part (a) of that problem.

7.5 EXAMPLES

EXAMPLE 1 Use matrix methods to solve

$$\begin{aligned} \dot{u}(t) &= u(t) + 2v(t) + 1 \\ \dot{v}(t) &= 4u(t) + 3v(t) - 1 \\ u(0) &= 1, \quad v(0) = 2. \end{aligned}$$

Solution This system can be put into fundamental form if we define $t_0 = 0$,

$$\mathbf{x}(t) = \begin{bmatrix} u(t) \\ v(t) \end{bmatrix}, \quad \mathbf{A} = \begin{bmatrix} 1 & 2 \\ 4 & 3 \end{bmatrix}, \quad \mathbf{f}(t) = \begin{bmatrix} 1 \\ -1 \end{bmatrix}, \quad \text{and} \quad \mathbf{c} = \begin{bmatrix} 1 \\ 2 \end{bmatrix}. \tag{46}$$

Since **A** is independent of t, this is a system with constant coefficients, and the solution is given by (40). For the **A** in (46), $e^{\mathbf{A}t}$ is found to be

$$e^{\mathbf{A}t} = \frac{1}{6}\begin{bmatrix} 2e^{5t} + 4e^{-t} & 2e^{5t} - 2e^{-t} \\ 4e^{5t} - 4e^{-t} & 4e^{5t} + 2e^{-t} \end{bmatrix}.$$

Hence,

$$e^{-\mathbf{A}s} = \frac{1}{6}\begin{bmatrix} 2e^{-5s} + 4e^{s} & 2e^{-5s} - 2e^{s} \\ 4e^{-5s} - 4e^{s} & 4e^{-5s} + 2e^{s} \end{bmatrix}$$

and

$$e^{\mathbf{A}(t-t_0)} = e^{\mathbf{A}t}, \qquad \text{since} \quad t_0 = 0.$$

Thus,

$$\begin{aligned} e^{\mathbf{A}(t-t_0)}\mathbf{c} &= \frac{1}{6}\begin{bmatrix} 2e^{5t} + 4e^{-t} & 2e^{5t} - 2e^{-t} \\ 4e^{5t} - 4e^{-t} & 4e^{5t} + 2e^{-t} \end{bmatrix}\begin{bmatrix} 1 \\ 2 \end{bmatrix} \\ &= \frac{1}{6}\begin{bmatrix} 1[2e^{5t} + 4e^{-t}] + 2[2e^{5t} - 2e^{-t}] \\ 1[4e^{5t} - 4e^{-t}] + 2[4e^{5t} + 2e^{-t}] \end{bmatrix} \\ &= \begin{bmatrix} e^{5t} \\ 2e^{5t} \end{bmatrix}, \end{aligned} \tag{47}$$

and

$$\begin{aligned} e^{-\mathbf{A}s}\mathbf{f}(s) &= \frac{1}{6}\begin{bmatrix} 2e^{-5s} + 4e^{s} & 2e^{-5s} - 2e^{s} \\ 4e^{-5s} - 4e^{s} & 4e^{-5s} + 2e^{s} \end{bmatrix}\begin{bmatrix} 1 \\ -1 \end{bmatrix} \\ &= \frac{1}{6}\begin{bmatrix} 1[2e^{-5s} + 4e^{s}] - 1[2e^{-5s} - 2e^{s}] \\ 1[4e^{-5s} - 4e^{s}] - 1[4e^{-5s} + 2e^{s}] \end{bmatrix} = \begin{bmatrix} e^{s} \\ -e^{s} \end{bmatrix}. \end{aligned}$$

Hence,

$$\int_{t_0}^{t} e^{-\mathbf{A}s}\mathbf{f}(s)\,ds = \begin{bmatrix} \displaystyle\int_0^t e^{s}\,ds \\ \displaystyle\int_0^t -e^{s}\,ds \end{bmatrix} = \begin{bmatrix} e^{s}|_0^t \\ -e^{s}|_0^t \end{bmatrix} = \begin{bmatrix} e^{t} - 1 \\ -e^{t} + 1 \end{bmatrix}$$

and

$$\begin{aligned} e^{\mathbf{A}t}\int_{t_0}^{t} e^{-\mathbf{A}s}\mathbf{f}(s)\,ds &= \frac{1}{6}\begin{bmatrix} 2e^{5t} + 4e^{-t} & 2e^{5t} - 2e^{-t} \\ 4e^{5t} - 4e^{-t} & 4e^{5t} + 2e^{-t} \end{bmatrix}\begin{bmatrix} (e^{t} - 1) \\ (1 - e^{t}) \end{bmatrix} \\ &= \frac{1}{6}\begin{bmatrix} [2e^{5t} + 4e^{-t}][e^{t} - 1] + [2e^{5t} - 2e^{-t}][1 - e^{t}] \\ [4e^{5t} - 4e^{-t}][e^{t} - 1] + [4e^{5t} + 2e^{-t}][1 - e^{t}] \end{bmatrix} \\ &= \begin{bmatrix} (1 - e^{-t}) \\ (-1 + e^{-t}) \end{bmatrix}. \end{aligned} \tag{48}$$

Substituting (47) and (48) into (40), we have

$$\begin{bmatrix} u(t) \\ v(t) \end{bmatrix} = \mathbf{x}(t) = \begin{bmatrix} e^{5t} \\ 2e^{5t} \end{bmatrix} + \begin{bmatrix} 1 - e^{-t} \\ -1 + e^{-t} \end{bmatrix} = \begin{bmatrix} e^{5t} + 1 - e^{-t} \\ 2e^{5t} - 1 + e^{-t} \end{bmatrix},$$

or

$$u(t) = e^{5t} - e^{-t} + 1,$$
$$v(t) = 2e^{5t} + e^{-t} - 1.$$

EXAMPLE 2 Use matrix methods to solve

$$\ddot{y} - 3\dot{y} + 2y = e^{-3t},$$
$$y(1) = 1, \quad \dot{y}(1) = 0.$$

Solution This system can be put into fundamental form if we define $t_0 = 1$;

$$\mathbf{x}(t) = \begin{bmatrix} x_1(t) \\ x_2(t) \end{bmatrix}, \qquad \mathbf{A} = \begin{bmatrix} 0 & 1 \\ -2 & 3 \end{bmatrix}, \qquad \mathbf{f}(t) = \begin{bmatrix} 0 \\ e^{-3t} \end{bmatrix}, \qquad \text{and} \qquad \mathbf{c} = \begin{bmatrix} 1 \\ 0 \end{bmatrix}.$$

The solution to this system is given by (40). For this **A**,

$$e^{\mathbf{A}t} = \begin{bmatrix} -e^{2t} + 2e^{t} & e^{2t} - e^{t} \\ -2e^{2t} + 2e^{t} & 2e^{2t} - e^{t} \end{bmatrix}.$$

Thus,

$$e^{\mathbf{A}(t-t_0)}\mathbf{c} = \begin{bmatrix} -e^{2(t-1)} + 2e^{(t-1)} & e^{2(t-1)} - e^{(t-1)} \\ -2e^{2(t-1)} + 2e^{(t-1)} & 2e^{2(t-1)} - e^{(t-1)} \end{bmatrix} \begin{bmatrix} 1 \\ 0 \end{bmatrix}$$
$$= \begin{bmatrix} -e^{2(t-1)} + 2e^{(t-1)} \\ -2e^{2(t-1)} + 2e^{(t-1)} \end{bmatrix}. \tag{49}$$

Now

$$\mathbf{f}(t) = \begin{bmatrix} 0 \\ e^{-3t} \end{bmatrix}, \qquad \mathbf{f}(s) = \begin{bmatrix} 0 \\ e^{-3s} \end{bmatrix},$$

and

$$e^{-\mathbf{A}s}\mathbf{f}(s) = \begin{bmatrix} -e^{-2s} + 2e^{-s} & e^{-2s} - e^{-s} \\ -2e^{-2s} + 2e^{-s} & 2e^{-2s} - e^{-s} \end{bmatrix} \begin{bmatrix} 0 \\ e^{-3s} \end{bmatrix}$$
$$= \begin{bmatrix} e^{-5s} - e^{-4s} \\ 2e^{-5s} - e^{-4s} \end{bmatrix}.$$

Hence,

$$\int_{t_0}^{t} e^{-\mathbf{A}s}\mathbf{f}(s)\,ds = \begin{bmatrix} \displaystyle\int_1^t (e^{-5s} - e^{-4s})\,ds \\ \displaystyle\int_1^t (2e^{-5s} - e^{-4s})\,ds \end{bmatrix}$$
$$= \begin{bmatrix} (-\tfrac{1}{5})e^{-5t} + (\tfrac{1}{4})e^{-4t} + (\tfrac{1}{5})e^{-5} - (\tfrac{1}{4})e^{-4} \\ (-\tfrac{2}{5})e^{-5t} + (\tfrac{1}{4})e^{-4t} + (\tfrac{2}{5})e^{-5} - (\tfrac{1}{4})e^{-4} \end{bmatrix},$$

and

$$e^{\mathbf{A}t}\int_{t_0}^{t} e^{-\mathbf{A}s}\mathbf{f}(s)\,ds$$

$$= \begin{bmatrix} (-e^{2t}+2e^{t}) & (e^{2t}-e^{t}) \\ (-2e^{2t}+2e^{t}) & (2e^{2t}-e^{t}) \end{bmatrix} \begin{bmatrix} (-\frac{1}{5}e^{-5t}+\frac{1}{4}e^{-4t}+\frac{1}{5}e^{-5}-\frac{1}{4}e^{-4}) \\ (-\frac{2}{5}e^{-5t}+\frac{1}{4}e^{-4t}+\frac{2}{5}e^{-5}-\frac{1}{4}e^{-4}) \end{bmatrix}$$

$$= \begin{bmatrix} \frac{1}{20}e^{-3t}+\frac{1}{5}e^{(2t-5)}-\frac{1}{4}e^{t-4} \\ -\frac{3}{20}e^{-3t}+\frac{2}{5}e^{(2t-5)}-\frac{1}{4}e^{t-4} \end{bmatrix}. \tag{50}$$

Substituting (49) and (50) into (40), we have that

$$\mathbf{x}(t) = \begin{bmatrix} x_1(t) \\ x_2(t) \end{bmatrix} = \begin{bmatrix} -e^{2(t-1)}+2e^{t-1} \\ -2e^{2(t-1)}+2e^{t-1} \end{bmatrix} + \begin{bmatrix} \frac{1}{20}e^{-3t}+\frac{1}{5}e^{(2t-5)}-\frac{1}{4}e^{t-4} \\ -\frac{3}{20}e^{-3t}+\frac{2}{5}e^{(2t-5)}-\frac{1}{4}e^{t-4} \end{bmatrix}$$

$$= \begin{bmatrix} -e^{2(t-1)}+2e^{t-1}+\frac{1}{20}e^{-3t}+\frac{1}{5}e^{(2t-5)}-\frac{1}{4}e^{t-4} \\ -2e^{2(t-1)}+2e^{t-1}-\frac{3}{20}e^{-3t}+\frac{2}{5}e^{(2t-5)}-\frac{1}{4}e^{t-4} \end{bmatrix}.$$

Thus, it follows that the solution to the initial-value problem is given by

$$y(t) = x_1(t) = -e^{2(t-1)}+2e^{t-1}+\tfrac{1}{20}e^{-3t}+\tfrac{1}{5}e^{(2t-5)}-\tfrac{1}{4}e^{t-4}.$$

The most tedious step in Example 2 was multiplying the matrix $e^{\mathbf{A}t}$ by the vector $\int_{t_0}^{t} e^{-\mathbf{A}s}\mathbf{f}(s)\,ds$. We could have eliminated this multiplication had we used (41) for the solution rather than (40). Of course, in using (41), we would have had to handle an integral rather more difficult than the one we encountered.

If $\mathbf{A}$ and $\mathbf{f}(t)$ are relatively simple (for instance, if $\mathbf{f}(t)$ is a constant vector), then the integral obtained in (41) may not be too difficult to evaluate, and its use can be a real savings in time and effort over the use of (40). We illustrate this point in the next example.

EXAMPLE 3 Use matrix methods to solve

$$\ddot{x}(t) + x(t) = 2,$$
$$x(\pi) = 0, \quad \dot{x}(\pi) = -1.$$

Solution This initial-valued problem can be put into fundamental form if we define $t_0 = \pi$,

$$\mathbf{x}(t) = \begin{bmatrix} x_1(t) \\ x_2(t) \end{bmatrix}, \quad \mathbf{A} = \begin{bmatrix} 0 & 1 \\ -1 & 0 \end{bmatrix}, \quad \mathbf{f}(t) = \begin{bmatrix} 0 \\ 2 \end{bmatrix}, \quad \text{and} \quad \mathbf{c} = \begin{bmatrix} 0 \\ -1 \end{bmatrix}. \tag{51}$$

Here, **A** is again independent of the variable t, hence, the solution is given by either (40) or (41). This time we elect to use (41). For the **A** given in (51), $e^{\mathbf{A}t}$ is found to be

$$e^{\mathbf{A}t} = \begin{bmatrix} \cos t & \sin t \\ -\sin t & \cos t \end{bmatrix}.$$

Thus,

$$\begin{aligned} e^{\mathbf{A}(t-t_0)}\mathbf{c} &= \begin{bmatrix} \cos(t-\pi) & \sin(t-\pi) \\ -\sin(t-\pi) & \cos(t-\pi) \end{bmatrix} \begin{bmatrix} 0 \\ -1 \end{bmatrix} \\ &= \begin{bmatrix} -\sin(t-\pi) \\ -\cos(t-\pi) \end{bmatrix}, \end{aligned} \tag{52}$$

and

$$\begin{aligned} e^{\mathbf{A}(t-s)}\mathbf{f}(s) &= \begin{bmatrix} \cos(t-s) & \sin(t-s) \\ -\sin(t-s) & \cos(t-s) \end{bmatrix} \begin{bmatrix} 0 \\ 2 \end{bmatrix} \\ &= \begin{bmatrix} 2\sin(t-s) \\ 2\cos(t-s) \end{bmatrix}. \end{aligned}$$

Hence,

$$\begin{aligned} \int_{t_0}^{t} e^{\mathbf{A}(t-s)}\mathbf{f}(s)\,ds &= \begin{bmatrix} \displaystyle\int_{\pi}^{t} 2\sin(t-s)\,ds \\ \displaystyle\int_{\pi}^{t} 2\cos(t-s)\,ds \end{bmatrix} \\ &= \begin{bmatrix} 2 - 2\cos(t-\pi) \\ 2\sin(t-\pi) \end{bmatrix}. \end{aligned} \tag{53}$$

Substituting (52) and (53) into (41) and using the trigonometric identities $\sin(t-\pi) = -\sin t$ and $\cos(t-\pi) = -\cos t$, we have

$$\begin{aligned} \begin{bmatrix} x_1(t) \\ x_2(t) \end{bmatrix} = \mathbf{x}(t) &= \begin{bmatrix} -\sin(t-\pi) \\ -\cos(t-\pi) \end{bmatrix} + \begin{bmatrix} 2 - 2\cos(t-\pi) \\ 2\sin(t-\pi) \end{bmatrix} \\ &= \begin{bmatrix} \sin t + 2\cos t + 2 \\ \cos t - 2\sin t \end{bmatrix}. \end{aligned}$$

Thus, since $x(t) = x_1(t)$, it follows that the solution to the intitial-value problem is given by

$$x(t) = \sin t + 2\cos t + 2.$$

EXAMPLE 4 Solve by matrix methods

$$\begin{aligned} \dot{u}(t) &= u(t) + 2v(t), \\ \dot{v}(t) &= 4u(t) + 3v(t). \end{aligned}$$

Solution This system can be put into fundamental form if we define

$$\mathbf{x}(t) = \begin{bmatrix} u(t) \\ v(t) \end{bmatrix}, \qquad \mathbf{A} = \begin{bmatrix} 1 & 2 \\ 4 & 3 \end{bmatrix}, \qquad \text{and} \qquad \mathbf{f}(t) = \begin{bmatrix} 0 \\ 0 \end{bmatrix}.$$

This is a homogeneous system with constant coefficients and no initial conditions specified; hence, the general solution is given by (45).

As in Example 1, for this **A**, we have

$$e^{\mathbf{A}t} = \frac{1}{6}\begin{bmatrix} 2e^{5t} + 4e^{-t} & 2e^{5t} - 2e^{-t} \\ 4e^{5t} - 4e^{-t} & 4e^{5t} + 2e^{-t} \end{bmatrix}.$$

Thus,

$$\begin{aligned} e^{\mathbf{A}t}\mathbf{k} &= \frac{1}{6}\begin{bmatrix} 2e^{5t} + 4e^{-t} & 2e^{5t} - 2e^{-t} \\ 4e^{5t} - 4e^{-t} & 4e^{5t} + 2e^{-t} \end{bmatrix}\begin{bmatrix} k_1 \\ k_2 \end{bmatrix} \\ &= \frac{1}{6}\begin{bmatrix} k_1[2e^{5t} + 4e^{-t}] + k_2[2e^{5t} - 2e^{-t}] \\ k_1[4e^{5t} - 4e^{-t}] + k_2[4e^{5t} + 2e^{-t}] \end{bmatrix} \\ &= \frac{1}{6}\begin{bmatrix} e^{5t}(2k_1 + 2k_2) + e^{-t}(4k_1 - 2k_2) \\ e^{5t}(4k_1 + 4k_2) + e^{-t}(-4k_1 + 2k_2) \end{bmatrix}. \end{aligned} \tag{54}$$

Substituting (54) into (45), we have that

$$\begin{bmatrix} u(t) \\ v(t) \end{bmatrix} = \mathbf{x}(t) = \frac{1}{6}\begin{bmatrix} e^{5t}(2k_1 + 2k_2) + e^{-t}(4k_1 - 2k_2) \\ e^{5t}(4k_1 + 4k_2) + e^{-t}(-4k_1 + 2k_2) \end{bmatrix}$$

or

$$\begin{aligned} u(t) &= \left(\frac{2k_1 + 2k_2}{6}\right)e^{5t} + \left(\frac{4k_1 - 2k_2}{6}\right)e^{-t} \\ v(t) &= 2\left(\frac{2k_1 + 2k_2}{6}\right)e^{5t} + \left(\frac{-4k_1 + 2k_2}{6}\right)e^{-t}. \end{aligned} \tag{55}$$

We can simplify the expressions for $u(t)$ and $v(t)$ if we introduce two new arbitrary constants k_3 and k_4 defined by

$$k_3 = \frac{2k_1 + 2k_2}{6}, \qquad k_4 = \frac{4k_1 - 2k_2}{6}. \tag{56}$$

Substituting these values into (55), we obtain

$$\begin{aligned} u(t) &= k_3 e^{5t} + k_4 e^{-t} \\ v(t) &= 2k_3 e^{5t} - k_4 e^{-t}. \end{aligned} \tag{57}$$

PROBLEMS 7.5

Solve each of the following systems by matrix methods. Note that Problems 1–4 have the same coefficient matrix.

1. $\dot{x}(t) = -2x(t) + 3y(t),$
   $\dot{y}(t) = -x(t) + 2y(t);$
   $x(2) = 2, \quad y(2) = 4.$

2. $\dot{x}(t) = -2x(t) + 3y(t) + 1,$
   $\dot{y}(t) = -x(t) + 2y(t) + 1;$
   $x(1) = 1, \quad y(1) = 1.$

3. $\dot{x}(t) = -2x(t) + 3y(t),$
   $\dot{y}(t) = -x(t) + 2y(t).$

4. $\dot{x}(t) = -2x(t) + 3y(t) + 1,$
   $\dot{y}(t) = -x(t) + 2y(t) + 1.$

5. $\ddot{x}(t) = -4x(t) + \sin t;$
   $x(0) = 1, \quad \dot{x}(0) = 0.$

6. $\dddot{x}(t) = t;$
   $x(1) = 1, \quad \dot{x}(1) = 2, \quad \ddot{x}(1) = 3$

7. $\ddot{x} - \dot{x} - 2x = e^{-t};$
   $x(0) = 1, \quad \dot{x}(0) = 0.$

8. $\ddot{x} = 2\dot{x} + 5y + 3,$
   $\dot{y} = -\dot{x} - 2y;$
   $x(0) = 0, \quad \dot{x}(0) = 0, \quad y(0) = 1.$

7.6 SOLUTIONS OF SYSTEMS—GENERAL CASE

Having completely solved systems of linear differential equations with constant coefficients, we now turn our attention to the solutions of systems of the form

$$\begin{aligned} \dot{\mathbf{x}}(t) &= \mathbf{A}(t)\mathbf{x}(t) + \mathbf{f}(t), \\ \mathbf{x}(t_0) &= \mathbf{c}. \end{aligned} \tag{58}$$

Note that $\mathbf{A}(t)$ may now depend on t, hence the analysis of Section 7.4 does not apply. However, since we still require both $\mathbf{A}(t)$ and $\mathbf{f}(t)$ to be continuous in some interval about $t = t_0$, Theorem 1 of Section 7.1 still guarantees that (58) has a unique solution. Our aim in this section is to obtain a representation for this solution.

Definition 1 The *transition* (or fundamental) *matrix* of the homogeneous equation $\dot{\mathbf{x}}(t) = \mathbf{A}(t)\mathbf{x}(t)$ is an $n \times n$ matrix $\mathbf{\Phi}(t, t_0)$ having the properties that

(a) $$\frac{d}{dt}\mathbf{\Phi}(t, t_0) = \mathbf{A}(t)\mathbf{\Phi}(t, t_0), \tag{59}$$

(b) $$\mathbf{\Phi}(t_0, t_0) = \mathbf{I}. \tag{60}$$

Here t_0 is the initial time given in (58). In the appendix to this chapter, we show that $\mathbf{\Phi}(t, t_0)$ exists and is unique.

EXAMPLE 1 Find $\mathbf{\Phi}(t, t_0)$ if $\mathbf{A}(t)$ is a constant matrix.

Solution Consider the matrix $e^{\mathbf{A}(t-t_0)}$. From Property 1 of Section 6.8, we have that $e^{\mathbf{A}(t_0-t_0)} = e^{\mathbf{0}} = \mathbf{I}$, while from Theorem 1 of Section 6.9, we have that

$$\begin{aligned}\frac{d}{dt} e^{\mathbf{A}(t-t_0)} &= \frac{d}{dt}(e^{\mathbf{A}t}e^{-\mathbf{A}t_0}) \\ &= \mathbf{A}e^{\mathbf{A}t}e^{-\mathbf{A}t_0} \\ &= \mathbf{A}e^{\mathbf{A}(t-t_0)}.\end{aligned}$$

Thus, $e^{\mathbf{A}(t-t_0)}$ satisfies (59) and (60). Since $\mathbf{\Phi}(t, t_0)$ is unique, it follows for the case where $\mathbf{A}$ is a *constant* matrix that

$$\mathbf{\Phi}(t, t_0) = e^{\mathbf{A}(t-t_0)}. \tag{61}$$

Caution: Although $\mathbf{\Phi}(t, t_0) = e^{\mathbf{A}(t-t_0)}$ if $\mathbf{A}$ is a constant matrix, this equality is not valid if $\mathbf{A}$ actually depends on t. In fact, it is usually impossible to explicitly find $\mathbf{\Phi}(t, t_0)$ in the general time varying case. Usually, the best we can say about the transition matrix is that it exists, it is unique, and, of course, it satisfies (59) and (60).

One immediate use of $\mathbf{\Phi}(t, t_0)$ is that it enables us to theoretically solve the general homogeneous initial-value problem

$$\begin{aligned}\dot{\mathbf{x}}(t) &= \mathbf{A}(t)\mathbf{x}(t) \\ \mathbf{x}(t_0) &= \mathbf{c}.\end{aligned} \tag{62}$$

Theorem 1 The unique solution to (62) is

$$\mathbf{x}(t) = \mathbf{\Phi}(t, t_0)\mathbf{c}. \tag{63}$$

Proof If $\mathbf{A}(t)$ is a constant matrix, (63) reduces to (43) (see (61)), hence Theorem 1 is valid. In general, however, we have that

$$\begin{aligned}\frac{d\mathbf{x}(t)}{dt} &= \frac{d}{dt}[\mathbf{\Phi}(t, t_0)\mathbf{c}] = \frac{d}{dt}[\mathbf{\Phi}(t, t_0)]\mathbf{c}, \\ &= \mathbf{A}(t)\mathbf{\Phi}(t, t_0)\mathbf{c} \quad \{\text{from (59)}, \\ &= \mathbf{A}(t)\mathbf{x}(t) \quad \{\text{from (63)},\end{aligned}$$

and

$$\begin{aligned}\mathbf{x}(t_0) &= \mathbf{\Phi}(t_0, t_0)\mathbf{c}, \\ &= \mathbf{Ic} \qquad \{\text{from (60)}, \\ &= \mathbf{c}.\end{aligned}$$

EXAMPLE 2 Find $x(t)$ and $y(t)$ if

$$\begin{aligned}\dot{x} &= ty \\ \dot{y} &= -tx \\ x(1) = 0, &\quad y(1) = 1,\end{aligned}$$

Solution Putting this system into fundamental form, we obtain

$$t_0 = 1, \qquad \mathbf{x}(t) = \begin{bmatrix} x(t) \\ y(t) \end{bmatrix}, \qquad \mathbf{A}(t) = \begin{bmatrix} 0 & t \\ -t & 0 \end{bmatrix}, \qquad \mathbf{f}(t) = \mathbf{0}, \qquad \mathbf{c} = \begin{bmatrix} 0 \\ 1 \end{bmatrix},$$

and

$$\begin{aligned}\dot{\mathbf{x}}(t) &= \mathbf{A}(t)\mathbf{x}(t), \\ \mathbf{x}(t_0) &= \mathbf{c}.\end{aligned}$$

The transition matrix for this system can be shown to be (see Problem 1)

$$\mathbf{\Phi}(t, t_0) = \begin{bmatrix} \cos\left(\dfrac{t^2 - t_0^2}{2}\right) & \sin\left(\dfrac{t^2 - t_0^2}{2}\right) \\ -\sin\left(\dfrac{t^2 - t_0^2}{2}\right) & \cos\left(\dfrac{t^2 - t_0^2}{2}\right) \end{bmatrix}.$$

Thus, from (63), we have

$$\begin{aligned}\mathbf{x}(t) &= \begin{bmatrix} \cos\left(\dfrac{t^2 - 1}{2}\right) & \sin\left(\dfrac{t^2 - 1}{2}\right) \\ -\sin\left(\dfrac{t^2 - 1}{2}\right) & \cos\left(\dfrac{t^2 - 1}{2}\right) \end{bmatrix} \begin{bmatrix} 0 \\ 1 \end{bmatrix} \\ &= \begin{bmatrix} \sin\left(\dfrac{t^2 - 1}{2}\right) \\ \cos\left(\dfrac{t^2 - 1}{2}\right) \end{bmatrix}.\end{aligned}$$

Consequently, the solution is

$$x(t) = \sin\left(\frac{t^2 - 1}{2}\right), \qquad y(t) = \cos\left(\frac{t^2 - 1}{2}\right).$$

The transition matrix also enables us to give a representation for the solution of the general time-varying initial-value problem

$$\begin{aligned} \dot{\mathbf{x}}(t) &= \mathbf{A}(t)\mathbf{x}(t) = \mathbf{f}(t), \\ \mathbf{x}(t_0) &= \mathbf{c}. \end{aligned} \tag{58}$$

Theorem 2 The unique solution to (58) is

$$\mathbf{x}(t) = \mathbf{\Phi}(t, t_0)\mathbf{c} + \int_{t_0}^{t} \mathbf{\Phi}(t, s)\mathbf{f}(s)\, ds. \tag{64}$$

Proof If $\mathbf{A}$ is a constant matrix, $\mathbf{\Phi}(t, t_0) = e^{\mathbf{A}(t-t_0)}$; hence $\mathbf{\Phi}(t, s) = e^{\mathbf{A}(t-s)}$ and (64) reduces to (41). We defer the proof of the general case, where $\mathbf{A}(t)$ depends on t, until the next section.

Equation (64) is the solution to the general initial-value problem given by (58). It should be noted, however, that since $\mathbf{\Phi}(t, t_0)$ is not explicitly known, $\mathbf{x}(t)$ will not be explicitly known either, hence, (64) is not as useful a formula as it might first appear. Unfortunately, (64) is the best solution that we can obtain for the general time varying problem. The student should not despair, though. It is often the case that by knowing enough properties of $\mathbf{\Phi}(t, t_0)$, we can extract a fair amount of information about the solution from (64). In fact, we can sometimes even obtain the exact solution! We shall consider some of these important properties of the transition matrix in the next section.

PROBLEMS 7.6

1. Use (59) and (60) to show that

$$\mathbf{\Phi}(t, t_0) = \begin{bmatrix} \cos\left(\dfrac{t^2 - t_0^2}{2}\right) & \sin\left(\dfrac{t^2 - t_0^2}{2}\right) \\ -\sin\left(\dfrac{t^2 - t_0^2}{2}\right) & \cos\left(\dfrac{t^2 - t_0^2}{2}\right) \end{bmatrix}$$

is a transition matrix for

$$\begin{aligned} \dot{x} &= ty, \\ \dot{y} &= -tx. \end{aligned}$$

2. As a generalization of Problem 1, use (59) and (60) to show that

$$\mathbf{\Phi}(t, t_0) = \begin{bmatrix} \cos \displaystyle\int_{t_0}^{t} g(s)\, ds & \sin \displaystyle\int_{t_0}^{t} g(s)\, ds \\ -\sin \displaystyle\int_{t_0}^{t} g(s)\, ds & \cos \displaystyle\int_{t_0}^{t} g(s)\, ds \end{bmatrix}$$

is a transition matrix for

$$\dot{x} = g(t)y,$$
$$\dot{y} = -g(t)x.$$

7.7 PROPERTIES OF THE TRANSITION MATRIX

Before considering some important properties of the transition matrix, we state one lemma that we ask the student to prove (see Problem 1).

Lemma 1 If $\mathbf{B}(t)$ is an $n \times n$ matrix having the property that $\mathbf{B}(t)\mathbf{c} = \mathbf{0}$ for every n-dimensional constant vector $\mathbf{c}$, then $\mathbf{B}(t)$ is the zero matrix.

For the remainder of this section we assume that $\mathbf{\Phi}(t, t_0)$ is the transition matrix for $\dot{\mathbf{x}}(t) = \mathbf{A}(t)\mathbf{x}(t)$.

Property 1 (The transition property)

$$\mathbf{\Phi}(t, \tau)\mathbf{\Phi}(\tau, t_0) = \mathbf{\Phi}(t, t_0). \tag{65}$$

Proof If $\mathbf{A}(t)$ is a constant matrix, $\mathbf{\Phi}(t, t_0) = e^{\mathbf{A}(t-t_0)}$ hence,

$$\begin{aligned}\mathbf{\Phi}(t, \tau)\mathbf{\Phi}(\tau, t_0) &= e^{\mathbf{A}(t-\tau)}e^{\mathbf{A}(\tau-t_0)} \\ &= e^{\mathbf{A}(t-\tau+\tau-t_0)} \\ &= e^{\mathbf{A}(t-t_0)} = \mathbf{\Phi}(t, t_0).\end{aligned}$$

Thus, Property 1 is immediate. For the more general case, that in which $\mathbf{A}(t)$ depends on t, the argument runs as follows: Consider the initial-value problem

$$\begin{aligned}\dot{\mathbf{x}}(t) &= \mathbf{A}(t)\mathbf{x}(t) \\ \mathbf{x}(t_0) &= \mathbf{c}.\end{aligned} \tag{66}$$

The unique solution of (66) is

$$\mathbf{x}(t) = \mathbf{\Phi}(t, t_0)\mathbf{c}. \tag{67}$$

Hence,

$$\mathbf{x}(t_1) = \mathbf{\Phi}(t_1, t_0)\mathbf{c} \tag{68}$$

and

$$\mathbf{x}(\tau) = \mathbf{\Phi}(\tau, t_0)\mathbf{c}, \tag{69}$$

where t_1 is any arbitrary time greater than τ. If we designate the vector $\mathbf{x}(t_1)$ by $\mathbf{d}$ and the vector $\mathbf{x}(\tau)$ by $\mathbf{b}$, then we can give the solution graphically by Fig. 1.

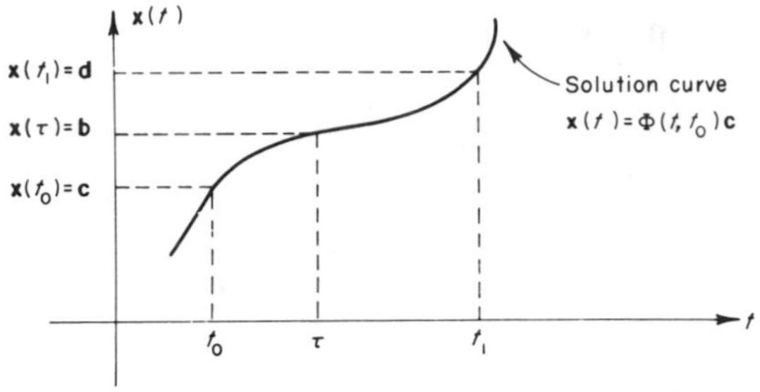

Figure 1

Consider an associated system governed by

$$\dot{\mathbf{x}}(t) = \mathbf{A}(t)\mathbf{x}(t), \tag{70}$$
$$\mathbf{x}(\tau) = \mathbf{b}.$$

We seek a solution to the above differential equation that has an initial value **b** at the initial time $t = \tau$. If we designate the solution by $\mathbf{y}(t)$, it follows from Theorem 1 of the previous section that

$$\mathbf{y}(t) = \mathbf{\Phi}(t, \tau)\mathbf{b}, \tag{71}$$

hence

$$\mathbf{y}(t_1) = \mathbf{\Phi}(t_1, \tau)\mathbf{b}. \tag{72}$$

But now we note that both $\mathbf{x}(t)$ and $\mathbf{y}(t)$ are governed by the same equation of motion, namely $\dot{\mathbf{x}}(t) = \mathbf{A}(t)\mathbf{x}(t)$, and both $\mathbf{x}(t)$ and $\mathbf{y}(t)$ go through the same point (τ, b). Thus, $\mathbf{x}(t)$ and $\mathbf{y}(t)$ must be the same solution. That is, the solution curve for $\mathbf{y}(t)$ looks exactly like that of $\mathbf{x}(t)$, shown in Fig. 1 except that it starts at $t = \tau$, while that of $\mathbf{x}(t)$ starts at $t = t_0$.
Hence,

$$\mathbf{x}(t) = \mathbf{y}(t), \qquad t \geq \tau,$$

and, in particular,

$$\mathbf{x}(t_1) = \mathbf{y}(t_1). \tag{73}$$

Thus, substituting (68) and (72) into (73) we obtain

$$\mathbf{\Phi}(t_1, t_0)\mathbf{c} = \mathbf{\Phi}(t_1, \tau)\mathbf{b}. \tag{74}$$

However, $\mathbf{x}(\tau) = \mathbf{b}$, thus (74) may be rewritten as

$$\mathbf{\Phi}(t_1, t_0)\mathbf{c} = \mathbf{\Phi}(t_1, \tau)\mathbf{x}(\tau). \tag{75}$$

Substituting (69) into (75), we have

$$\mathbf{\Phi}(t_1, t_0)\mathbf{c} = \mathbf{\Phi}(t_1, \tau)\mathbf{\Phi}(\tau, t_0)\mathbf{c}$$

or

$$[\boldsymbol{\Phi}(t_1, t_0) - \boldsymbol{\Phi}(t_1, \tau)\boldsymbol{\Phi}(\tau, t_0)]\mathbf{c} = \mathbf{0}. \tag{76}$$

Since $\mathbf{c}$ may represent any n-dimensional initial state, it follows from Lemma 1 that

$$\boldsymbol{\Phi}(t_1, t_0) - \boldsymbol{\Phi}(t_1, \tau)\boldsymbol{\Phi}(\tau, t_0) = \mathbf{0}$$

or

$$\boldsymbol{\Phi}(t_1, t_0) = \boldsymbol{\Phi}(t_1, \tau)\boldsymbol{\Phi}(\tau, t_0). \tag{77}$$

Since t_1 is arbitrary, it can be replaced by t; Eq. (77) therefore implies Eq. (65).

Property 2 $\boldsymbol{\Phi}(t, t_0)$ is invertible and

$$[\boldsymbol{\Phi}(t, t_0)]^{-1} = \boldsymbol{\Phi}(t_0, t). \tag{78}$$

Proof This result is obvious if $\mathbf{A}(t)$ is a constant matrix. We know from Section 6.8 that the inverse of $e^{\mathbf{A}t}$ is $e^{-\mathbf{A}t}$, hence,

$$\begin{aligned}[\boldsymbol{\Phi}(t, t_0)]^{-1} &= [e^{\mathbf{A}(t-t_0)}]^{-1} \\ &= e^{-\mathbf{A}(t-t_0)} \\ &= e^{\mathbf{A}(t_0-t)} \\ &= \boldsymbol{\Phi}(t_0, t).\end{aligned}$$

In order to prove Property 2 for any $\mathbf{A}(t)$, we note that (65) is valid for any t, hence it must be valid for $t = t_0$. Thus

$$\boldsymbol{\Phi}(t_0, \tau)\boldsymbol{\Phi}(\tau, t_0) = \boldsymbol{\Phi}(t_0, t_0).$$

It follows from (60) that

$$\boldsymbol{\Phi}(t_0, \tau)\boldsymbol{\Phi}(\tau, t_0) = \mathbf{I}.$$

Thus, from the definition of the inverse, we have

$$[\boldsymbol{\Phi}(\tau, t_0)]^{-1} = \boldsymbol{\Phi}(t_0, \tau)$$

which implies (78).

EXAMPLE 1 Find the inverse of

$$\begin{bmatrix} \cos\left(\dfrac{t^2 - t_0^2}{2}\right) & \sin\left(\dfrac{t^2 - t_0^2}{2}\right) \\ -\sin\left(\dfrac{t^2 - t_0^2}{2}\right) & \cos\left(\dfrac{t^2 - t_0^2}{2}\right) \end{bmatrix}.$$

Solution From Problem 1 of Section 7.6, we know that the above matrix is a transition matrix. Hence using (78) we find the inverse to be

$$\begin{bmatrix} \cos\left(\dfrac{t_0^2 - t^2}{2}\right) & \sin\left(\dfrac{t_0^2 - t^2}{2}\right) \\ -\sin\left(\dfrac{t_0^2 - t^2}{2}\right) & \cos\left(\dfrac{t_0^2 - t^2}{2}\right) \end{bmatrix} = \begin{bmatrix} \cos\left(\dfrac{t^2 - t_0^2}{2}\right) & -\sin\left(\dfrac{t^2 - t_0^2}{2}\right) \\ \sin\left(\dfrac{t^2 - t_0^2}{2}\right) & \cos\left(\dfrac{t^2 - t_0^2}{2}\right) \end{bmatrix}.$$

Here we have used the identities $\sin(-\theta) = -\sin\theta$ and $\cos(-\theta) = \cos\theta$.

Properties 1 and 2 now enable us to prove Theorem 2 of Section 7.6, namely, that the solution of

$$\dot{\mathbf{x}}(t) = \mathbf{A}(t)\mathbf{x}(t) + \mathbf{f}(t),$$
$$\mathbf{x}(t_0) = \mathbf{c}$$

is

$$\mathbf{x}(t) = \mathbf{\Phi}(t, t_0)\mathbf{c} + \int_{t_0}^{t} \mathbf{\Phi}(t, s)\mathbf{f}(s)\, ds. \tag{79}$$

Using Property 1, we have that $\mathbf{\Phi}(t, s) = \mathbf{\Phi}(t, t_0)\mathbf{\Phi}(t_0, s)$; hence, (79) may be rewritten as

$$\mathbf{x}(t) = \mathbf{\Phi}(t, t_0)\mathbf{c} + \mathbf{\Phi}(t, t_0)\int_{t_0}^{t} \mathbf{\Phi}(t_0, s)\mathbf{f}(s)\, ds. \tag{80}$$

Now

$$\begin{aligned} \mathbf{x}(t_0) &= \mathbf{\Phi}(t_0, t_0)\mathbf{c} + \mathbf{\Phi}(t_0, t_0)\int_{t_0}^{t_0} \mathbf{\Phi}(t_0, s)\mathbf{f}(s)\, ds \\ &= \mathbf{Ic} + \mathbf{I0} = \mathbf{c}. \end{aligned}$$

Thus, the initial condition is satisfied by (80). To show that the differential equation is also satisfied, we differentiate (80) and obtain

$$\begin{aligned} \frac{d\mathbf{x}(t)}{dt} &= \frac{d}{dt}\left[\mathbf{\Phi}(t, t_0)\mathbf{c} + \mathbf{\Phi}(t, t_0)\int_{t_0}^{t} \mathbf{\Phi}(t_0, s)\mathbf{f}(s)\, ds\right] \\ &= \left[\frac{d}{dt}\mathbf{\Phi}(t, t_0)\right]\mathbf{c} + \left[\frac{d}{dt}\mathbf{\Phi}(t, t_0)\right]\int_{t_0}^{t} \mathbf{\Phi}(t_0, s)\mathbf{f}(s)\, ds \\ &\quad + \mathbf{\Phi}(t, t_0)\left[\frac{d}{dt}\int_{t_0}^{t} \mathbf{\Phi}(t_0, s)\mathbf{f}(s)\, ds\right] \\ &= \mathbf{A}(t)\mathbf{\Phi}(t, t_0)\mathbf{c} + \mathbf{A}(t)\mathbf{\Phi}(t, t_0)\int_{t_0}^{t} \mathbf{\Phi}(t_0, s)\mathbf{f}(s)\, ds \\ &\quad + \mathbf{\Phi}(t, t_0)\mathbf{\Phi}(t_0, t)\mathbf{f}(t) \\ &= \mathbf{A}(t)\left[\mathbf{\Phi}(t, t_0)\mathbf{c} + \mathbf{\Phi}(t, t_0)\int_{t_0}^{t} \mathbf{\Phi}(t_0, s)\mathbf{f}(s)\, ds\right] \\ &\quad + \mathbf{\Phi}(t, t_0)\mathbf{\Phi}^{-1}(t, t_0)\mathbf{f}(t). \end{aligned}$$

However, the quantity inside the bracket is given by (80) to be $\mathbf{x}(t)$; hence,

$$\frac{d\mathbf{x}(t)}{dt} = \mathbf{A}(t)\mathbf{x}(t) + \mathbf{f}(t).$$

We conclude this section with one final property of the transition matrix, the proof of which is beyond the scope of this book.[21]

Property 3

$$\det \boldsymbol{\Phi}(t, t_0) = \exp\left\{\int_{t_0}^{t} \operatorname{tr}[\mathbf{A}(t)]\, dt\right\}. \tag{81}$$

Since the exponential is never zero, (81) establishes that $\det \boldsymbol{\Phi}(t, t_0) \neq 0$, hence, we have an alternate proof that $\boldsymbol{\Phi}(t, t_0)$ is invertible.

PROBLEMS 7.7

1. Prove Lemma 1. (Hint: Consider the product $\mathbf{B}(t)\mathbf{c}$ where

$$\text{first, } \mathbf{c} = \begin{bmatrix} 1 \\ 0 \\ 0 \\ \vdots \\ 0 \end{bmatrix}, \qquad \text{second, } \mathbf{c} = \begin{bmatrix} 0 \\ 1 \\ 0 \\ \vdots \\ 0 \end{bmatrix}, \qquad \text{etc.)}$$

2. If $\boldsymbol{\Phi}(t, t_0)$ is a transition matrix, prove that

$$\boldsymbol{\Phi}'(t_1, t_0)\left[\int_{t_0}^{t_1} \boldsymbol{\Phi}(t_1, s)\boldsymbol{\Phi}'(t_1, s)\, ds\right]^{-1} \boldsymbol{\Phi}(t_1, t_0)$$
$$= \left[\int_{t_0}^{t_1} \boldsymbol{\Phi}(t_0, s)\boldsymbol{\Phi}'(t_0, s)\, ds\right]^{-1}.$$

7.8 THE ADJOINT SYSTEM

Often, when discussing the initial-value problem given by (58), it is advantageous to consider the related system

$$\dot{\mathbf{y}}(t) = -\mathbf{A}'(t)\mathbf{y}(t), \tag{82}$$

which is called the *adjoint system* of the homogeneous differential equation

$$\dot{\mathbf{x}}(t) = \mathbf{A}(t)\mathbf{x}(t). \tag{83}$$

If we define

$$\mathbf{B}(t) = -\mathbf{A}'(t), \tag{84}$$

[21] See E. A. Coddington and N. Levinson, "Theory of Ordinary Differential Equations," p. 67. McGraw-Hill, New York, 1955.

then the differential equation in (82) can be rewritten as

$$\dot{\mathbf{y}}(t) = \mathbf{B}(t)\mathbf{y}(t). \tag{85}$$

The transition matrix for this system is the matrix $\mathbf{\Psi}(t, t_0)$ having the properties (see Definition 1 of Section 7.6) that

$$\frac{d}{dt}\mathbf{\Psi}(t, t_0) = \mathbf{B}(t)\mathbf{\Psi}(t, t_0)$$

and

$$\mathbf{\Psi}(t_0, t_0) = \mathbf{I}. \tag{86}$$

Using (84) and (85), it follows that $\mathbf{\Psi}(t, t_0)$ is a transition matrix for (82) and $\mathbf{\Psi}(t, t_0)$ satisfies the properties (see 86)

$$\begin{aligned} \frac{d\mathbf{\Psi}(t, t_0)}{dt} &= -\mathbf{A}'(t)\mathbf{\Psi}(t, t_0) \\ \mathbf{\Psi}(t_0, t_0) &= \mathbf{I}. \end{aligned} \tag{87}$$

Define $\mathbf{\Phi}(t, t_0)$ to be the transition matrix of (83). Thus, $\mathbf{\Phi}(t, t_0)$ satisfies the properties

$$\begin{aligned} \frac{d\mathbf{\Phi}(t, t_0)}{dt} &= \mathbf{A}(t)\mathbf{\Phi}(t, t_0) \\ \mathbf{\Phi}(t_0, t_0) &= \mathbf{I}. \end{aligned} \tag{88}$$

Since the differential equations are so closely related, one might expect the transition matrices of (82) and (83) also to be closely related. This is fact turns out to be the case. However, in order to show the relationship between $\mathbf{\Phi}(t, t_0)$ and $\mathbf{\Psi}(t, t_0)$ we will need the following lemma.

Lemma 1 Let $\mathbf{B}(t)$ be a differentiable nonsingular matrix. Then

$$\frac{d\mathbf{B}^{-1}(t)}{dt} = -\mathbf{B}^{-1}(t)\left[\frac{d\mathbf{B}(t)}{dt}\right]\mathbf{B}^{-1}(t).$$

Proof $\mathbf{B}(t)\mathbf{B}^{-1}(t) = \mathbf{I}$.

Differentiating both sides of this equation, we obtain

$$\left[\frac{d\mathbf{B}(t)}{dt}\right]\mathbf{B}^{-1}(t) + \mathbf{B}(t)\frac{d}{dt}[\mathbf{B}^{-1}(t)] = \frac{d\mathbf{I}}{dt} = \mathbf{0}.$$

Thus,

$$\mathbf{B}(t)\left[\frac{d\mathbf{B}^{-1}(t)}{dt}\right] = -\left[\frac{d\mathbf{B}(t)}{dt}\right]\mathbf{B}^{-1}(t)$$

or

$$\frac{d\mathbf{B}^{-1}(t)}{dt} = -\mathbf{B}^{-1}(t)\left[\frac{d\,\mathbf{B}(t)}{dt}\right]\mathbf{B}^{-1}(t).$$

Theorem 1 $\mathbf{\Psi}(t, t_0) = [\mathbf{\Phi}'(t, t_0)]^{-1}$. (89)

Proof

$$[\mathbf{\Phi}'(t_0, t_0)]^{-1} = [\mathbf{I}']^{-1} = \mathbf{I}^{-1} = \mathbf{I}.$$

From Lemma 1,

$$\begin{aligned}\frac{d}{dt}[\mathbf{\Phi}'(t, t_0)]^{-1} &= -[\mathbf{\Phi}'(t, t_0)]^{-1}\left[\frac{d\mathbf{\Phi}'(t, t_0)}{dt}\right][\mathbf{\Phi}'(t, t_0)]^{-1}\\ &= -[\mathbf{\Phi}'(t, t_0)]^{-1}\left[\frac{d\mathbf{\Phi}(t, t_0)}{dt}\right]'[\mathbf{\Phi}'(t, t_0)]^{-1}\\ &= -[\mathbf{\Phi}'(t, t_0)]^{-1}[\mathbf{A}(t)\mathbf{\Phi}(t, t_0)]'[\mathbf{\Phi}'(t, t_0)]^{-1}\\ &= -[\mathbf{\Phi}'(t, t_0)]^{-1}\mathbf{\Phi}'(t, t_0)\mathbf{A}'(t)[\mathbf{\Phi}'(t, t_0)]^{-1}\\ &= -\mathbf{A}'(t)[\mathbf{\Phi}'(t, t_0)]^{-1}.\end{aligned}$$

Thus, $[\mathbf{\Phi}'(t, t_0)]^{-1}$ satisfies (87) and hence must be a transition matrix for (82). However, since the transition matrix is unique, it follows that

$$\mathbf{\Psi}(t, t_0) = [\mathbf{\Phi}'(t, t_0)]^{-1}.$$

Recall that the inverse of the transpose is the transpose of the inverse. Thus, we have that

$$[\mathbf{\Phi}'(t, t_0)]^{-1} = [\mathbf{\Phi}^{-1}(t, t_0)]'.$$

Using Property 2 of the previous section, it follows that (89) can be rewritten as

$$\mathbf{\Psi}(t, t_0) = \mathbf{\Phi}'(t_0, t). \tag{90}$$

Then from Theorem 1 of Section 7.6 and Eq. (90), it follows that the unique solution to (82) satisfying the initial condition $\mathbf{y}(t_0) = \mathbf{c}$ is

$$\mathbf{y}(t) = \mathbf{\Phi}'(t_0, t)\mathbf{c}.$$

PROBLEMS 7.8

1. Show that the solution to the initial-value problem

$$\dot{\mathbf{y}}(t) = -\mathbf{A}'(t)\mathbf{y}(t) + \mathbf{f}(t),$$

$$\mathbf{y}(t_0) = \mathbf{c}$$

is $\mathbf{y}(t) = \mathbf{\Phi}'(t_0, t)\mathbf{c} + \int_{t_0}^{t} \mathbf{\Phi}'(s, t)\mathbf{f}(s)\, ds$

where $\mathbf{\Phi}(t, t_0)$ is the transition matrix for $\dot{\mathbf{x}}(t) = \mathbf{A}(t)\mathbf{x}(t)$.

2. Verify (89) for

$$\mathbf{A}(t) = \begin{bmatrix} 0 & 1 \\ 1 & 0 \end{bmatrix}.$$

That is, first find $\mathbf{\Phi}(t, t_0)$ and $\mathbf{\Psi}(t, t_0)$ directly and then check whether or not (89) is valid. (Hint: $\mathbf{A}(t)$ is a constant matrix, hence the transition matrices can be found explicitly.)

APPENDIX TO CHAPTER 7

We now prove that there exists a unique matrix $\mathbf{\Phi}(t, t_0)$ having properties (59) and (60).

Define n-dimensional unit vectors $\mathbf{e}_1, \mathbf{e}_2, \ldots, \mathbf{e}_n$ by

$$\mathbf{e}_1 = \begin{bmatrix} 1 \\ 0 \\ 0 \\ 0 \\ \vdots \\ 0 \end{bmatrix}, \quad \mathbf{e}_2 = \begin{bmatrix} 0 \\ 1 \\ 0 \\ 0 \\ \vdots \\ 0 \end{bmatrix}, \quad \mathbf{e}_3 = \begin{bmatrix} 0 \\ 0 \\ 1 \\ 0 \\ \vdots \\ 0 \end{bmatrix}, \quad \ldots, \quad \mathbf{e}_n = \begin{bmatrix} 0 \\ 0 \\ 0 \\ 0 \\ \vdots \\ 1 \end{bmatrix}. \tag{91}$$

Thus,

$$[\mathbf{e}_1\ \mathbf{e}_2\ \mathbf{e}_3 \cdots \mathbf{e}_n] = \begin{bmatrix} 1 & 0 & 0 & \cdots & 0 \\ 0 & 1 & 0 & & 0 \\ 0 & 0 & 0 & & 0 \\ \vdots & \vdots & \vdots & & \vdots \\ 0 & 0 & 0 & \cdots & 1 \end{bmatrix} = \mathbf{I}. \tag{92}$$

Consider the homogeneous systems given by

$$\begin{aligned} \dot{\mathbf{x}}(t) &= \mathbf{A}(t)\mathbf{x}(t) \\ \mathbf{x}(t_0) &= \mathbf{e}_j \qquad (j = 1, 2, \ldots, n), \end{aligned} \tag{93}$$

where $\mathbf{A}(t)$ and t_0 are taken from (58).

For each j ($j = 1, 2, \ldots, n$), Theorem 1 of Section 7.1 guarantees the existence of a unique solution of (93); denote this solution by $\mathbf{x}_j(t)$. Thus, $\mathbf{x}_1(t)$ solves the system

$$\begin{aligned} \dot{\mathbf{x}}_1(t) &= \mathbf{A}(t)\mathbf{x}_1(t) \\ \mathbf{x}_1(t_0) &= \mathbf{e}_1, \end{aligned} \tag{94}$$

$\mathbf{x}_2(t)$ satisfies the system

$$\begin{aligned} \dot{\mathbf{x}}_2(t) &= \mathbf{A}(t)\mathbf{x}_2(t) \\ \mathbf{x}_2(t_0) &= \mathbf{e}_2, \end{aligned} \tag{95}$$

and $\mathbf{x}_n(t)$ satisfies the system

$$\begin{aligned} \dot{\mathbf{x}}_n(t) &= \mathbf{A}(t)\mathbf{x}_n(t) \\ \mathbf{x}_n(t_0) &= \mathbf{e}_n. \end{aligned} \tag{96}$$

Define the matrix $\mathbf{\Phi}(t, t_0) = [\mathbf{x}_1(t)\ \mathbf{x}_2(t) \cdots \mathbf{x}_n(t)]$. Then,

$$\begin{aligned} \mathbf{\Phi}(t_0, t_0) &= [x_1(t_0)\ \mathbf{x}_2(t_0) \cdots \mathbf{x}_n(t_0)] \\ &= [\mathbf{e}_1\ \mathbf{e}_2 \cdots \mathbf{e}_n] \qquad \{\text{from (94)–(96)} \\ &= \mathbf{I} \qquad \{\text{from (92)} \end{aligned}$$

and

$$\begin{aligned} \frac{d\mathbf{\Phi}(t, t_0)}{dt} &= \frac{d}{dt}[\mathbf{x}_1(t)\ \mathbf{x}_2(t) \cdots \mathbf{x}_n(t)] \\ &= [\dot{\mathbf{x}}_1(t)\ \dot{\mathbf{x}}_2(t) \cdots \dot{\mathbf{x}}_n(t)] \\ &= [\mathbf{A}(t)\mathbf{x}_1(t)\ \mathbf{A}(t)\mathbf{x}_2(t) \cdots \mathbf{A}(t)\mathbf{x}_n(t)] \quad \{\text{from (94)–(96)} \\ &= \mathbf{A}(t)[\mathbf{x}_1(t)\ \mathbf{x}_2(t) \cdots \mathbf{x}_n(t)] \\ &= \mathbf{A}(t)\mathbf{\Phi}(t, t_0). \end{aligned}$$

Thus, $\mathbf{\Phi}(t, t_0)$, as defined above, is a matrix that satisfies (59) and (60). Since this $\mathbf{\Phi}(t, t_0)$ always exists, it follows that there will always exist a matrix that satisfies these equations.

It only remains to be shown that $\mathbf{\Phi}(t, t_0)$ is unique. Let $\mathbf{\Psi}(t, t_0)$ be any matrix satisfying (59) and (60). Then the jth column of $\mathbf{\Psi}(t, t_0)$ must satisfy the initial-valued problem given by (93). However, the solution to (93) is unique by Theorem 1 of Section 7.1, hence, the jth column of $\mathbf{\Psi}(t, t_0)$ must be $\mathbf{x}_j(t)$. Thus,

$$\mathbf{\Psi}(t, t_0) = [\mathbf{x}_1(t)\ \mathbf{x}_2(t) \cdots \mathbf{x}_n(t)] = \mathbf{\Phi}(t, t_0).$$

From this equation, it follows that the transition matrix is unique.

Chapter 8 JORDAN CANONICAL FORMS

8.1 SIMILAR MATRICES

Definition 1 A matrix $\mathbf{A}$ is *similar* to a matrix $\mathbf{B}$ if there exists an invertible matrix $\mathbf{P}$ such that

$$\mathbf{A} = \mathbf{P}^{-1}\mathbf{B}\mathbf{P}. \tag{1}$$

If we premultiply (1) by $\mathbf{P}$, it follows that $\mathbf{A}$ is similar to $\mathbf{B}$ if and only if there exists a nonsingular matrix $\mathbf{P}$ such that

$$\mathbf{P}\mathbf{A} = \mathbf{B}\mathbf{P}. \tag{2}$$

Furthermore, if we postmultiply (2) by $\mathbf{P}^{-1}$, we see that **A** is similar to **B** if and only if **B** is similar to **A**.

EXAMPLE 1 Determine whether

$$\mathbf{A}=\begin{bmatrix}4 & 3\\ -2 & -1\end{bmatrix} \quad \text{is similar to } \mathbf{B}=\begin{bmatrix}5 & -4\\ 3 & -2\end{bmatrix}.$$

Solution **A** will be similar to **B** if and only if there exists a nonsingular matrix **P** such that (2) is satisfied. Designate **P** by

$$\begin{bmatrix}a & b\\ c & d\end{bmatrix}.$$

Then **PA** = **BP** implies that

$$\begin{bmatrix}(4a-2b) & (3a-b)\\ (4c-2d) & (3c-d)\end{bmatrix}=\begin{bmatrix}(5a-4c) & (5b-4d)\\ (3a-2c) & (3b-2d)\end{bmatrix}.$$

Equating corresponding elements, we find that in order for **A** to be similar to **B**, the elements of **P** must satisfy the following set of equations:

$$\begin{aligned}-a-2b+4c\quad &=0,\\ 3a-6b+4d&=0,\\ -3a+6c-2d&=0,\\ -3b+3c+\quad d&=0.\end{aligned}$$

A solution to this set of equations is $a=-\frac{2}{3}d$, $b=\frac{1}{3}d$, with $c=0$, d arbitrary. Thus,

$$\mathbf{P}=\begin{bmatrix}a & b\\ c & d\end{bmatrix}=\frac{d}{3}\begin{bmatrix}-2 & 1\\ 0 & 3\end{bmatrix}.$$

In order for **P** to be invertible, we must have $d \neq 0$. Thus, by choosing $d \neq 0$, we obtain an invertible matrix **P** that satisfies (2), which implies that **A** is similar to **B**.

The following theorem reveals the basic relationship between similar matrices.

Theorem 1 Similar matrices have the same characteristic equation (and, therefore, the same eigenvalues).

Proof Let **A** and **B** be similar matrices. The characteristic equation of **A** is $\det(\mathbf{A}-\lambda\mathbf{I})=0$, while the characteristic equation of **B** is $\det(\mathbf{B}-\lambda\mathbf{I})=0$. Thus, it is sufficient to show that $\det(\mathbf{A}-\lambda\mathbf{I})=\det(\mathbf{B}-\lambda\mathbf{I})$. We first note that since **A** is similar to **B**, there must exist a nonsingular matrix **P** such that (1) is satisfied, and since **P** is invertible, we can write

$$\lambda\mathbf{I} = \lambda\mathbf{P}^{-1}\mathbf{P} = \mathbf{P}^{-1}\lambda\mathbf{P} = \mathbf{P}^{-1}\lambda\mathbf{I}\mathbf{P}$$

Then,

$$\begin{aligned}|\mathbf{A} - \lambda\mathbf{I}| = |\mathbf{P}^{-1}\mathbf{B}\mathbf{P} - \lambda\mathbf{I}| &= |\mathbf{P}^{-1}\mathbf{B}\mathbf{P} - \mathbf{P}^{-1}\lambda\mathbf{I}\mathbf{P}| \\ &= |\mathbf{P}^{-1}(\mathbf{B} - \lambda\mathbf{I})\mathbf{P}| \\ &= |\mathbf{P}^{-1}|\,|\mathbf{B} - \lambda\mathbf{I}|\,|\mathbf{P}| \quad \{\text{by Property 8 of Section 2.3} \\ &= \frac{1}{|\mathbf{P}|}|\mathbf{B} - \lambda\mathbf{I}|\,|\mathbf{P}| \quad \{\text{by Problem 10 of Section 3.3} \\ &= |\mathbf{B} - \lambda\mathbf{I}|.\end{aligned}$$

EXAMPLE 2 Determine whether

$$\mathbf{A} = \begin{bmatrix} 1 & 2 \\ 4 & 3 \end{bmatrix} \quad \text{is similar to } \mathbf{B} = \begin{bmatrix} 1 & 4 \\ 3 & 2 \end{bmatrix}.$$

Solution The characteristic equation of **A** is $\lambda^2 - 4\lambda - 5 = 0$, while that of **B** is $\lambda^2 - 3\lambda - 10 = 0$. Since these equations are not identical, it follows from Theorem 1 that **A** is not similar to *B*.

Warning: Theorem I does *not* imply that two matrices are similar if their characteristic equations are the same. This statement is, in fact, false. Consider the matrices

$$\mathbf{A} = \begin{bmatrix} 2 & 0 \\ 0 & 2 \end{bmatrix} \quad \text{and } \mathbf{B} = \begin{bmatrix} 2 & 1 \\ 0 & 2 \end{bmatrix}.$$

Although both matrices have the same characteristic equation, namely $(\lambda - 2)^2 = 0$, they are not similar (see Problem 2). This example serves to emphasize that Theorem 1 can only be used to prove that two matrices are not similar; it *cannot* be used to prove that two matrices are similar. In other words, if two matrices do not have the same characteristic equation, then we can state, categorically, that they are not similar. However, if two matrices have the same characteristic equations, they may or may not be similar, and for the present, an analysis similar to that employed in Example 1 must be used in order to reach a conclusion. In later sections, we will develop more sophisticated methods to determine when two matrices are similar.

PROBLEMS 8.1

1. Prove that **A** is similar to **B** if and only if **B** is similar to **A**.

2. Prove that

$$\mathbf{A} = \begin{bmatrix} 2 & 0 \\ 0 & 2 \end{bmatrix} \quad \text{is not similar to } \mathbf{B} = \begin{bmatrix} 2 & 1 \\ 0 & 2 \end{bmatrix}.$$

3. For the following, determine whether or not **A** is similar to **B**:

(a) $$\mathbf{A} = \begin{bmatrix} 3 & 5 \\ 3 & 1 \end{bmatrix}, \qquad \mathbf{B} = \begin{bmatrix} 2 & 4 \\ 4 & 2 \end{bmatrix};$$

(b) $$\mathbf{A} = \begin{bmatrix} 3 & 0 \\ 0 & 3 \end{bmatrix}, \qquad \mathbf{B} = \begin{bmatrix} 6 & 3 \\ 1 & 4 \end{bmatrix};$$

(c) $$\mathbf{A} = \begin{bmatrix} 1 & 1 \\ 4 & 2 \end{bmatrix}, \qquad \mathbf{B} = \begin{bmatrix} 1 & 1 \\ 2 & 4 \end{bmatrix};$$

(d) $$\mathbf{A} = \begin{bmatrix} 2 & 1 & 2 \\ 1 & -2 & 1 \\ 0 & 0 & 1 \end{bmatrix}, \qquad \mathbf{B} = \begin{bmatrix} 3 & 1 & 0 \\ 4 & 1 & 0 \\ 2 & 1 & 1 \end{bmatrix}.$$

8.2 DIAGONALIZABLE MATRICES

Definition 1 A matrix is *diagonalizable* if it is similar to a diagonal matrix.

Diagonalizable matrices are of particular interest since matrix functions of them can be computed easily. As such we devote this section to determining which matrices are diagonalizable and to finding those matrices **P** which will perform the similarity transformations. We note by Theorem 1 of the previous section that if a matrix is similar to a diagonal matrix **D**, then the form of **D** is known. Since the eigenvalues of **D** are precisely the elements on the main diagonal of **D**, it follows that the main diagonal of **D** must consist of the eigenvalues of **A**. For example, if

$$\mathbf{A} = \begin{bmatrix} 1 & 2 \\ 4 & 3 \end{bmatrix},$$

having eigenvalues -1 and 5, is diagonalizable, then it must be similar to either

$$\begin{bmatrix} -1 & 0 \\ 0 & 5 \end{bmatrix} \quad \text{or} \quad \begin{bmatrix} 5 & 0 \\ 0 & -1 \end{bmatrix}.$$

Before continuing our discussion of diagonalizable matrices, we first review two important properties of matrix factoring which we will have occasion to use. The verification of these properties is left as an exercise for the student.

Property 1 Let $\mathbf{B} = [\mathbf{b}_1\ \mathbf{b}_2 \ldots \mathbf{b}_n]$ be an $n \times n$ matrix, where $\mathbf{b}_j$ $(j = 1, 2, \ldots, n)$ is the jth column of $\mathbf{B}$ considered as a vector. Then for any $n \times n$ matrix $\mathbf{A}$,

$$\mathbf{AB} = \mathbf{A}[\mathbf{b}_1\ \mathbf{b}_2 \ldots \mathbf{b}_n] = [\mathbf{Ab}_1\ \mathbf{Ab}_2 \ldots . \mathbf{Ab}_n].$$

EXAMPLE 1 Verify Property 1 for

$$\mathbf{A} = \begin{bmatrix} 1 & 2 & -1 \\ 1 & -2 & 3 \\ -2 & 1 & 1 \end{bmatrix} \quad \text{and} \quad \mathbf{B} = \begin{bmatrix} 1 & 2 & 3 \\ 4 & 5 & 6 \\ 7 & 8 & 9 \end{bmatrix}.$$

Solution In this case,

$$\mathbf{b}_1 = \begin{bmatrix} 1 \\ 4 \\ 7 \end{bmatrix}, \quad \mathbf{b}_2 = \begin{bmatrix} 2 \\ 5 \\ 8 \end{bmatrix}, \quad \mathbf{b}_3 = \begin{bmatrix} 3 \\ 6 \\ 9 \end{bmatrix}.$$

Thus

$$\mathbf{Ab}_1 = \begin{bmatrix} 1 & 2 & -1 \\ 1 & -2 & 3 \\ -2 & 1 & 1 \end{bmatrix} \begin{bmatrix} 1 \\ 4 \\ 7 \end{bmatrix} = \begin{bmatrix} 2 \\ 14 \\ 9 \end{bmatrix},$$

$$\mathbf{Ab}_2 = \begin{bmatrix} 1 & 2 & -1 \\ 1 & -2 & 3 \\ -2 & 1 & 1 \end{bmatrix} \begin{bmatrix} 2 \\ 5 \\ 8 \end{bmatrix} = \begin{bmatrix} 4 \\ 16 \\ 9 \end{bmatrix},$$

$$\mathbf{Ab}_3 = \begin{bmatrix} 1 & 2 & -1 \\ 1 & -2 & 3 \\ -2 & 1 & 1 \end{bmatrix} \begin{bmatrix} 3 \\ 6 \\ 9 \end{bmatrix} = \begin{bmatrix} 6 \\ 18 \\ 9 \end{bmatrix},$$

and

$$[\mathbf{Ab}_1\ \mathbf{Ab}_2\ \mathbf{Ab}_3] = \begin{bmatrix} 2 & 4 & 6 \\ 14 & 16 & 18 \\ 9 & 9 & 9 \end{bmatrix},$$

which is exactly $\mathbf{AB}$.

Property 2[22] Designate the $n \times n$ matrix $\mathbf{B}$ by $[\mathbf{b}_1\ \mathbf{b}_2 \ldots \mathbf{b}_n]$ as in Property 1 and let $\lambda_1, \lambda_2, \ldots, \lambda_n$ represent scalars. Then:

$$[\lambda_1\mathbf{b}_1\ \lambda_2\mathbf{b}_2 \ldots \lambda_n\mathbf{b}_n] = [\mathbf{b}_1\ \mathbf{b}_2 \ldots \mathbf{b}_n] \begin{bmatrix} \lambda_1 & 0 & \ldots & 0 \\ 0 & \lambda_2 & \ldots & 0 \\ \vdots & \vdots & & \vdots \\ 0 & 0 & \ldots & \lambda_n \end{bmatrix}$$

$$= \mathbf{B} \begin{bmatrix} \lambda_1 & 0 & \ldots & 0 \\ 0 & \lambda_2 & \ldots & 0 \\ \vdots & \vdots & \ddots & \vdots \\ 0 & 0 & \ldots & \lambda_n \end{bmatrix}.$$

[22] See Problem 6 of Section 1.5.

Now let $\mathbf{A}$ be an $n \times n$ matrix that has *n linearly independent eigenvectors* $\mathbf{x}_1, \mathbf{x}_2, \ldots, \mathbf{x}_n$ which correspond to the eigenvalues $\lambda_1, \lambda_2, \ldots, \lambda_n$. (Recall from Section 5.5 that a matrix will have n linearly independent eigenvectors if all the eigenvalues are distinct or, depending upon the matrix, even if some or all of the eigenvalues are equal. *A priori*, therefore, we place no restrictions on the multiplicities of the eigenvalues.) Define

$$\mathbf{M} = [\mathbf{x}_1\ \mathbf{x}_2 \ldots \mathbf{x}_n] \qquad \text{and} \qquad \mathbf{D} = \begin{bmatrix} \lambda_1 & 0 & \ldots & 0 \\ 0 & \lambda_2 & \ldots & 0 \\ \vdots & \vdots & & \vdots \\ 0 & 0 & \ldots & \lambda_n \end{bmatrix}.$$

Here $\mathbf{M}$ is called a *modal matrix* for $\mathbf{A}$, and $\mathbf{D}$ is called a *spectral matrix* for $\mathbf{A}$. Note that since eigenvectors themselves are not unique, and since the columns of both $\mathbf{M}$ and $\mathbf{D}$ may be interchanged (although the jth column of $\mathbf{M}$ must still correspond to the jth column of $\mathbf{D}$; that is, the jth column of $\mathbf{M}$ must be the eigenvector of $\mathbf{A}$ associated with the eigenvalue located in the (j, j) position of $\mathbf{D}$), it follows that both $\mathbf{M}$ and $\mathbf{D}$ are not unique. Using Properties 1 and 2 and the fact that $\mathbf{x}_j$ $(j = 1, 2, \ldots, n)$ is an eigenvector of $\mathbf{A}$, we have that

$$\begin{aligned} \mathbf{AM} &= \mathbf{A}[\mathbf{x}_1\ \mathbf{x}_2 \ldots \mathbf{x}_n] \\ &= [\mathbf{Ax}_1\ \mathbf{Ax}_2 \ldots \mathbf{Ax}_n] \\ &= [\lambda_1\mathbf{x}_1\ \lambda_2\mathbf{x}_2 \ldots \lambda_n\mathbf{x}_n] \\ &= [\mathbf{x}_1\ \mathbf{x}_2 \ldots \mathbf{x}_n]\mathbf{D} = \mathbf{MD} \end{aligned} \tag{3}$$

(Since the columns of $\mathbf{M}$ are linearly independent, it follows that the column rank of $\mathbf{M}$ is n, the rank of $\mathbf{M}$ is n, the determinant of $\mathbf{M}$ is nonzero, and $\mathbf{M}^{-1}$ exists.) Premultiplying (3) by $\mathbf{M}^{-1}$ we obtain,

$$\mathbf{D} = \mathbf{M}^{-1}\mathbf{AM}, \tag{4}$$

which implies that $\mathbf{D}$ is similar to $\mathbf{A}$. Furthermore, by defining $\mathbf{P} = \mathbf{M}^{-1}$, it follows that

$$\mathbf{A} = \mathbf{P}^{-1}\mathbf{DP} = \mathbf{MDM}^{-1}, \tag{5}$$

which implies that $\mathbf{A}$ is similar to $\mathbf{D}$. Since we can retrace our steps and show that if (5) is satisfied, then $\mathbf{P}$ must be $\mathbf{M}^{-1}$, we have proved the following theorem.

Theorem 1 An $n \times n$ matrix $\mathbf{A}$ is diagonalizable if and only if it possesses n linearly independent eigenvectors. The inverse of the matrix $\mathbf{P}$ is a modal matrix of $\mathbf{A}$.

EXAMPLE 2 Determine whether

$$\mathbf{A} = \begin{bmatrix} 1 & 2 \\ 4 & 3 \end{bmatrix}$$

is diagonalizable.

Solution The eigenvalues of **A** are -1 and 5. Since the eigenvalues are distinct, their respective eigenvectors

$$\mathbf{x}_1 = \begin{bmatrix} 1 \\ -1 \end{bmatrix} \quad \text{and} \quad \mathbf{x}_2 = \begin{bmatrix} 1 \\ 2 \end{bmatrix}$$

are linearly independent, hence the matrix is diagonalizable. We can choose either

$$\mathbf{M} = \begin{bmatrix} 1 & 1 \\ -1 & 2 \end{bmatrix} \quad \text{or} \quad \mathbf{M} = \begin{bmatrix} 1 & 1 \\ 2 & -1 \end{bmatrix}.$$

Making the first choice, we find

$$\mathbf{D} = \mathbf{M}^{-1}\mathbf{A}\mathbf{M} = \frac{1}{3}\begin{bmatrix} 2 & -1 \\ 1 & 1 \end{bmatrix}\begin{bmatrix} 1 & 2 \\ 4 & 3 \end{bmatrix}\begin{bmatrix} 1 & 1 \\ -1 & 2 \end{bmatrix} = \begin{bmatrix} -1 & 0 \\ 0 & 5 \end{bmatrix}.$$

Making the second choice, we obtain

$$\mathbf{D} = \mathbf{M}^{-1}\mathbf{A}\mathbf{M} = \frac{1}{3}\begin{bmatrix} 1 & 1 \\ 2 & -1 \end{bmatrix}\begin{bmatrix} 1 & 2 \\ 4 & 3 \end{bmatrix}\begin{bmatrix} 1 & 1 \\ 2 & -1 \end{bmatrix} = \begin{bmatrix} 5 & 0 \\ 0 & -1 \end{bmatrix}.$$

Example 2 illustrates a point we made previously that neither **M** nor **D** is unique. However, the columns of **M** must still correspond to the columns of **D**; that is, once **M** is chosen, then **D** is uniquely determined. For example, if we choose $\mathbf{M} = [\mathbf{x}_2\ \mathbf{x}_1\ \mathbf{x}_3 \ldots \mathbf{x}_n]$, then **D** must be

$$\begin{bmatrix} \lambda_2 & & & & \\ & \lambda_1 & & & \\ & & \lambda_3 & & \\ & & & \ddots & \\ & & & & \lambda_n \end{bmatrix}$$

while if we choose $\mathbf{M} = [\mathbf{x}_n\ \mathbf{x}_{n-1} \ldots \mathbf{x}_1]$, then **D** must be

$$\begin{bmatrix} \lambda_n & & & \\ & \lambda_{n-1} & & \\ & & \ddots & \\ & & & \lambda_1 \end{bmatrix}.$$

EXAMPLE 3 Is

$$\mathbf{A} = \begin{bmatrix} 3 & 2 & 1 \\ 0 & 2 & 0 \\ 1 & 2 & 3 \end{bmatrix}$$

diagonalizable?

Solution The eigenvalues of **A** are 2, 2, and 4. Even though the eigenvalues of **A** are not all distinct, **A** still possesses three linearly independent eigenvectors, namely

$$\mathbf{x}_1 = \begin{bmatrix} -2 \\ 1 \\ 0 \end{bmatrix}, \quad \mathbf{x}_2 = \begin{bmatrix} -1 \\ 0 \\ 1 \end{bmatrix}, \quad \text{and} \quad \mathbf{x}_3 = \begin{bmatrix} 1 \\ 0 \\ 1 \end{bmatrix},$$

hence it is diagonalizable. If we choose

$$\mathbf{M} = \begin{bmatrix} -2 & -1 & 1 \\ 1 & 0 & 0 \\ 0 & 1 & 1 \end{bmatrix},$$

we find that

$$\mathbf{M}^{-1}\mathbf{A}\mathbf{M} = \begin{bmatrix} 2 & 0 & 0 \\ 0 & 2 & 0 \\ 0 & 0 & 4 \end{bmatrix}.$$

EXAMPLE 4 Is

$$\mathbf{A} = \begin{bmatrix} 2 & 1 \\ 0 & 2 \end{bmatrix}$$

diagonalizable?

Solution The eigenvalues of **A** are 2 and 2. **A** has only one linearly independent eigenvector associated with it, namely

$$\begin{bmatrix} 1 \\ 0 \end{bmatrix},$$

hence it is *not* diagonalizable. (See Problem 2 of the Section 8.1).

PROBLEMS 8.2

Determine whether or not the following matrices are diagonalizable. If they are, determine **M** and compute $\mathbf{M}^{-1}\mathbf{A}\mathbf{M}$.

1. $\mathbf{A} = \begin{bmatrix} 2 & -3 \\ 1 & -2 \end{bmatrix}$.

2. $\mathbf{A} = \begin{bmatrix} 2 & -5 \\ 1 & -2 \end{bmatrix}$.

3. $\mathbf{A} = \begin{bmatrix} 1 & 1 & 1 \\ 0 & 1 & 0 \\ 0 & 0 & 1 \end{bmatrix}$.

4. $\mathbf{A} = \begin{bmatrix} 1 & 0 & 0 \\ 2 & -3 & 3 \\ 1 & 2 & 2 \end{bmatrix}$.

5. $\mathbf{A} = \begin{bmatrix} 5 & 1 & 2 \\ 0 & 3 & 0 \\ 2 & 1 & 5 \end{bmatrix}$.

6. $\mathbf{A} = \begin{bmatrix} 7 & 3 & 3 \\ 0 & 1 & 0 \\ -3 & -3 & 1 \end{bmatrix}$.

8.3 FUNCTIONS OF MATRICES—DIAGONALIZABLE MATRICES

By utilizing (5), we can develop a simple procedure for computing functions of a diagonalizable matrix. We begin by directing our attention to those matrices that are already in diagonal form. In particular, we have from Section 6.1 (Eq. (7) and Problem 2) that if

$$\mathbf{D} = \begin{bmatrix} \lambda_1 & & & \\ & \lambda_2 & & \\ & & \ddots & \\ & & & \lambda_n \end{bmatrix} \tag{6}$$

then

$$\mathbf{D}^n = \begin{bmatrix} \lambda_1^n & & & \\ & \lambda_2^n & & \\ & & \ddots & \\ & & & \lambda_n^n \end{bmatrix}, \tag{7}$$

$$p_k(\mathbf{D}) = \begin{bmatrix} p_k(\lambda_1) & & & \\ & p_k(\lambda_2) & & \\ & & \ddots & \\ & & & p_k(\lambda_n) \end{bmatrix}, \tag{8}$$

and

$$e^{\mathbf{D}} = \begin{bmatrix} e^{\lambda_1} & & & \\ & e^{\lambda_2} & & \\ & & \ddots & \\ & & & e^{\lambda_n} \end{bmatrix}. \tag{9}$$

EXAMPLE 1 Find $\mathbf{D}^5$, $\mathbf{D}^3 + 2\mathbf{D} - 3\mathbf{I}$, and $e^{\mathbf{D}}$ for

$$\mathbf{D} = \begin{bmatrix} 1 & 0 & 0 \\ 0 & 2 & 0 \\ 0 & 0 & 2 \end{bmatrix}.$$

Solution From (7), (8), and (9) we obtain

$$\mathbf{D}^5 = \begin{bmatrix} (1)^5 & 0 & 0 \\ 0 & (2)^5 & 0 \\ 0 & 0 & (2)^5 \end{bmatrix} = \begin{bmatrix} 1 & 0 & 0 \\ 0 & 32 & 0 \\ 0 & 0 & 32 \end{bmatrix}.$$

$$\mathbf{D}^3 + 2\mathbf{D} - 3\mathbf{I} = \begin{bmatrix} (1)^3 + 2(1) - 3 & 0 & 0 \\ 0 & (2)^3 + 2(2) - 3 & 0 \\ 0 & 0 & (2)^3 + 2(2) - 3 \end{bmatrix}$$

$$= \begin{bmatrix} 0 & 0 & 0 \\ 0 & 9 & 0 \\ 0 & 0 & 9 \end{bmatrix}.$$

$$e^{\mathbf{D}} = \begin{bmatrix} e^1 & 0 & 0 \\ 0 & e^2 & 0 \\ 0 & 0 & e^2 \end{bmatrix}.$$

Now assume that a matrix **A** is diagonalizable. Then it follows from (5) that

$$\mathbf{A} = \mathbf{M}\mathbf{D}\mathbf{M}^{-1}.$$

Thus,

$$\begin{aligned} \mathbf{A}^2 = \mathbf{A} \cdot \mathbf{A} &= (\mathbf{M}\mathbf{D}\mathbf{M}^{-1})(\mathbf{M}\mathbf{D}\mathbf{M}^{-1}) \\ &= (\mathbf{M}\mathbf{D})(\mathbf{M}^{-1}\mathbf{M})(\mathbf{D}\mathbf{M}^{-1}) \\ &= \mathbf{M}\mathbf{D}(\mathbf{I})\mathbf{D}\mathbf{M}^{-1} = \mathbf{M}\mathbf{D}^2\mathbf{M}^{-1}, \\ \mathbf{A}^3 = \mathbf{A}^2 \cdot \mathbf{A} &= (\mathbf{M}\mathbf{D}^2\mathbf{M}^{-1})(\mathbf{M}\mathbf{D}\mathbf{M}^{-1}) \\ &= (\mathbf{M}\mathbf{D}^2)(\mathbf{M}^{-1}\mathbf{M})(\mathbf{D}\mathbf{M}^{-1}) = \mathbf{M}\mathbf{D}^3\mathbf{M}^{-1}, \end{aligned}$$

and, in general,

$$\mathbf{A}^n = \mathbf{M}\mathbf{D}^n\mathbf{M}^{-1}. \tag{10}$$

Therefore, to obtain any power of a diagonalizable matrix **A**, we need only compute **D** to that power (this, in itself, is easily done using (7)), premultiply $\mathbf{D}^n$ by **M**, and postmultiply the result by $\mathbf{M}^{-1}$.

EXAMPLE 2 Find $\mathbf{A}^{915}$ for

$$\mathbf{A} = \begin{bmatrix} 1 & 2 \\ 4 & 3 \end{bmatrix}.$$

Solution The eigenvalues for **A** are -1, 5 and a set of linearly independent eigenvectors corresponding to these eigenvalues are

$$\begin{bmatrix} 1 \\ -1 \end{bmatrix} \quad \text{and} \quad \begin{bmatrix} 1 \\ 2 \end{bmatrix}.$$

Hence

$$\mathbf{M} = \begin{bmatrix} 1 & 1 \\ -1 & 2 \end{bmatrix}, \quad \mathbf{M}^{-1} = \frac{1}{3}\begin{bmatrix} 2 & -1 \\ 1 & 1 \end{bmatrix}, \quad \text{and} \quad \mathbf{D} = \begin{bmatrix} -1 & 0 \\ 0 & 5 \end{bmatrix}.$$

It follows from (10) that

$$\mathbf{A}^{915} = \begin{bmatrix} 1 & 2 \\ 4 & 3 \end{bmatrix}^{915} = \mathbf{M}\mathbf{D}^{915}\mathbf{M}^{-1} = \begin{bmatrix} 1 & 1 \\ -1 & 2 \end{bmatrix} \begin{bmatrix} (-1)^{915} & 0 \\ 0 & (5)^{915} \end{bmatrix} \frac{1}{3} \begin{bmatrix} 2 & -1 \\ 1 & 1 \end{bmatrix}$$

$$= \frac{1}{3} \begin{bmatrix} -2 + (5)^{915} & 1 + (5)^{915} \\ 2 + 2(5)^{915} & -1 + 2(5)^{915} \end{bmatrix}.$$

We may also use (10) to find a simplified expression for $p_k(\mathbf{A})$ where $p_k(x)$ is a kth degree polynomial in x. For instance, suppose that $p_5(x) = 5x^5 - 3x^3 + 2x^2 + 4$ and $p_5(\mathbf{A})$ is to be calculated. Making repeated use of (10), we have

$$\begin{aligned} p_5(\mathbf{A}) &= 5\mathbf{A}^5 - 3\mathbf{A}^3 + 2\mathbf{A}^2 + 4\mathbf{I} \\ &= 5\mathbf{M}\mathbf{D}^5\mathbf{M}^{-1} - 3\mathbf{M}\mathbf{D}^3\mathbf{M}^{-1} + 2\mathbf{M}\mathbf{D}^2\mathbf{M}^{-1} + 4\mathbf{M}\mathbf{I}\mathbf{M}^{-1} \\ &= \mathbf{M}[5\mathbf{D}^5 - 3\mathbf{D}^3 + 2\mathbf{D}^2 + 4\mathbf{I}]\mathbf{M}^{-1} \\ &= \mathbf{M}p_5(\mathbf{D})\mathbf{M}^{-1}. \end{aligned}$$

Thus, to calculate $p_5(\mathbf{A})$, we need compute only $p_5(\mathbf{D})$, which can easily be done by using (8), premultiply this result by $\mathbf{M}$, and postmultiply by $\mathbf{M}^{-1}$. We can extend this reasoning in a straightforward manner (see Problem 7) to prove that if $p_k(\mathbf{A})$ is any kth degree polynomial of $\mathbf{A}$, then

$$p_k(\mathbf{A}) = \mathbf{M}p_k(\mathbf{D})\mathbf{M}^{-1}. \tag{11}$$

EXAMPLE 3 Find $4\mathbf{A}^{15} - 2\mathbf{A}^7 + \mathbf{I}$ for

$$\mathbf{A} = \begin{bmatrix} 1 & 2 \\ 4 & 3 \end{bmatrix}.$$

Solution From Example 2, we have $\mathbf{M}$, $\mathbf{M}^{-1}$, and $\mathbf{D}$. Therefore, it follows from (11) and (8) that

$$4\mathbf{A}^{15} - 2\mathbf{A}^7 + \mathbf{I} = \mathbf{M}(4\mathbf{D}^{15} - 2\mathbf{D}^7 + \mathbf{I})\mathbf{M}^{-1}$$

$$= \begin{bmatrix} 1 & 1 \\ -1 & 2 \end{bmatrix} \begin{bmatrix} 4(-1)^{15} - 2(-1)^7 + 1 & 0 \\ 0 & 4(5)^{15} - 2(5)^7 + 1 \end{bmatrix}$$

$$\times \frac{1}{3} \begin{bmatrix} 2 & -1 \\ 1 & 1 \end{bmatrix}$$

$$= \frac{1}{3} \begin{bmatrix} 4(5)^{15} - 2(5)^7 - 1 & 4(5)^{15} - 2(5)^7 + 2 \\ 8(5)^{15} - 4(5)^7 + 4 & 8(5)^{15} - 4(5)^7 + 1 \end{bmatrix}.$$

The student should realize from Chapter 7, that a function of great interest is the exponential. If the matrix $\mathbf{A}$ is diagonalizable, then we can use (10) to obtain a useful representation for $e^{\mathbf{A}}$.

$$e^{\mathbf{A}} = \sum_{k=0}^{\infty} \frac{\mathbf{A}^k}{k!} = \sum_{k=0}^{\infty} \frac{\mathbf{M}\mathbf{D}^k\mathbf{M}^{-1}}{k!} = \mathbf{M}\left(\sum_{k=0}^{\infty} \frac{\mathbf{D}^k}{k!}\right)\mathbf{M}^{-1} = \mathbf{M}e^{\mathbf{D}}\mathbf{M}^{-1}. \tag{12}$$

Thus, to calculate $e^{\mathbf{A}}$, we need only compute $e^{\mathbf{D}}$, which can be done easily by using (9), and then premultiply this result by $\mathbf{M}$ and postmultiply by $\mathbf{M}^{-1}$.

EXAMPLE 4 Find $e^{\mathbf{A}}$ for

$$\mathbf{A} = \begin{bmatrix} 1 & 2 \\ 4 & 3 \end{bmatrix}.$$

Solution Once again $\mathbf{M}$, $\mathbf{M}^{-1}$, and $\mathbf{D}$ are known from Example 2. It follows, therefore, from (12) and (9) that

$$\begin{aligned} e^{\mathbf{A}} = \mathbf{M}e^{\mathbf{D}}\mathbf{M}^{-1} &= \begin{bmatrix} 1 & 1 \\ -1 & 2 \end{bmatrix}\begin{bmatrix} e^{-1} & 0 \\ 0 & e^{5} \end{bmatrix}\frac{1}{3}\begin{bmatrix} 2 & -1 \\ 1 & 1 \end{bmatrix} \\ &= \frac{1}{3}\begin{bmatrix} 2e^{-1} + e^{5} & -e^{-1} + e^{5} \\ -2e^{-1} + 2e^{5} & e^{-1} + 2e^{5} \end{bmatrix}. \end{aligned}$$

(Check this result against Example 1 of Section 6.5.)

EXAMPLE 5 Find $e^{\mathbf{A}}$ for

$$\mathbf{A} = \begin{bmatrix} 3 & 2 & 1 \\ 0 & 2 & 0 \\ 1 & 2 & 3 \end{bmatrix}.$$

Solution Using the results of Example 3 of Section 8.2, we have

$$\mathbf{D} = \begin{bmatrix} 2 & 0 & 0 \\ 0 & 2 & 0 \\ 0 & 0 & 4 \end{bmatrix}, \quad \mathbf{M} = \begin{bmatrix} -2 & -1 & 1 \\ 1 & 0 & 0 \\ 0 & 1 & 1 \end{bmatrix}, \quad \mathbf{M}^{-1} = \frac{1}{2}\begin{bmatrix} 0 & 2 & 0 \\ -1 & -2 & 1 \\ 1 & 2 & 1 \end{bmatrix}.$$

Thus,

$$\begin{aligned} e^{\mathbf{A}} = \mathbf{M}e^{\mathbf{D}}\mathbf{M}^{-1} &= \begin{bmatrix} -2 & -1 & 1 \\ 1 & 0 & 0 \\ 0 & 1 & 1 \end{bmatrix}\begin{bmatrix} e^{2} & 0 & 0 \\ 0 & e^{2} & 0 \\ 0 & 0 & e^{4} \end{bmatrix}\frac{1}{2}\begin{bmatrix} 0 & 2 & 0 \\ -1 & -2 & 1 \\ 1 & 2 & 1 \end{bmatrix} \\ &= \frac{1}{2}\begin{bmatrix} e^{4} + e^{2} & 2e^{4} - 2e^{2} & e^{4} - e^{2} \\ 0 & 2e^{2} & 0 \\ e^{4} - e^{2} & 2e^{4} - 2e^{2} & e^{4} + e^{2} \end{bmatrix}. \end{aligned}$$

By employing the same reasoning as that used in obtaining (12), we can prove the following theorem (see Problem 8):

Theorem 1 If $\mathbf{A}$ and $f(z)$ satisfy the hypothesis of Theorem 1 of Section 6.1, and if $\mathbf{A}$ is diagonalizable, then

$$f(\mathbf{A}) = \mathbf{M}f(\mathbf{D})\mathbf{M}^{-1}. \tag{13}$$

EXAMPLE 6 Find cos **A** for

$$\mathbf{A} = \begin{bmatrix} 4\pi & 2\pi & 0 \\ -\pi & \pi & 0 \\ 3\pi & -2\pi & \pi \end{bmatrix}.$$

Solution The eigenvalues of **A** are π, 2π, 3π, hence

$$\mathbf{D} = \begin{bmatrix} \pi & 0 & 0 \\ 0 & 2\pi & 0 \\ 0 & 0 & 3\pi \end{bmatrix}.$$

An appropriate **M** is found to be

$$\mathbf{M} = \begin{bmatrix} 0 & 1 & -2 \\ 0 & -1 & 1 \\ 1 & 5 & -4 \end{bmatrix}, \qquad \text{hence} \qquad \mathbf{M}^{-1} = \begin{bmatrix} 1 & 6 & 1 \\ -1 & -2 & 0 \\ -1 & -1 & 0 \end{bmatrix}.$$

Thus, it follows from (13) that

$$\begin{aligned} \cos(\mathbf{A}) &= \mathbf{M}\cos(\mathbf{D})\mathbf{M}^{-1} \\ &= \begin{bmatrix} 0 & 1 & -2 \\ 0 & -1 & 1 \\ 1 & 5 & -4 \end{bmatrix} \begin{bmatrix} \cos\pi & 0 & 0 \\ 0 & \cos 2\pi & 0 \\ 0 & 0 & \cos 3\pi \end{bmatrix} \begin{bmatrix} 1 & 6 & 1 \\ -1 & -2 & 0 \\ -1 & -1 & 0 \end{bmatrix} \\ &= \begin{bmatrix} -3 & -4 & 0 \\ 2 & 3 & 0 \\ -10 & -20 & -1 \end{bmatrix}. \end{aligned}$$

PROBLEMS 8.3

1. Find $\mathbf{A}^{27}$ for

$$\mathbf{A} = \begin{bmatrix} 0 & 1 \\ -2 & 3 \end{bmatrix}.$$

2. Find $\mathbf{A}^{17} - 3\mathbf{A}^5 + 2\mathbf{A}^2 + \mathbf{I}$ for the **A** of Problem 1.

3. Find $e^{\mathbf{A}}$ for the **A** of Problem 1.

4. Find $e^{\mathbf{A}}$ for

$$\mathbf{A} = \begin{bmatrix} 2 & 1 \\ -5 & 2 \end{bmatrix}.$$

5. Find $e^{\mathbf{A}}$ for

$$\mathbf{A} = \begin{bmatrix} 1 & 0 & 0 \\ 2 & -3 & 3 \\ 1 & 2 & 2 \end{bmatrix}.$$

6. Find $e^{\mathbf{A}}$ for

$$\mathbf{A} = \begin{bmatrix} 5 & 1 & 2 \\ 0 & 3 & 0 \\ 2 & 1 & 5 \end{bmatrix}.$$

7. Prove that (11) is valid for any $p_k(\mathbf{A})$.

8. Prove Theorem 1 for $f(\mathbf{A}) = \sin \mathbf{A}$.

9. Find $\sin \mathbf{A}$ for the $\mathbf{A}$ of Example 6.

8.4 GENERALIZED EIGENVECTORS

In the previous section, we showed that if a matrix $\mathbf{A}$ has n linearly independent eigenvectors associated with it, and hence is diagonalizable, then certain matrix functions of $\mathbf{A}$ can be computed quite easily. We now can generalize our analysis and obtain similar results for matrices which are not diagonalizable. We begin by generalizing the concept of the eigenvector. It then will follow that every matrix $\mathbf{A}$ has n linearly independent generalized eigenvectors and hence is similar to an "almost diagonal" matrix. These results, in turn, will provide us with a straightforward method for computing certain matrix functions of $\mathbf{A}$.

Definition 1 A vector $\mathbf{x}_m$ is a *generalized eigenvector of rank m* corresponding to the matrix $\mathbf{A}$ and the eigenvalue λ if $(\mathbf{A} - \lambda\mathbf{I})^m\mathbf{x}_m = \mathbf{0}$ but $(\mathbf{A} - \lambda\mathbf{I})^{m-1}\mathbf{x}_m \neq \mathbf{0}$.

For example, if

$$\mathbf{A} = \begin{bmatrix} 2 & 1 & -1 \\ 0 & 2 & 1 \\ 0 & 0 & 2 \end{bmatrix}, \qquad \text{then} \qquad \mathbf{x}_3 = \begin{bmatrix} 0 \\ 0 \\ 1 \end{bmatrix}$$

is a generalized eigenvector of rank 3 corresponding to $\lambda = 2$ since

$$(\mathbf{A} - 2\mathbf{I})^3\mathbf{x}_3 = \begin{bmatrix} 0 & 0 & 0 \\ 0 & 0 & 0 \\ 0 & 0 & 0 \end{bmatrix}\begin{bmatrix} 0 \\ 0 \\ 1 \end{bmatrix} = \begin{bmatrix} 0 \\ 0 \\ 0 \end{bmatrix}$$

but

$$(\mathbf{A} - 2\mathbf{I})^2\mathbf{x}_3 = \begin{bmatrix} 0 & 0 & 1 \\ 0 & 0 & 0 \\ 0 & 0 & 0 \end{bmatrix}\begin{bmatrix} 0 \\ 0 \\ 1 \end{bmatrix} = \begin{bmatrix} 1 \\ 0 \\ 0 \end{bmatrix} \neq \mathbf{0}.$$

Also,

$$\mathbf{x}_2 = \begin{bmatrix} -1 \\ 1 \\ 0 \end{bmatrix}$$

is a generalized eigenvector of rank 2 corresponding to $\lambda = 2$ since

$$(\mathbf{A} - 2\mathbf{I})^2\mathbf{x}_2 = \begin{bmatrix} 0 & 0 & 1 \\ 0 & 0 & 0 \\ 0 & 0 & 0 \end{bmatrix}\begin{bmatrix} -1 \\ 1 \\ 0 \end{bmatrix} = \begin{bmatrix} 0 \\ 0 \\ 0 \end{bmatrix}$$

but

$$(\mathbf{A} - 2\mathbf{I})^1\mathbf{x}_2 = \begin{bmatrix} 0 & 1 & -1 \\ 0 & 0 & 1 \\ 0 & 0 & 0 \end{bmatrix}\begin{bmatrix} -1 \\ 1 \\ 0 \end{bmatrix} = \begin{bmatrix} 1 \\ 0 \\ 0 \end{bmatrix} \neq \mathbf{0},$$

and

$$\mathbf{x}_1 = \begin{bmatrix} 1 \\ 0 \\ 0 \end{bmatrix}$$

is a generalized eigenvector of rank 1 corresponding to $\lambda = 2$ since $(\mathbf{A} - \lambda\mathbf{I})^1\mathbf{x}_1 = \mathbf{0}$ but $(\mathbf{A} - \lambda\mathbf{I})^0\ \mathbf{x}_1 = \mathbf{I}\mathbf{x}_1 = \mathbf{x}_1 \neq \mathbf{0}$.

We note for reference that a generalized eigenvector of rank 1 is, in fact, an eigenvector (see Problem 8).

EXAMPLE 1 It is known, and we shall see why in Section 8.6, that the matrix

$$\mathbf{A} = \begin{bmatrix} 5 & 1 & -2 & 4 \\ 0 & 5 & 2 & 2 \\ 0 & 0 & 5 & 3 \\ 0 & 0 & 0 & 4 \end{bmatrix}$$

has a generalized eigenvector of rank 3 corresponding to $\lambda = 5$. Find it.

Solution We seek a vector $\mathbf{x}_3$ such that $(\mathbf{A} - 5\mathbf{I})^3\mathbf{x}_3 = \mathbf{0}$ and $(\mathbf{A} - 5\mathbf{I})^2\mathbf{x}_3 \neq \mathbf{0}$. Designate $\mathbf{x}_3$ by

$$\begin{bmatrix} w \\ x \\ y \\ z \end{bmatrix}.$$

Then

$$(\mathbf{A} - 5\mathbf{I})^3\mathbf{x}_3 = \begin{bmatrix} 0 & 0 & 0 & 14 \\ 0 & 0 & 0 & -4 \\ 0 & 0 & 0 & 3 \\ 0 & 0 & 0 & -1 \end{bmatrix}\begin{bmatrix} w \\ x \\ y \\ z \end{bmatrix} = \begin{bmatrix} 14z \\ -4z \\ 3z \\ -z \end{bmatrix}$$

and

$$(\mathbf{A} - 5\mathbf{I})^2\mathbf{x}_3 = \begin{bmatrix} 0 & 0 & 2 & -8 \\ 0 & 0 & 0 & 4 \\ 0 & 0 & 0 & -3 \\ 0 & 0 & 0 & 1 \end{bmatrix} \begin{bmatrix} w \\ x \\ y \\ z \end{bmatrix} = \begin{bmatrix} 2y - 8z \\ 4z \\ -3z \\ z \end{bmatrix}.$$

Thus, in order to satisfy the conditions that $(\mathbf{A} - 5\mathbf{I})^3\mathbf{x}_3 = \mathbf{0}$, and $(\mathbf{A} - 5\mathbf{I})^2\mathbf{x}_3 \neq \mathbf{0}$, we must have $z = 0$ and $y \neq 0$. No restrictions are placed on w and x. By choosing $w = x = z = 0$, $y = 1$, we obtain

$$x_3 = \begin{bmatrix} 0 \\ 0 \\ 1 \\ 0 \end{bmatrix}$$

as a generalized eigenvector of rank 3 corresponding to $\lambda = 5$. Note that it is possible to obtain infinitely many other generalized eigenvectors of rank 3 by choosing different values of w, x, and y $(y \neq 0)$. For instance, if we had picked $w = -1$, $x = 2$ and $y = 15$ (z must be zero), we would have found

$$\mathbf{x}_3 = \begin{bmatrix} -1 \\ 2 \\ 15 \\ 0 \end{bmatrix}.$$

Our first choice, however, is the simplest.

EXAMPLE 2 It is known that the matrix

$$\mathbf{A} = \begin{bmatrix} 5 & 0 & 2 \\ 2 & 1 & 1 \\ -5 & 1 & -1 \end{bmatrix}$$

has a generalized eigenvector of rank 2 corresponding to $\lambda = 2$. Find it.

Solution We seek a vector $\mathbf{x}_2$ such that $(\mathbf{A} - 2\mathbf{I})^2\mathbf{x}_2 = \mathbf{0}$ and $(\mathbf{A} - 2\mathbf{I})\mathbf{x}_2 \neq \mathbf{0}$. Designate $\mathbf{x}_2$ by

$$\begin{bmatrix} x \\ y \\ z \end{bmatrix}.$$

Then

$$(\mathbf{A} - 2\mathbf{I})^2\mathbf{x}_2 = \begin{bmatrix} -1 & 2 & 0 \\ -1 & 2 & 0 \\ 2 & -4 & 0 \end{bmatrix} \begin{bmatrix} x \\ y \\ z \end{bmatrix} = \begin{bmatrix} -x + 2y \\ -x + 2y \\ 2x - 4y \end{bmatrix}$$

and

$$(\mathbf{A} - 2\mathbf{I})\mathbf{x}_2 = \begin{bmatrix} 3 & 0 & 2 \\ 2 & -1 & 1 \\ -5 & 1 & -3 \end{bmatrix} \begin{bmatrix} x \\ y \\ z \end{bmatrix} = \begin{bmatrix} 3x + 2z \\ 2x - y + z \\ -5x + y - 3z \end{bmatrix}.$$

In order to have $(\mathbf{A} - 2\mathbf{I})^2\mathbf{x}_2 = \mathbf{0}$, it follows that $x = 2y$. Using this result, we obtain

$$(\mathbf{A} - 2\mathbf{I})\mathbf{x}_2 = \begin{bmatrix} 6y + 2z \\ 3y + z \\ -9y - 3z \end{bmatrix}.$$

Since this vector must not be zero, it follows that $z \neq -3y$. There are infinitely many values of x, y, z that simultaneously satisfy the requirements $x = 2y$ and $z \neq -3y$ (for instance, $x = 2$, $y = 1$, $z = 4$); the simplest choice is $x = y = 0$, $z = 1$. Thus,

$$\mathbf{x}_2 = \begin{bmatrix} 0 \\ 0 \\ 1 \end{bmatrix}$$

is a generalized eigenvector of rank two corresponding to $\lambda = 2$.

EXAMPLE 3 It is known that the matrix

$$\mathbf{A} = \begin{bmatrix} 4 & 1 \\ 0 & 4 \end{bmatrix}$$

has a generalized eigenvector of rank 2 corresponding to $\lambda = 4$. Find it.

Solution We seek a vector $\mathbf{x}_2$ such that $(\mathbf{A} - 4\mathbf{I})^2\mathbf{x}_2 = \mathbf{0}$ but such that $(\mathbf{A} - 4\mathbf{I})\mathbf{x}_2 \neq \mathbf{0}$. Designate $\mathbf{x}_2$ by

$$\begin{bmatrix} x \\ y \end{bmatrix}.$$

Then

$$(\mathbf{A} - 4\mathbf{I})^2\mathbf{x}_2 = \begin{bmatrix} 0 & 0 \\ 0 & 0 \end{bmatrix}\begin{bmatrix} x \\ y \end{bmatrix} = \begin{bmatrix} 0 \\ 0 \end{bmatrix}.$$

Thus, we see that every vector has the property that $(\mathbf{A} - 4\mathbf{I})^2\mathbf{x}_2 = \mathbf{0}$; hence, we need place no restrictions on either x or y to achieve this result. However, since

$$(\mathbf{A} - 4\mathbf{I})\mathbf{x}_2 = \begin{bmatrix} 0 & 1 \\ 0 & 0 \end{bmatrix}\begin{bmatrix} x \\ y \end{bmatrix} = \begin{bmatrix} y \\ 0 \end{bmatrix}$$

cannot be the zero vector, it must be the case that $y \neq 0$. Thus, by choosing $x = 0$ and $y = 1$ (once again there are infinitely many other choices), we obtain

$$\mathbf{x}_2 = \begin{bmatrix} 0 \\ 1 \end{bmatrix}$$

as a generalized eigenvector of rank 2 corresponding to $\lambda = 4$.

PROBLEMS 8.4

1. Determine whether the following vectors are generalized eigenvectors of rank 3 corresponding to $\lambda = 2$ for the matrix

$$\mathbf{A} = \begin{bmatrix} 2 & 2 & 1 & 1 \\ 0 & 2 & -1 & 0 \\ 0 & 0 & 2 & 0 \\ 0 & 0 & 0 & 1 \end{bmatrix}.$$

(a) $\begin{bmatrix} 1 \\ 1 \\ 1 \\ 0 \end{bmatrix}$ (b) $\begin{bmatrix} 0 \\ 1 \\ 0 \\ 0 \end{bmatrix}$ (c) $\begin{bmatrix} 0 \\ 0 \\ 1 \\ 0 \end{bmatrix}$ (d) $\begin{bmatrix} 2 \\ 0 \\ 3 \\ 0 \end{bmatrix}$ (e) $\begin{bmatrix} 0 \\ 0 \\ 0 \\ 1 \end{bmatrix}$

(f) $\begin{bmatrix} 0 \\ 0 \\ 0 \\ 0 \end{bmatrix}.$

For the following matrices find a generalized eigenvector of rank 2 corresponding to the eigenvalue $\lambda = -1$:

2. $\begin{bmatrix} -1 & 1 \\ 0 & 1 \end{bmatrix}$

3. $\begin{bmatrix} -1 & 1 & 0 \\ 0 & 1 & 1 \\ 0 & 0 & 1 \end{bmatrix}$

4. $\begin{bmatrix} 0 & 4 & 2 \\ -1 & 4 & 1 \\ -1 & -7 & -4 \end{bmatrix}$

5. $\begin{bmatrix} 3 & -2 & 2 \\ 2 & -2 & 1 \\ -9 & 9 & -4 \end{bmatrix}$

6. $\begin{bmatrix} 2 & 0 & 3 \\ 2 & -1 & 1 \\ -1 & 0 & -2 \end{bmatrix}.$

7. Find a generalized eigenvector of rank 3 corresponding to $\lambda = 3$ and a generalized eigenvector of rank 2 corresponding to $\lambda = 4$ for

$$\mathbf{A} = \begin{bmatrix} 4 & 1 & 0 & 0 & 1 \\ 0 & 4 & 0 & 0 & 0 \\ 0 & 0 & 3 & 1 & 0 \\ 0 & 0 & 0 & 3 & 2 \\ 0 & 0 & 0 & 0 & 3 \end{bmatrix}.$$

8. Prove that a generalized eigenvector of rank 1 is an eigenvector.

8.5 CHAINS

Definition 1 Let $\mathbf{x}_m$ be a generalized eigenvector of rank m corresponding to the matrix $\mathbf{A}$ and the eigenvalue λ. The *chain generated by* $\mathbf{x}_m$ is a set of vectors $\{\mathbf{x}_m\ \mathbf{x}_{m-1} \dots \mathbf{x}_1\}$ given by

$$\begin{aligned} \mathbf{x}_{m-1} &= (\mathbf{A} - \lambda\mathbf{I})\mathbf{x}_m \\ \mathbf{x}_{m-2} &= (\mathbf{A} - \lambda\mathbf{I})^2\mathbf{x}_m = (\mathbf{A} - \lambda\mathbf{I})\mathbf{x}_{m-1} \\ \mathbf{x}_{m-3} &= (\mathbf{A} - \lambda\mathbf{I})^3\mathbf{x}_m = (\mathbf{A} - \lambda\mathbf{I})\mathbf{x}_{m-2} \\ &\vdots \\ \mathbf{x}_1 &= (\mathbf{A} - \lambda\mathbf{I})^{m-1}\mathbf{x}_m = (\mathbf{A} - \lambda\mathbf{I})\mathbf{x}_2 . \end{aligned} \tag{14}$$

Thus, in general,

$$\mathbf{x}_j = (\mathbf{A} - \lambda\mathbf{I})^{m-j}\mathbf{x}_m = (\mathbf{A} - \lambda\mathbf{I})\mathbf{x}_{j+1} \quad (j = 1, 2, \dots, m-1). \tag{15}$$

Theorem 1 $\mathbf{x}_j$ (given by (15)) is a generalized eigenvector of rank j corresponding to the eigenvalue λ.

Proof Since $\mathbf{x}_m$ is a generalized eigenvector of rank m, $(\mathbf{A} - \lambda\mathbf{I})^m\mathbf{x}_m = \mathbf{0}$ and $(\mathbf{A} - \lambda\mathbf{I})^{m-1}\mathbf{x}_m \neq \mathbf{0}$. Thus, using (15), we find that

$$(\mathbf{A} - \lambda\mathbf{I})^j\mathbf{x}_j = (\mathbf{A} - \lambda\mathbf{I})^j(\mathbf{A} - \lambda\mathbf{I})^{m-j}\mathbf{x}_m = (\mathbf{A} - \lambda\mathbf{I})^m\mathbf{x}_m = \mathbf{0}$$

and

$$(\mathbf{A} - \lambda\mathbf{I})^{j-1}\mathbf{x}_j = (\mathbf{A} - \lambda\mathbf{I})^{j-1}(\mathbf{A} - \lambda\mathbf{I})^{m-j}\mathbf{x}_m = (\mathbf{A} - \lambda\mathbf{I})^{m-1}\mathbf{x}_m \neq \mathbf{0}$$

which together imply Theorem 1.

It follows from (14) and Theorem 1 that once we have found a generalized eigenvector of rank m, it is simple to obtain a generalized eigenvector of any rank less than m. For example, we found in the previous section that

$$\mathbf{x}_3 = \begin{bmatrix} 0 \\ 0 \\ 1 \\ 0 \end{bmatrix}$$

is a generalized eigenvector of rank three for

$$\mathbf{A} = \begin{bmatrix} 5 & 1 & -2 & 4 \\ 0 & 5 & 2 & 2 \\ 0 & 0 & 5 & 3 \\ 0 & 0 & 0 & 4 \end{bmatrix}$$

corresponding to $\lambda = 5$ (see Example 1). Using Theorem 1, we now can state that

$$\mathbf{x}_2 = (\mathbf{A} - 5\mathbf{I})\mathbf{x}_3 = \begin{bmatrix} 0 & 1 & -2 & 4 \\ 0 & 0 & 2 & 2 \\ 0 & 0 & 0 & 3 \\ 0 & 0 & 0 & -1 \end{bmatrix} \begin{bmatrix} 0 \\ 0 \\ 1 \\ 0 \end{bmatrix} = \begin{bmatrix} -2 \\ 2 \\ 0 \\ 0 \end{bmatrix}$$

is a generalized eigenvector of rank 2 corresponding to $\lambda = 5$, while

$$\mathbf{x}_1 = (\mathbf{A} - 5\mathbf{I})^2\mathbf{x}_3 = (\mathbf{A} - 5\mathbf{I})\mathbf{x}_2 = \begin{bmatrix} 0 & 1 & -2 & 4 \\ 0 & 0 & 2 & 2 \\ 0 & 0 & 0 & 3 \\ 0 & 0 & 0 & -1 \end{bmatrix} \begin{bmatrix} -2 \\ 2 \\ 0 \\ 0 \end{bmatrix} = \begin{bmatrix} 2 \\ 0 \\ 0 \\ 0 \end{bmatrix}$$

is a generalized eigenvector of rank 1, hence an eigenvector, corresponding to $\lambda = 5$. The set

$$\{\mathbf{x}_3, \mathbf{x}_2, \mathbf{x}_1\} = \left\{ \begin{bmatrix} 0 \\ 0 \\ 1 \\ 0 \end{bmatrix}, \begin{bmatrix} -2 \\ 2 \\ 0 \\ 0 \end{bmatrix}, \begin{bmatrix} 2 \\ 0 \\ 0 \\ 0 \end{bmatrix} \right\}$$

is the chain generated by $\mathbf{x}_3$.

The value of chains is hinted at by the following theorem.

Theorem 2 A chain is a linearly independent set of vectors.

Proof Let $\{\mathbf{x}_m, \mathbf{x}_{m-1}, \ldots, \mathbf{x}_1\}$ be a chain generated from $\mathbf{x}_m$, a generalized eigenvector of rank m corresponding to the eigenvalue λ of a matrix $\mathbf{A}$, and consider the vector equation

$$c_m\mathbf{x}_m + c_{m-1}\mathbf{x}_{m-1} + \cdots + c_1\mathbf{x}_1 = \mathbf{0}. \tag{16}$$

In order to prove that this chain is a linearly independent set, we must show that the only constants satisfying (16) are $c_m = c_{m-1} = \cdots = c_1 = 0$. Multiply (16) by $(\mathbf{A} - \lambda\mathbf{I})^{m-1}$, and note that for $j = 1, 2, \ldots, (m-1)$

$$\begin{aligned}(\mathbf{A} - \lambda\mathbf{I})^{m-1}c_j\mathbf{x}_j &= c_j(\mathbf{A} - \lambda\mathbf{I})^{m-j-1}(\mathbf{A} - \lambda\mathbf{I})^j\mathbf{x}_j \\ &= c_j(\mathbf{A} - \lambda\mathbf{I})^{m-j-1}\mathbf{0} \quad \{\text{since } \mathbf{x}_j \text{ is a generalized eigenvector of rank } j \\ &= \mathbf{0}.\end{aligned}$$

Thus, (16) becomes $c_m(\mathbf{A} - \lambda\mathbf{I})^{m-1}\mathbf{x}_m = \mathbf{0}$. However, since $\mathbf{x}_m$ is a generalized eigenvector of rank m, $(\mathbf{A} - \lambda\mathbf{I})^{m-1}\mathbf{x}_m \neq \mathbf{0}$, from which it follows that $c_m = 0$. Substituting $c_m = 0$ into (16) and then multiplying (16) by $(\mathbf{A} - \lambda\mathbf{I})^{m-2}$, we find by similar reasoning that $c_{m-1} = 0$. Continuing this process, we finally obtain $c_m = c_{m-1} = \cdots = c_1 = 0$, which implies that the chain is linearly independent.

PROBLEMS 8.5

1. Find a generalized eigenvector of rank 4 corresponding to $\lambda = 2$ for

$$\mathbf{A} = \begin{bmatrix} 2 & 1 & 3 & -1 \\ 0 & 2 & -1 & 4 \\ 0 & 0 & 2 & 1 \\ 0 & 0 & 0 & 2 \end{bmatrix}$$

   and construct a chain from this vector. Check to see whether each vector in the chain has the correct rank.
2. Find a generalized eigenvector of rank 3 corresponding to $\lambda = 3$ and a generalized eigenvector of rank 2 corresponding to $\lambda = 4$ for the matrix of Problem 7 of Section 8.4 and construct chains from these vectors. Check to see whether or not each vector in the chains has the correct rank.

8.6 CANONICAL BASIS

As the reader might suspect from our work with diagonalizable matrices, we are interested only in sets of linearly independent generalized eigenvectors. The following theorem, the proof of which is beyond the scope of this book,[23] answers many of the questions regarding the number of such vectors.

Theorem 1 Every $n \times n$ matrix $\mathbf{A}$ possesses n linearly independent generalized eigenvectors, henceforth abbreviated *liges*. Generalized eigenvectors corresponding to distinct eigenvalues are linearly independent. If λ is an eigenvalue of $\mathbf{A}$ of multiplicity v, then $\mathbf{A}$ will have v liges corresponding to λ.

For any given matrix $\mathbf{A}$, there are infinitely many ways to pick the n liges. If they are chosen in a particularly judicious manner, we can use these vectors to show that $\mathbf{A}$ is similar to an "almost diagonal matrix." In particular,

Definition 1 A set of n liges (linearly independent generalized eigenvectors) is a *canonical basis* if it is composed entirely of chains.

[23] See B. Friedman, "Principles and Techniques of Applied Mathematics." Wiley, New York, 1956.

Thus, once we have determined that a generalized eigenvector of rank m is in a canonical basis, it follows that the $m - 1$ vectors $\mathbf{x}_{m-1}, \mathbf{x}_{m-2}, \ldots, \mathbf{x}_1$ that are in the chain generated by $\mathbf{x}_m$ given by (14) are also in the canonical basis.

For the remainder of this section we concern ourselves with determining a canonical basis for an arbitrary $n \times n$ matrix $\mathbf{A}$.

Let λ_i be an eigenvalue of $\mathbf{A}$ of multiplicity v. First, find the ranks of the matrices $(\mathbf{A} - \lambda_i\mathbf{I})$, $(\mathbf{A} - \lambda_i\mathbf{I})^2$, $(\mathbf{A} - \lambda_i\mathbf{I})^3, \ldots, (\mathbf{A} - \lambda_i\mathbf{I})^m$. The integer m is determined to be the first integer for which $(\mathbf{A} - \lambda_i\mathbf{I})^m$ has rank $n - v$ (n being the number of rows or columns of $\mathbf{A}$, that is, $\mathbf{A}$ is $n \times n$).

EXAMPLE 1 Determine m corresponding to $\lambda_i = 2$ for

$$\mathbf{A} = \begin{bmatrix} 2 & 1 & -1 & 0 & 0 & 0 \\ 0 & 2 & 1 & 0 & 0 & 0 \\ 0 & 0 & 2 & 0 & 0 & 0 \\ 0 & 0 & 0 & 2 & 1 & 0 \\ 0 & 0 & 0 & 0 & 2 & 1 \\ 0 & 0 & 0 & 0 & 0 & 4 \end{bmatrix}.$$

Solution $n = 6$ and the eigenvalue $\lambda_i = 2$ has multiplicity $v = 5$, hence $n - v = 1$.

$$(\mathbf{A} - 2\mathbf{I}) = \begin{bmatrix} 0 & 1 & -1 & 0 & 0 & 0 \\ 0 & 0 & 1 & 0 & 0 & 0 \\ 0 & 0 & 0 & 0 & 0 & 0 \\ 0 & 0 & 0 & 0 & 1 & 0 \\ 0 & 0 & 0 & 0 & 0 & 1 \\ 0 & 0 & 0 & 0 & 0 & 2 \end{bmatrix} \tag{17}$$

has rank 4.

$$(\mathbf{A} - 2\mathbf{I})^2 = \begin{bmatrix} 0 & 0 & 1 & 0 & 0 & 0 \\ 0 & 0 & 0 & 0 & 0 & 0 \\ 0 & 0 & 0 & 0 & 0 & 0 \\ 0 & 0 & 0 & 0 & 0 & 1 \\ 0 & 0 & 0 & 0 & 0 & 2 \\ 0 & 0 & 0 & 0 & 0 & 4 \end{bmatrix} \tag{18}$$

has rank 2.

$$(\mathbf{A} - 2\mathbf{I})^3 = \begin{bmatrix} 0 & 0 & 0 & 0 & 0 & 0 \\ 0 & 0 & 0 & 0 & 0 & 0 \\ 0 & 0 & 0 & 0 & 0 & 0 \\ 0 & 0 & 0 & 0 & 0 & 2 \\ 0 & 0 & 0 & 0 & 0 & 4 \\ 0 & 0 & 0 & 0 & 0 & 8 \end{bmatrix} \tag{19}$$

has rank $1 = n - v$. Therefore, corresponding to $\lambda_i = 2$, we have $m = 3$.

Now define

$$\rho_k = r(\mathbf{A} - \lambda_i \mathbf{I})^{k-1} - r(\mathbf{A} - \lambda_i \mathbf{I})^k \qquad (k = 1, 2, \ldots, m). \tag{20}$$

ρ_k designates the number of liges of rank k corresponding to the eigenvalue λ_i that will appear in a canonical basis for $\mathbf{A}$. Note that $r(\mathbf{A} - \lambda_i \mathbf{I})^0 = r(\mathbf{I}) = n$.

EXAMPLE 2 Determine how many generalized eigenvectors of each rank corresponding to $\lambda_1 = 2$ will appear in a canonical basis for the $\mathbf{A}$ of Example 1.

Solution Using the results of Example 1, we have that

$$\begin{aligned} \rho_3 &= r(\mathbf{A} - 2\mathbf{I})^2 - r(\mathbf{A} - 2\mathbf{I})^3 = 2 - 1 = 1 \\ \rho_2 &= r(\mathbf{A} - 2\mathbf{I})^1 - r(\mathbf{A} - 2\mathbf{I})^2 = 4 - 2 = 2 \\ \rho_1 &= r(\mathbf{A} - 2\mathbf{I})^0 - r(\mathbf{A} - 2\mathbf{I})^1 = 6 - 4 = 2. \end{aligned}$$

Thus, a canonical basis for the matrix given in Example 1 will have, corresponding to $\lambda_1 = 2$, one generalized eigenvector of rank 3, two liges of rank 2, and two liges of rank 1.

If in the previous example the question had been how many liges of each rank corresponding to $\lambda_2 = 4$ will appear in a canonical basis for $\mathbf{A}$, we would have found $m = 1$ and $\rho_1 = 1$; hence, a canonical basis would have contained one generalized eigenvector of rank 1 corresponding to $\lambda_2 = 4$. This is, of course, the eigenvector corresponding to $\lambda_2 = 4$ (see Problem 8 of Section 8.4).

Once we have determined the number of generalized eigenvectors of each rank that a canonical basis has, we can use the techniques of Section 8.4 (see Examples 1 and 2 of that section) together with (14) to obtain the vectors explicitly.

EXAMPLE 3 Find a canonical basis for the $\mathbf{A}$ given in Example 1.

Solution We first find the liges corresponding to $\lambda_1 = 2$. From Example 2, we know that there is one generalized eigenvector of rank 3; using the methods of Section 8.4, we find this vector to be

$$\mathbf{x}_3 = \begin{bmatrix} 0 \\ 0 \\ 1 \\ 0 \\ 0 \\ 0 \end{bmatrix}.$$

Then using (14), we obtain $\mathbf{x}_2$ and $\mathbf{x}_1$ as generalized eigenvectors of rank 2 and 1 respectively, where

$$\mathbf{x}_2 = (\mathbf{A} - 2\mathbf{I})\mathbf{x}_3 = \begin{bmatrix} -1 \\ 1 \\ 0 \\ 0 \\ 0 \\ 0 \end{bmatrix}, \qquad \text{and } \mathbf{x}_1 = (\mathbf{A} - 2\mathbf{I})\mathbf{x}_2 = \begin{bmatrix} 1 \\ 0 \\ 0 \\ 0 \\ 0 \\ 0 \end{bmatrix}.$$

From Example 2, we know that a canonical basis for $\mathbf{A}$ also has two liges of rank 2 corresponding to $\lambda_1 = 2$. We already found one of these vectors to be $\mathbf{x}_2$; therefore, we seek a generalized eigenvector $\mathbf{y}_2$ of rank 2 that is linearly independent of $\{\mathbf{x}_3, \mathbf{x}_2, \mathbf{x}_1\}$. Designate

$$\mathbf{y}_2 = \begin{bmatrix} u_2 \\ v_2 \\ w_2 \\ x_2 \\ y_2 \\ z_2 \end{bmatrix}.$$

Using the methods of Section 8.4, we find that in order for $\mathbf{y}_2$ to be a generalized eigenvector of rank 2, $w_2 = z_2 = 0$, v_2 or y_2 must be unequal to zero, and u_2 and x_2 are arbitrary. If we pick $u_2 = w_2 = x_2 = y_2 = z_2 = 0$, $v_2 = 1$, we obtain

$$\mathbf{y}_2 = \begin{bmatrix} 0 \\ 1 \\ 0 \\ 0 \\ 0 \\ 0 \end{bmatrix}$$

as a generalized eigenvector of rank 2. This vector, however, is not linearly independent of $\{\mathbf{x}_3, \mathbf{x}_2, \mathbf{x}_1\}$ since $\mathbf{y}_2 = \mathbf{x}_2 + \mathbf{x}_1$. If instead we choose $u_2 = v_2 = w_2 = x_2 = z_2 = 0$, $y_2 = 1$, we obtain

$$\mathbf{y}_2 = \begin{bmatrix} 0 \\ 0 \\ 0 \\ 0 \\ 1 \\ 0 \end{bmatrix},$$

which satisfies all the necessary requirements. (Note that there are many other adequate choices for $\mathbf{y}_2$. In particular, we could have chosen $u_2 = w_2 = x_2 = z_2 = 0$, $v_2 = y_2 = 1$.) Using (14) again, we find that

$$\mathbf{y}_1 = (\mathbf{A} - 2\mathbf{I})\mathbf{y}_2 = \begin{bmatrix} 0 \\ 0 \\ 0 \\ 1 \\ 0 \\ 0 \end{bmatrix}$$

is a generalized eigenvector of rank 1.

From Example 2, we know that a canonical basis for $\mathbf{A}$ has two liges of rank 1 corresponding to $\lambda_1 = 2$. However, we have determined these vectors already to be $\mathbf{x}_1$ and $\mathbf{y}_1$.

Having found all the liges corresponding to $\lambda_1 = 2$, we direct our attention to the liges corresponding to $\lambda_2 = 4$. From our previous discussion, we know that the only generalized eigenvector corresponding to $\lambda_2 = 4$ is the eigenvector itself, which we determine to be

$$\mathbf{z}_1 = \begin{bmatrix} 0 \\ 0 \\ 0 \\ 1 \\ 2 \\ 4 \end{bmatrix}.$$

Thus, a canonical basis for $\mathbf{A}$ is $\{\mathbf{x}_3, \mathbf{x}_2, \mathbf{x}_1, \mathbf{y}_2, \mathbf{y}_1, \mathbf{z}_1\}$. Note that due to Theorem 1, we do not have to check whether $\mathbf{z}_1$ is linearly independent of $\{\mathbf{x}_3, \mathbf{x}_2, \mathbf{x}_1, \mathbf{y}_2, \mathbf{y}_1\}$. Since $\mathbf{z}_1$ corresponds to λ_2 and all the other vectors correspond to λ_1 where $\lambda_1 \neq \lambda_2$, linear independence is guaranteed.

For future reference, we note that this canonical basis consists of one chain containing three vectors $\{\mathbf{x}_3, \mathbf{x}_2, \mathbf{x}_1\}$, one chain containing two vectors $\{\mathbf{y}_2, \mathbf{y}_1\}$, and one chain containing one vector $\{\mathbf{z}_1\}$.

EXAMPLE 4 Find a canonical basis for

$$\mathbf{A} = \begin{bmatrix} 1 & 1 & 0 & -1 \\ 0 & 1 & 0 & 0 \\ 0 & 0 & 1 & 1 \\ 0 & 0 & 0 & 1 \end{bmatrix}.$$

Solution $\mathbf{A}$ is a 4×4 and $\lambda_1 = 1$ is an eigenvalue of multiplicity 4; hence, $n = 4$, $v = 4$ and $n - v = 0$.

$$(\mathbf{A} - 1\mathbf{I}) = \begin{bmatrix} 0 & 1 & 0 & -1 \\ 0 & 0 & 0 & 0 \\ 0 & 0 & 0 & 1 \\ 0 & 0 & 0 & 0 \end{bmatrix}$$

has rank 2, and

$$(\mathbf{A} - 1\mathbf{I})^2 = \begin{bmatrix} 0 & 0 & 0 & 0 \\ 0 & 0 & 0 & 0 \\ 0 & 0 & 0 & 0 \\ 0 & 0 & 0 & 0 \end{bmatrix}$$

has rank $0 = n - \nu$. Thus, $m = 2$, $\rho_2 = r(\mathbf{A} - 1\mathbf{I}) - r(\mathbf{A} - 1\mathbf{I})^2 = 2 - 0 = 2$ and $\rho_1 = r(\mathbf{A} - 1\mathbf{I})^0 - r(\mathbf{A} - 1\mathbf{I})^1 = 4 2 = -2$; hence, a canonical basis for $\mathbf{A}$ will have two liges of rank 2 and two liges of rank 1. In order for a vector

$$\begin{bmatrix} w \\ x \\ y \\ z \end{bmatrix}$$

to be a generalized eigenvector of rank 2, either x or z must be nonzero and w and y arbitrary (see Section 8.4). If we first choose $x = 1$, $w = y = z = 0$, and then choose $z = 1$, $w = x = y = 0$, we obtain two liges of rank 2 to be

$$\mathbf{x}_2 = \begin{bmatrix} 0 \\ 1 \\ 0 \\ 0 \end{bmatrix}, \qquad \mathbf{y}_2 = \begin{bmatrix} 0 \\ 0 \\ 0 \\ 1 \end{bmatrix}.$$

Note that we could have chosen w, x, y, z in such a manner as to generate *four* linearly independent generalized eigenvectors of rank 2. The vectors

$$\begin{bmatrix} 1 \\ 1 \\ 0 \\ 0 \end{bmatrix} \quad \text{and} \quad \begin{bmatrix} 0 \\ 1 \\ 1 \\ 0 \end{bmatrix}$$

together with $\mathbf{x}_2$ and $\mathbf{y}_2$ form such a set. Thus, we immediately have found a set of four liges corresponding to $\lambda_1 = 1$. This set, however is *not* a canonical basis for $\mathbf{A}$, since it is not composed of chains. In order to obtain a canonical basis for $\mathbf{A}$, we use only two of these vectors (we will in particular use $\mathbf{x}_2$ and $\mathbf{y}_2$) and form chains from them.

Using (14), we obtain the two liges of rank 1 to be

$$\mathbf{x}_1 = (\mathbf{A} - \mathbf{I})\mathbf{x}_2 = \begin{bmatrix} 1 \\ 0 \\ 0 \\ 0 \end{bmatrix} \quad \text{and} \quad \mathbf{y}_1 = (\mathbf{A} - \mathbf{I})\mathbf{y}_2 = \begin{bmatrix} -1 \\ 0 \\ 1 \\ 0 \end{bmatrix}.$$

Thus, a canonical basis for $\mathbf{A}$ is $\{\mathbf{x}_2, \mathbf{x}_1, \mathbf{y}_2, \mathbf{y}_1\}$, which consists of the two chains each containing two vectors $\{\mathbf{x}_2, \mathbf{x}_1\}$ and $\{\mathbf{y}_2, \mathbf{y}_1\}$.

EXAMPLE 5 Find a canonical basis for

$$\mathbf{A} = \begin{bmatrix} 4 & 0 & 1 & 0 \\ 2 & 2 & 3 & 0 \\ -1 & 0 & 2 & 0 \\ 4 & 0 & 1 & 2 \end{bmatrix}.$$

Solution The characteristic equation for $\mathbf{A}$ is $(\lambda - 3)^2(\lambda - 2)^2 = 0$; hence, $\lambda_1 = 3$ and $\lambda_2 = 2$ are both eigenvalues of multiplicity 2. For $\lambda_1 = 3$, we find that $n - v = 2$, $m = 2$, $\rho_2 = 1$, and $\rho_1 = 1$, so that a canonical basis for $\mathbf{A}$ has one generalized eigenvector of rank 2 and one generalized eigenvector of rank 1 corresponding to $\lambda_1 = 3$. Using the methods of Section 8.4, we find that a generalized eigenvector of rank 2 is

$$\mathbf{x}_2 = \begin{bmatrix} 1 \\ 3 \\ 0 \\ 1 \end{bmatrix}.$$

By (14), therefore, we have that

$$\mathbf{x}_1 = (\mathbf{A} - 3\mathbf{I})\mathbf{x}_2 = \begin{bmatrix} 1 \\ -1 \\ -1 \\ 3 \end{bmatrix}$$

is a generalized eigenvector of rank 1.

For $\lambda_2 = 2$, we find that $n - v = 2$, $m = 1$, and $\rho_1 = 2$; hence, there are two generalized eigenvectors of rank 1 corresponding to $\lambda_2 = 2$. Using the methods of Section 8.4 (or, equivalently, the methods of Section 5.5 since generalized eigenvectors of rank 1 are themselves eigenvectors), we obtain

$$\mathbf{y}_1 = \begin{bmatrix} 0 \\ 1 \\ 0 \\ 0 \end{bmatrix} \quad \text{and} \quad \mathbf{z}_1 = \begin{bmatrix} 0 \\ 0 \\ 0 \\ 1 \end{bmatrix}$$

as the required vectors. Thus, a canonical basis for $\mathbf{A}$ is $\{\mathbf{x}_2, \mathbf{x}_1, \mathbf{y}_1, \mathbf{z}_1\}$ which consists of one chain containing two vectors $\{\mathbf{x}_2, \mathbf{x}_1\}$ and two chains containing one vector apiece $\{\mathbf{y}_1\}$ and $\{\mathbf{z}_1\}$. Note that once again, due to Theorem 1, we are guaranteed that $\{\mathbf{x}_2, \mathbf{x}_1\}$ are linearly independent of $\{\mathbf{y}_1, \mathbf{z}_1\}$ since they correspond to different eigenvalues.

PROBLEMS 8.6

Find a canonical basis for the following matrices:

1. $\begin{bmatrix} 3 & 1 \\ -1 & 1 \end{bmatrix}$

2. $\begin{bmatrix} 7 & 3 & 3 \\ 0 & 1 & 0 \\ -3 & -3 & 1 \end{bmatrix}$

3. $\begin{bmatrix} 5 & 1 & -1 \\ 0 & 5 & 2 \\ 0 & 0 & 5 \end{bmatrix}$

4. $\begin{bmatrix} 5 & 1 & 2 \\ 0 & 3 & 0 \\ 2 & 1 & 5 \end{bmatrix}$

5. $\begin{bmatrix} 2 & 1 & 0 & -1 \\ 0 & 2 & 1 & 1 \\ 0 & 0 & 2 & 0 \\ 0 & 0 & 0 & 2 \end{bmatrix}$

6. $\begin{bmatrix} 3 & 1 & 0 & -1 \\ 0 & 3 & 1 & 0 \\ 0 & 0 & 4 & 1 \\ 0 & 0 & 0 & 4 \end{bmatrix}$

7. $\begin{bmatrix} 4 & 1 & 1 & 0 & 0 & -1 \\ 0 & 4 & 2 & 0 & 0 & 1 \\ 0 & 0 & 4 & 1 & 0 & 0 \\ 0 & 0 & 0 & 5 & 1 & 0 \\ 0 & 0 & 0 & 0 & 5 & 2 \\ 0 & 0 & 0 & 0 & 0 & 4 \end{bmatrix}.$

8.7 JORDAN CANONICAL FORMS

In this section, we will show that every matrix is similar to an "almost diagonal" matrix, or in more precise terminology, a matrix in Jordan canonical form. We start by defining a square matrix $\mathbf{S}_k$ (k represents some positive integer and has *no* direct bearing on the order of $\mathbf{S}_k$) given by

$$\mathbf{S}_k = \begin{bmatrix} \lambda_k & 1 & 0 & 0 & \cdots & 0 & 0 \\ 0 & \lambda_k & 1 & 0 & \cdots & 0 & 0 \\ 0 & 0 & \lambda_k & 1 & \cdots & 0 & 0 \\ \vdots & \vdots & \vdots & \vdots & \ddots & \vdots & \vdots \\ 0 & 0 & 0 & 0 & \cdots & \lambda_k & 1 \\ 0 & 0 & 0 & 0 & \cdots & 0 & \lambda_k \end{bmatrix}. \tag{21}$$

Thus, $\mathbf{S}_k$ is a matrix that has all of its diagonal elements equal to λ_k, all of its superdiagonal elements equal to 1, and all of its other elements equal to zero.

Definition 1 A square matrix $\mathbf{A}$ is in *Jordan canonical form* if it is a diagonal matrix or can be expressed in either one of the following two partitioned diagonal forms:

$$\begin{bmatrix} \mathbf{D} & & & & \\ & \mathbf{S}_1 & & & \\ & & \cdot & & \\ & & & \cdot & \\ & & & & \cdot & \\ & & & & & \mathbf{S}_r \end{bmatrix} \tag{22}$$

or

$$\begin{bmatrix} \mathbf{S}_1 & & & \\ & \cdot & & \\ & & \cdot & \\ & & & \cdot & \\ & & & & \mathbf{S}_r \end{bmatrix} \tag{23}$$

Here $\mathbf{D}$ is a diagonal matrix and $\mathbf{S}_k$ $(k = 1, 2, \ldots, r)$ is defined by (21).

Consider the following matrices:

$$\begin{bmatrix} 2 & 1 & 0 & 0 \\ 0 & 2 & 0 & 0 \\ 0 & 0 & 3 & 1 \\ 0 & 0 & 0 & 3 \end{bmatrix} \tag{24}$$

$$\begin{bmatrix} 2 & 1 & 0 & 0 & 0 \\ 0 & 2 & 1 & 0 & 0 \\ 0 & 0 & 2 & 0 & 0 \\ 0 & 0 & 0 & 2 & 1 \\ 0 & 0 & 0 & 0 & 2 \end{bmatrix} \tag{25}$$

$$\begin{bmatrix} 0 & 0 & 0 & 0 & 0 & 0 \\ 0 & 2 & 0 & 0 & 0 & 0 \\ 0 & 0 & 2 & 1 & 0 & 0 \\ 0 & 0 & 0 & 2 & 0 & 0 \\ 0 & 0 & 0 & 0 & 2 & 1 \\ 0 & 0 & 0 & 0 & 0 & 2 \end{bmatrix} \tag{26}$$

$$\begin{bmatrix} 2 & 1 & 0 & 1 \\ 0 & 2 & 1 & 0 \\ 0 & 0 & 2 & 1 \\ 0 & 0 & 0 & 2 \end{bmatrix} \tag{27}$$

$$\begin{bmatrix} 2 & 2 & 0 \\ 0 & 2 & 2 \\ 0 & 0 & 2 \end{bmatrix}. \tag{28}$$

Matrix (24) is in Jordan canonical form since it can be written

$$\begin{bmatrix} \mathbf{S}_1 & \mathbf{0} \\ \mathbf{0} & \mathbf{S}_2 \end{bmatrix}, \quad \text{where} \quad \mathbf{S}_1 = \begin{bmatrix} 2 & 1 \\ 0 & 2 \end{bmatrix} \quad \text{and} \quad \mathbf{S}_2 = \begin{bmatrix} 3 & 1 \\ 0 & 3 \end{bmatrix}.$$

Matrix (25) is in Jordan canonical form since it can be expressed as

$$\begin{bmatrix} \mathbf{S}_1 & \mathbf{0} \\ \mathbf{0} & \mathbf{S}_2 \end{bmatrix}, \quad \text{where} \quad \mathbf{S}_1 = \begin{bmatrix} 2 & 1 & 0 \\ 0 & 2 & 1 \\ 0 & 0 & 2 \end{bmatrix} \quad \text{and} \quad \mathbf{S}_2 = \begin{bmatrix} 2 & 1 \\ 0 & 2 \end{bmatrix},$$

while (26) is in Jordan canonical form since it can be written

$$\begin{bmatrix} \mathbf{D} & 0 & 0 \\ 0 & \mathbf{S}_1 & 0 \\ 0 & 0 & \mathbf{S}_2 \end{bmatrix}, \quad \text{where} \quad \mathbf{D} = \begin{bmatrix} 0 & 0 \\ 0 & 2 \end{bmatrix} \quad \text{and} \quad \mathbf{S}_1 = \mathbf{S}_2 = \begin{bmatrix} 2 & 1 \\ 0 & 2 \end{bmatrix}.$$

Matrices (27) and (28) are not in Jordan canonical form; the first because of the nonzero term in the (1, 4) position and the second due to the 2's on the superdiagonal.

Note that a matrix in Jordan canonical form has nonzero elements only on the main diagonal and superdiagonal, and that the elements on the super diagonal are restricted to be either zero or one. In particular, a diagonal matrix is a matrix in Jordan canonical form (by definition) that has all its superdiagonal elements equal to zero.

In order to prove that every matrix $\mathbf{A}$ is similar to a matrix in Jordan canonical form, we must first generalize the concept of the modal matrix (see Section 8.2).

Definition 2 Let $\mathbf{A}$ be an $n \times n$ matrix. A *generalized modal matrix* $\mathbf{M}$ for $\mathbf{A}$ is an $n \times n$ matrix whose columns, considered as vectors, form a canonical basis for $\mathbf{A}$ and appear in $\mathbf{M}$ according to the following rules:

(M1) All chains consisting of one vector (that is, one vector in length) appear in the first columns of $\mathbf{M}$.

(M2) All vectors of one chain appear together in adjacent columns of $\mathbf{M}$.

(M3) Each chain appears in $\mathbf{M}$ in order of increasing rank (that is, the generalized eigenvector of rank 1 appears before the generalized eigenvector of rank 2 of the same chain, which appears before the generalized eigenvector of rank 3 of the same chain, etc.).

EXAMPLE 1 Find a generalized modal matrix $\mathbf{M}$ corresponding to the $\mathbf{A}$ given in Example 5 of Section 8.6.

Solution In that example, we found that a canonical basis for $\mathbf{A}$ has one chain of two vectors $\{\mathbf{x}_2, \mathbf{x}_1\}$ and two chains of one vector each $\{\mathbf{y}_1\}$ and $\{\mathbf{z}_1\}$. Thus, the first two columns of $\mathbf{M}$ must be $\mathbf{y}_1$ and $\mathbf{z}_1$ (however, in any order) due to (M1) while the third and fourth columns must be $\mathbf{x}_1$ and $\mathbf{x}_2$ respectively due to (M3). Hence,

$$\mathbf{M} = [\mathbf{y}_1\ \mathbf{z}_1\ \mathbf{x}_1\ \mathbf{x}_2] = \begin{bmatrix} 0 & 0 & 1 & 1 \\ 1 & 0 & -1 & 3 \\ 0 & 0 & -1 & 0 \\ 0 & 1 & 3 & 1 \end{bmatrix}$$

or

$$\mathbf{M} = [\mathbf{z}_1\ \mathbf{y}_1\ \mathbf{x}_1\ \mathbf{x}_2] = \begin{bmatrix} 0 & 0 & 1 & 1 \\ 0 & 1 & -1 & 3 \\ 0 & 0 & -1 & 0 \\ 1 & 0 & 3 & 1 \end{bmatrix}.$$

EXAMPLE 2 Find a generalized modal matrix **M** corresponding to the **A** given in Example 4 of Section 8.6.

Solution In that example, we found that a canonical basis for **A** has two chains consisting of two vectors apiece $\{\mathbf{x}_2, \mathbf{x}_1\}$ and $\{\mathbf{y}_2, \mathbf{y}_1\}$. Since this canonical basis has no chain consisting of one vector, (M1) does not apply. From (M2), we assign either $\mathbf{x}_2$ and $\mathbf{x}_1$ to the first two columns of **M** and $\mathbf{y}_2$ and $\mathbf{y}_1$ to the last two columns of **M** or, alternatively, $\mathbf{y}_2$ and $\mathbf{y}_1$ to the first two columns of **M** and $\mathbf{x}_2$ and $\mathbf{x}_1$ to the last two columns of **M**. We can not, however, define $\mathbf{M} = [\mathbf{x}_1\ \mathbf{y}_1\ \mathbf{y}_2\ \mathbf{x}_2]$ since this alignment would split the $\{\mathbf{x}_2, \mathbf{x}_1\}$ chain and violate (M2). Due to (M3), $\mathbf{x}_1$ must precede $\mathbf{x}_2$ and $\mathbf{y}_1$ must precede $\mathbf{y}_2$; hence

$$\mathbf{M} = [\mathbf{x}_1\ \mathbf{x}_2\ \mathbf{y}_1\ \mathbf{y}_2] = \begin{bmatrix} 1 & 0 & -1 & 0 \\ 0 & 1 & 0 & 0 \\ 0 & 0 & 1 & 0 \\ 0 & 0 & 0 & 1 \end{bmatrix}$$

or

$$\mathbf{M} = [\mathbf{y}_1\ \mathbf{y}_2\ \mathbf{x}_1\ \mathbf{x}_2] = \begin{bmatrix} -1 & 0 & 1 & 0 \\ 0 & 0 & 0 & 1 \\ 1 & 0 & 0 & 0 \\ 0 & 1 & 0 & 0 \end{bmatrix}.$$

Examples 1 and 2 show that **M** is not unique. The important fact, however, is that for any arbitrary $n \times n$ matrix **A**, there does exist at least one generalized modal matrix **M** corresponding to it. Furthermore, since the columns of **M** considered as vectors form a linearly independent set, it follows that the column rank of **M** is n, the rank of **M** is n, the determinant of **M** is nonzero, **M** is invertible (that is, $\mathbf{M}^{-1}$ exists).

Now let **A** represent any $n \times n$ matrix and let **M** be a generalized modal matrix for **A**. Then, one can show (see the appendix to this chapter for a proof) that

$$\mathbf{AM} = \mathbf{MJ}, \tag{29}$$

where **J** is a matrix in Jordan canonical form. By either premultiplying or postmultiplying (29) by $\mathbf{M}^{-1}$, we obtain either

$$\mathbf{J} = \mathbf{M}^{-1}\mathbf{A}\mathbf{M} \tag{30}$$

or

$$\mathbf{A} = \mathbf{M}\mathbf{J}\mathbf{M}^{-1}. \tag{31}$$

Using (30), we see how to find a matrix similar to **A** that is in Jordan canonical form. It is possible, however, to write **J** immediately from just a knowledge of the numbers ρ_k (see (20)) associated with **A**. We refer the reader to the appendix of this chapter for the method. Equation (31) provides us with a proof of

Theorem 1 Every $n \times n$ matrix **A** is similar to a matrix in Jordan canonical form.

Note the resemblance of Theorem 1, (30) and (31) above to Theorem 1, (4) and (5) of Section 8.2.

EXAMPLE 3 Verify (30) for the **A** of Example 1.

Solution In that example, we found a generalized modal matrix for **A** to be

$$\mathbf{M} = \begin{bmatrix} 0 & 0 & 1 & 1 \\ 1 & 0 & -1 & 3 \\ 0 & 0 & -1 & 0 \\ 0 & 1 & 3 & 1 \end{bmatrix}.$$

We compute

$$\mathbf{M}^{-1} = \begin{bmatrix} -3 & 1 & -4 & 0 \\ -1 & 0 & 2 & 1 \\ 0 & 0 & -1 & 0 \\ 1 & 0 & 1 & 0 \end{bmatrix}.$$

Thus,

$$\mathbf{M}^{-1}\mathbf{A}\mathbf{M} = \begin{bmatrix} -3 & 1 & -4 & 0 \\ -1 & 0 & 2 & 1 \\ 0 & 0 & -1 & 0 \\ 1 & 0 & 1 & 0 \end{bmatrix} \begin{bmatrix} 4 & 0 & 1 & 0 \\ 2 & 2 & 3 & 0 \\ -1 & 0 & 2 & 0 \\ 4 & 0 & 1 & 2 \end{bmatrix} \begin{bmatrix} 0 & 0 & 1 & 1 \\ 1 & 0 & -1 & 3 \\ 0 & 0 & -1 & 0 \\ 0 & 1 & 3 & 1 \end{bmatrix}$$

$$= \begin{bmatrix} 2 & 0 & 0 & 0 \\ 0 & 2 & 0 & 0 \\ 0 & 0 & 3 & 1 \\ 0 & 0 & 0 & 3 \end{bmatrix} = \mathbf{J},$$

a matrix in Jordan canonical form.

EXAMPLE 4 Find a matrix in Jordan canonical form that is similar to

$$\mathbf{A} = \begin{bmatrix} 0 & 4 & 2 \\ -3 & 8 & 3 \\ 4 & -8 & -2 \end{bmatrix}.$$

Solution The characteristic equation of **A** is $(\lambda - 2)^3 = 0$, hence, $\lambda = 2$ is an eigenvalue of multiplicity three. Following the procedures of the previous sections, we find that $r(\mathbf{A} - 2\mathbf{I}) = 1$ and $r(\mathbf{A} - 2\mathbf{I})^2 = 0 = n - \nu$. Thus, $\rho_2 = 1$ and $\rho_1 = 2$, which implies that a canonical basis for **A** will contain one lige of rank 2 and two liges of rank 1, or equivalently, one chain of two vectors $\{\mathbf{x}_2, \mathbf{x}_1\}$ and one chain of one vector $\{\mathbf{y}_1\}$. Designating $\mathbf{M} = [\mathbf{y}_1\ \mathbf{x}_1\ \mathbf{x}_2]$, we find that

$$\mathbf{M} = \begin{bmatrix} 2 & 2 & 0 \\ 1 & 3 & 0 \\ 0 & -4 & 1 \end{bmatrix}.$$

Thus,

$$\mathbf{M}^{-1} = \frac{1}{4}\begin{bmatrix} 3 & -2 & 0 \\ -1 & 2 & 0 \\ -4 & 8 & 4 \end{bmatrix}$$

and

$$\mathbf{J} = \mathbf{M}^{-1}\mathbf{A}\mathbf{M} = \frac{1}{4}\begin{bmatrix} 8 & 0 & 0 \\ 0 & 8 & 4 \\ 0 & 0 & 8 \end{bmatrix} = \begin{bmatrix} 2 & 0 & 0 \\ 0 & 2 & 1 \\ 0 & 0 & 2 \end{bmatrix}.$$

EXAMPLE 5 Find a matrix in Jordan canonical form that is similar to

$$\mathbf{A} = \begin{bmatrix} -1 & 0 & -1 & 1 & 1 & 3 & 0 \\ 0 & 1 & 0 & 0 & 0 & 0 & 0 \\ 2 & 1 & 2 & -1 & -1 & -6 & 0 \\ -2 & 0 & -1 & 2 & 1 & 3 & 0 \\ 0 & 0 & 0 & 0 & 1 & 0 & 0 \\ 0 & 0 & 0 & 0 & 0 & 1 & 0 \\ -1 & -1 & 0 & 1 & 2 & 4 & 1 \end{bmatrix}.$$

Solution The characteristic equation of **A** is $(\lambda - 1)^7 = 0$, hence, $\lambda = 1$ is an eigenvalue of multiplicity 7. Following the procedures of the previous sections, we find that $r(\mathbf{A} - 1\mathbf{I}) = 3$, $r(\mathbf{A} - 1\mathbf{I})^2 = 1$, $r(\mathbf{A} - 1\mathbf{I})^3 = 0 = n - \nu$. Thus $\rho_3 = 1$, $\rho_2 = 2$, and $\rho_1 = 4$ which implies that a canonical basis for **A** will consist of one lige of rank 3, two liges of rank 2 and four liges of rank 1, or equivalently, one chain of three vectors $\{\mathbf{x}_3, \mathbf{x}_2, \mathbf{x}_1\}$ one chain of two vectors $\{\mathbf{y}_2, \mathbf{y}_1\}$, and two chains of one vector $\{\mathbf{z}_1\}$, $\{\mathbf{w}_1\}$. Designating

$$\mathbf{M} = [\mathbf{z}_1\ \mathbf{w}_1\ \mathbf{x}_1\ \mathbf{x}_2\ \mathbf{x}_3\ \mathbf{y}_1\ \mathbf{y}_2],$$

we find that

$$\mathbf{M} = \begin{bmatrix} 0 & 1 & -1 & 0 & 0 & -2 & 1 \\ 0 & 3 & 0 & 0 & 1 & 0 & 0 \\ -1 & 1 & 1 & 1 & 0 & 2 & 0 \\ -2 & 0 & -1 & 0 & 0 & -2 & 0 \\ 1 & 0 & 0 & 0 & 0 & 0 & 0 \\ 0 & 1 & 0 & 0 & 0 & 0 & 0 \\ 0 & 0 & 0 & -1 & 0 & -1 & 0 \end{bmatrix}.$$

Thus,

$$\mathbf{M}^{-1} = \begin{bmatrix} 0 & 0 & 0 & 0 & 1 & 0 & 0 \\ 0 & 0 & 0 & 0 & 0 & 1 & 0 \\ 0 & 0 & 2 & 1 & 4 & -2 & 2 \\ 0 & 0 & 1 & 1 & 3 & -1 & 0 \\ 0 & 1 & 0 & 0 & 0 & -3 & 0 \\ 0 & 0 & -1 & -1 & -3 & 1 & -1 \\ 1 & 0 & 0 & -1 & -2 & -1 & 0 \end{bmatrix}$$

and

$$\mathbf{J} = \mathbf{M}^{-1}\mathbf{A}\mathbf{M} = \begin{bmatrix} 1 & 0 & 0 & 0 & 0 & 0 & 0 \\ 0 & 1 & 0 & 0 & 0 & 0 & 0 \\ 0 & 0 & 1 & 1 & 0 & 0 & 0 \\ 0 & 0 & 0 & 1 & 1 & 0 & 0 \\ 0 & 0 & 0 & 0 & 1 & 0 & 0 \\ 0 & 0 & 0 & 0 & 0 & 1 & 1 \\ 0 & 0 & 0 & 0 & 0 & 0 & 1 \end{bmatrix}.$$

PROBLEMS 8.7

In the following problems find a matrix in Jordan canonical form that is similar to the given matrix in

1. Problem 1 of Section 8.6
2. Problem 2 of Section 8.6
3. Problem 3 of Section 8.6
4. Problem 5 of Section 8.6
5. Problem 7 of Section 8.6.

8.8 FUNCTIONS OF MATRICES—GENERAL CASE

By using (31), we now are able to generalize the results of Section 8.3 and develop a straightforward method for computing functions of nondiagonalizable matrices. We begin by directing our attention to those matrices that are already in Jordan canonical form.

Consider any arbitrary $n \times n$ matrix $\mathbf{J}$ in the Jordan canonical form[24]

$$\mathbf{J} = \begin{bmatrix} \mathbf{D} & & & & \\ & \mathbf{S}_1 & & & \\ & & \mathbf{S}_2 & & \\ & & & \ddots & \\ & & & & \mathbf{S}_r \end{bmatrix}$$

Using the methods of Section 1.6 for multiplying together partitioned matrices, it follows that

$$\mathbf{J}^2 = \begin{bmatrix} \mathbf{D} & & & \\ & \mathbf{S}_1 & & \\ & & \ddots & \\ & & & \mathbf{S}_r \end{bmatrix} \begin{bmatrix} \mathbf{D} & & & \\ & \mathbf{S}_1 & & \\ & & \ddots & \\ & & & \mathbf{S}_r \end{bmatrix} = \begin{bmatrix} \mathbf{D}^2 & & & \\ & \mathbf{S}_1^2 & & \\ & & \ddots & \\ & & & \mathbf{S}_r^2 \end{bmatrix},$$

$$\mathbf{J}^3 = \mathbf{J} \cdot \mathbf{J}^2 = \begin{bmatrix} \mathbf{D}^3 & & & \\ & \mathbf{S}_1^3 & & \\ & & \ddots & \\ & & & \mathbf{S}_r^3 \end{bmatrix}, \quad .$$

and, in general,

$$\mathbf{J}^n = \begin{bmatrix} \mathbf{D}^n & & & \\ & \mathbf{S}_1^n & & \\ & & \ddots & \\ & & & \mathbf{S}_r^n \end{bmatrix}. \quad n = 0, 1, 2, 3, \ldots .$$

Furthermore, if $f(z)$ is a function that together with $\mathbf{J}$, or equivalently, $\mathbf{D}, \mathbf{S}_1, \ldots, \mathbf{S}_r$, satisfies the requirements of Theorem 1 of Section 6.1, then we can use the procedures developed in Section 6.1 to show that[25]

$$f(\mathbf{J}) = \begin{bmatrix} f(\mathbf{D}) & & & \\ & f(\mathbf{S}_1) & & \\ & & \ddots & \\ & & & f(\mathbf{S}_r) \end{bmatrix}. \tag{32}$$

[24] If instead the matrix has a form given by (23), the same analysis may be carried over in total by suppressing the D term.

[25] For a proof of (32), see D. T. Finkbeiner, "Introduction to Matrices and Linear Algebra," p. 196. Freeman, San Francisco, 1960.

Since $f(\mathbf{D})$ has already been determined in Section 8.3, we only need develop a method for calculating $f(\mathbf{S}_k)$ in order to have $f(\mathbf{J})$ determined completely.

From (21), we have the $(p+1) \times (p+1)$ matrix $\mathbf{S}_k$ defined by

$$\mathbf{S}_k = \begin{bmatrix} \lambda_k & 1 & 0 & \cdots & 0 & 0 \\ 0 & \lambda_k & 1 & \cdots & 0 & 0 \\ 0 & 0 & \lambda_k & \cdots & 1 & 0 \\ \vdots & \vdots & \vdots & \ddots & & \\ 0 & 0 & 0 & \cdots & \lambda_k & 1 \\ 0 & 0 & 0 & \cdots & 0 & \lambda_k \end{bmatrix}.$$

It can be shown that [26]

$$f(\mathbf{S}_k) = \begin{bmatrix} f(\lambda_k) & \dfrac{f'(\lambda_k)}{1!} & \dfrac{f''(\lambda_k)}{2!} & \cdots & \dfrac{f^{(p)}(\lambda_k)}{p!} \\ 0 & f(\lambda_k) & \dfrac{f'(\lambda_k)}{1!} & \cdots & \dfrac{f^{(p-1)}(\lambda_k)}{(p-1)!} \\ 0 & 0 & f(\lambda_k) & \cdots & \dfrac{f^{(p-2)}(\lambda_k)}{(p-2)!} \\ \vdots & \vdots & \vdots & \ddots & \vdots \\ 0 & 0 & 0 & \cdots & f(\lambda_k) \end{bmatrix}. \tag{33}$$

Caution: The derivatives in (33) are taken with respect to λ_k. For instance, if $\lambda_k = 3t$ and $f(\lambda_k) = e^{\lambda_k} = e^{3t}$, then $f''(\lambda_k)$ is not equal to $9e^{3t}$ but rather e^{3t}. That is, the derivative must be taken with respect to $\lambda_k = 3t$ and not with respect to t. Perhaps the safest way to make the necessary computations in (33) without incurring an error is to first keep $f(\lambda_k)$ in terms of λ_k (do not substitute a numerical value for λ_k such as $3t$), then take the derivative of $f(\lambda_k)$ with respect to λ_k (the second derivative of e^{λ_k} with respect to λ_k is e^{λ_k}), and finally, as the last step, substitute in the correct value for λ_k where needed.

EXAMPLE 1 Find $e^{\mathbf{S}_k}$ if

$$\mathbf{S}_k = \begin{bmatrix} 2t & 1 & 0 \\ 0 & 2t & 1 \\ 0 & 0 & 2t \end{bmatrix}.$$

Solution In this case, $\lambda_k = 2t$, $f(\mathbf{S}_k) = e^{\mathbf{S}_k}$, and $f(\lambda_k) = e^{\lambda_k}$.

[26] D. T. Finkbeiner, "Introduction to Matrices and Linear Algebra," p. 197. Freeman San Francisco, 1960.

$$e^{\mathbf{S}_k} = f(\mathbf{S}_k) = \begin{bmatrix} f(\lambda_k) & f'(\lambda_k) & \dfrac{f''(\lambda_k)}{2} \\ 0 & f(\lambda_k) & f'(\lambda_k) \\ 0 & 0 & f(\lambda_k) \end{bmatrix} = \begin{bmatrix} e^{\lambda_k} & e^{\lambda_k} & \dfrac{e^{\lambda_k}}{2} \\ 0 & e^{\lambda_k} & e^{\lambda_k} \\ 0 & 0 & e^{\lambda_k} \end{bmatrix}$$

$$= \begin{bmatrix} e^{2t} & e^{2t} & \dfrac{e^{2t}}{2} \\ 0 & e^{2t} & e^{2t} \\ 0 & 0 & e^{2t} \end{bmatrix} = e^{2t}\begin{bmatrix} 1 & 1 & \frac{1}{2} \\ 0 & 1 & 1 \\ 0 & 0 & 1 \end{bmatrix}.$$

EXAMPLE 2 Find $\mathbf{J}^6$ if

$$\mathbf{J} = \begin{bmatrix} 2 & 0 & 0 & 0 & 0 & 0 \\ 0 & 3 & 0 & 0 & 0 & 0 \\ 0 & 0 & 1 & 1 & 0 & 0 \\ 0 & 0 & 0 & 1 & 1 & 0 \\ 0 & 0 & 0 & 0 & 1 & 1 \\ 0 & 0 & 0 & 0 & 0 & 1 \end{bmatrix}.$$

Solution $\mathbf{J}$ is in the Jordan canonical form

$$\mathbf{J} = \begin{bmatrix} \mathbf{D} & \mathbf{0} \\ \mathbf{0} & \mathbf{S}_1 \end{bmatrix}.$$

In this case, $f(\mathbf{J}) = \mathbf{J}^6$. It follows, therefore, from (32) that

$$\mathbf{J}^6 = \begin{bmatrix} \mathbf{D}^6 & \mathbf{0} \\ \mathbf{0} & \mathbf{S}_1^6 \end{bmatrix}. \tag{34}$$

Using (7) and (33), we find that

$$\mathbf{D}^6 = \begin{bmatrix} 2^6 & 0 \\ 0 & 3^6 \end{bmatrix} = \begin{bmatrix} 64 & 0 \\ 0 & 729 \end{bmatrix} \tag{35}$$

and

$$\mathbf{S}_1^6 = \begin{bmatrix} f(\lambda_1) & \dfrac{f'(\lambda_1)}{1} & \dfrac{f''(\lambda_1)}{2} & \dfrac{f'''(\lambda_1)}{6} \\ 0 & f(\lambda_1) & f'(\lambda_1) & \dfrac{f''(\lambda_1)}{2} \\ 0 & 0 & f(\lambda_1) & f'(\lambda_1) \\ 0 & 0 & 0 & f(\lambda_1) \end{bmatrix}$$

$$= \begin{bmatrix} \lambda_1^6 & 6\lambda_1^5 & 15\lambda_1^4 & 20\lambda_1^3 \\ 0 & \lambda_1^6 & 6\lambda_1^5 & 15\lambda_1^4 \\ 0 & 0 & \lambda_1^6 & 6\lambda_1^5 \\ 0 & 0 & 0 & \lambda_1^6 \end{bmatrix}$$

$$= \begin{bmatrix} (1)^6 & 6(1)^5 & 15(1)^4 & 20(1)^3 \\ 0 & (1)^6 & 6(1)^5 & 15(1)^4 \\ 0 & 0 & (1)^6 & 6(1)^5 \\ 0 & 0 & 0 & (1)^6 \end{bmatrix} = \begin{bmatrix} 1 & 6 & 15 & 20 \\ 0 & 1 & 6 & 15 \\ 0 & 0 & 1 & 6 \\ 0 & 0 & 0 & 1 \end{bmatrix}. \tag{36}$$

Substituting (35) and (36) to (34), we obtain

$$\mathbf{J}^6 = \begin{bmatrix} 64 & 0 & 0 & 0 & 0 & 0 \\ 0 & 729 & 0 & 0 & 0 & 0 \\ 0 & 0 & 1 & 6 & 15 & 20 \\ 0 & 0 & 0 & 1 & 6 & 15 \\ 0 & 0 & 0 & 0 & 1 & 6 \\ 0 & 0 & 0 & 0 & 0 & 1 \end{bmatrix}.$$

Now let $\mathbf{A}$ be any arbitrary matrix. We know from the previous section that there exists a matrix $\mathbf{J}$ in Jordan canonical form and an invertible generalized modal matrix $\mathbf{M}$ such that

$$\mathbf{A} = \mathbf{MJM}^{-1}. \tag{37}$$

By an analysis identical to that used in Section 8.3 to obtain (10), (11), and (13), we have

$$f(\mathbf{A}) = \mathbf{M}f(\mathbf{J})\mathbf{M}^{-1},$$

providing, of course, that $f(z)$ and $\mathbf{A}$ satisfy the requirements of Theorem 1 of Section 6.1. Thus, $f(\mathbf{A})$ is obtained simply by first calculating $f(\mathbf{J})$, which in view of (32) can be done quite easily, then premultiplying $f(\mathbf{J})$ by $\mathbf{M}$, and finally postmultiplying this result by $\mathbf{M}^{-1}$.

EXAMPLE 3 Find $e^{\mathbf{A}}$ if

$$\mathbf{A} = \begin{bmatrix} 0 & 4 & 2 \\ -3 & 8 & 3 \\ 4 & -8 & -2 \end{bmatrix}.$$

Solution From Example 4 of Section 8.7 we have that a modal matrix for $\mathbf{A}$ is

$$\mathbf{M} = \begin{bmatrix} 2 & 2 & 0 \\ 1 & 3 & 0 \\ 0 & -4 & 1 \end{bmatrix} \quad \text{and} \quad \mathbf{J} = \begin{bmatrix} 2 & 0 & 0 \\ 0 & 2 & 1 \\ 0 & 0 & 2 \end{bmatrix}.$$

Thus, $e^{\mathbf{A}} = \mathbf{M}e^{\mathbf{J}}\mathbf{M}^{-1}$. In order to calculate $e^{\mathbf{J}}$, we note that

$$\mathbf{J} = \begin{bmatrix} \mathbf{D} & \mathbf{0} \\ \mathbf{0} & \mathbf{S}_1 \end{bmatrix},$$

where $\mathbf{D}$ is the 1×1 matrix $[2]$ and $\mathbf{S}_1$ is the 2×2 matrix

$$\begin{bmatrix} 2 & 1 \\ 0 & 2 \end{bmatrix}.$$

Using (7) and (33), we find that

$$e^{\mathbf{D}} = [e^2]$$

and

$$e^{\mathbf{S}_1} = \begin{bmatrix} e^2 & e^2 \\ 0 & e^2 \end{bmatrix};$$

hence,

$$e^{\mathbf{J}} = \begin{bmatrix} e^{\mathbf{D}} & \mathbf{0} \\ \mathbf{0} & e^{\mathbf{S}_1} \end{bmatrix} = \begin{bmatrix} e^2 & 0 & 0 \\ 0 & e^2 & e^2 \\ 0 & 0 & e^2 \end{bmatrix}.$$

Thus,

$$e^{\mathbf{A}} = \mathbf{M}e^{\mathbf{J}}\mathbf{M}^{-1} = \begin{bmatrix} 2 & 2 & 0 \\ 1 & 3 & 0 \\ 0 & -4 & 1 \end{bmatrix} \begin{bmatrix} e^2 & 0 & 0 \\ 0 & e^2 & e^2 \\ 0 & 0 & e^2 \end{bmatrix} \begin{bmatrix} 3 & -2 & 0 \\ -1 & 2 & 0 \\ -4 & 8 & 4 \end{bmatrix} \frac{1}{4}$$

$$= e^2 \begin{bmatrix} -1 & 4 & 2 \\ -3 & 7 & 3 \\ 4 & -8 & -3 \end{bmatrix}.$$

EXAMPLE 4 Find $\sin \mathbf{A}$ if

$$\mathbf{A} = \begin{bmatrix} \pi & \pi/3 & -\pi \\ 0 & \pi & \pi/2 \\ 0 & 0 & \pi \end{bmatrix}$$

Solution By the methods of the previous section, we find that $\mathbf{A}$ is similar to

$$\mathbf{J} = \begin{bmatrix} \pi & 1 & 0 \\ 0 & \pi & 1 \\ 0 & 0 & \pi \end{bmatrix}$$

with a corresponding generalized modal matrix given by

$$\mathbf{M} = \begin{bmatrix} \pi^2/6 & -\pi & 0 \\ 0 & \pi/2 & 0 \\ 0 & 0 & 1 \end{bmatrix}.$$

Now $\sin \mathbf{A} = \mathbf{M}(\sin \mathbf{J})\mathbf{M}^{-1}$. If we note that $\mathbf{J} = \mathbf{S}_1$, then $\sin \mathbf{J}$ can be calculated from (33). Here, $f(\lambda_1) = \sin \lambda_1$, $f'(\lambda_1) = \cos \lambda_1$, $f''(\lambda_1) = -\sin \lambda_1$, and $\lambda_1 = \pi$; hence,

$$\sin \mathbf{J} = \begin{bmatrix} \sin \pi & \cos \pi & \dfrac{-\sin \pi}{2!} \\ 0 & \sin \pi & \cos \pi \\ 0 & 0 & \sin \pi \end{bmatrix},$$

and

$$\sin \mathbf{A} = \begin{bmatrix} \pi^2/6 & -\pi & 0 \\ 0 & \pi/2 & 0 \\ 0 & 0 & 1 \end{bmatrix} \begin{bmatrix} \sin \pi & \cos \pi & \dfrac{-\sin \pi}{2} \\ 0 & \sin \pi & \cos \pi \\ 0 & 0 & \sin \pi \end{bmatrix} \begin{bmatrix} 6/\pi^2 & 12/\pi^2 & 0 \\ 0 & 2/\pi & 0 \\ 0 & 0 & 1 \end{bmatrix}$$

$$= \begin{bmatrix} 0 & -\pi/3 & \pi \\ 0 & 0 & -\pi/2 \\ 0 & 0 & 0 \end{bmatrix}.$$

PROBLEMS 8.8

1. Find $\mathbf{A}^4$ if

$$\mathbf{A} = \begin{bmatrix} 2 & 1 & 0 \\ 0 & 2 & 1 \\ 0 & 0 & 2 \end{bmatrix}.$$

2. Find $\mathbf{A}^{10}$ if

$$\mathbf{A} = \begin{bmatrix} 1 & 0 & 0 & 0 & 0 & 0 \\ 0 & -1 & 0 & 0 & 0 & 0 \\ 0 & 0 & -1 & 1 & 0 & 0 \\ 0 & 0 & 0 & -1 & 1 & 0 \\ 0 & 0 & 0 & 0 & -1 & 1 \\ 0 & 0 & 0 & 0 & 0 & -1 \end{bmatrix}.$$

3. Find $e^{\mathbf{A}}$ if

$$\mathbf{A} = \begin{bmatrix} 4 & 0 & 0 \\ 0 & 4 & 1 \\ 0 & 0 & 4 \end{bmatrix}.$$

4. Find $e^{\mathbf{A}}$ if

$$\mathbf{A} = \begin{bmatrix} 2 & 1 & 0 \\ 0 & 2 & 2 \\ 0 & 0 & 2 \end{bmatrix}.$$

5. Find cos $\mathbf{A}$ for the $\mathbf{A}$ of Example 4.
6. Find $e^{\mathbf{A}}$ for the $\mathbf{A}$ of Example 4.
7. Find $e^{\mathbf{A}}$ for

$$\mathbf{A} = \begin{bmatrix} 1 & 0 & 0 \\ 2 & 3 & -1 \\ 1 & 1 & 1 \end{bmatrix}.$$

8. Find $e^{\mathbf{A}}$ for

$$\mathbf{A} = \begin{bmatrix} 2 & 1 & 0 & 0 \\ 0 & 2 & 0 & 0 \\ 0 & 0 & 2 & 7 \\ 0 & 0 & -1 & 1 \end{bmatrix}.$$

8.9 THE FUNCTION $e^{\mathbf{A}t}$

As was the case in Chapter 6, once we have developed a procedure to find functions of arbitrary matrices, no new difficulties arise in the calculation of $e^{\mathbf{A}t}$. Again we simply define $\mathbf{B} = \mathbf{A}t$ and use the methods of the previous sections to compute $e^{\mathbf{B}}$.

EXAMPLE 1 Find $e^{\mathbf{A}t}$ if

$$\mathbf{A} = \begin{bmatrix} 3 & 1 & 0 \\ 0 & 3 & 1 \\ 0 & 0 & 3 \end{bmatrix}.$$

Solution Define

$$\mathbf{B} = \mathbf{A}t = \begin{bmatrix} 3t & t & 0 \\ 0 & 3t & t \\ 0 & 0 & 3t \end{bmatrix}.$$

Note that although $\mathbf{A}$ is in Jordan canonical form, $\mathbf{B}$, due to the t terms on the superdiagonal, is not. We find a generalized modal matrix $\mathbf{M}$ to be

$$\mathbf{M} = \begin{bmatrix} t^2 & 0 & 0 \\ 0 & t & 0 \\ 0 & 0 & 1 \end{bmatrix}, \qquad \text{hence} \qquad \mathbf{J} = \mathbf{M}^{-1}\mathbf{B}\mathbf{M} = \begin{bmatrix} 3t & 1 & 0 \\ 0 & 3t & 1 \\ 0 & 0 & 3t \end{bmatrix}.$$

Now

$$e^{\mathbf{A}t} = e^{\mathbf{B}} = \mathbf{M}e^{\mathbf{J}}\mathbf{M}^{-1}.$$

Using the techniques of Section 8.8, we find that

$$e^{\mathbf{J}} = \begin{bmatrix} e^{3t} & e^{3t} & e^{3t}/2 \\ 0 & e^{3t} & e^{3t} \\ 0 & 0 & e^{3t} \end{bmatrix}.$$

Hence,

$$e^{\mathbf{A}t} = \begin{bmatrix} t^2 & 0 & 0 \\ 0 & t & 0 \\ 0 & 0 & 1 \end{bmatrix} e^{3t} \begin{bmatrix} 1 & 1 & 1/2 \\ 0 & 1 & 1 \\ 0 & 0 & 1 \end{bmatrix} \begin{bmatrix} 1/t^2 & 0 & 0 \\ 0 & 1/t & 0 \\ 0 & 0 & 1 \end{bmatrix}$$

$$= e^{3t} \begin{bmatrix} 1 & t & t^2/2 \\ 0 & 1 & t \\ 0 & 0 & 1 \end{bmatrix}. \tag{38}$$

Note that this derivation of $e^{\mathbf{A}t}$ is not valid for $t = 0$ since $\mathbf{M}^{-1}$ is undefined there. Considering the case $t = 0$ separately, we find that $e^{\mathbf{A}0} = e^0 = \mathbf{I}$. However, since (38) also reduces to the identity matrix at $t = 0$, we are justified in using (38) for all t.

The student is urged to compare (38) with the result obtained in Example 2 of Section 6.6.

EXAMPLE 2 Find $e^{\mathbf{A}t}$ if

$$\mathbf{A} = \begin{bmatrix} 3 & 0 & 4 \\ 1 & 2 & 1 \\ -1 & 0 & -2 \end{bmatrix}.$$

Solution Define

$$\mathbf{B} = \mathbf{A}t = \begin{bmatrix} 3t & 0 & 4t \\ t & 2t & t \\ -t & 0 & -2t \end{bmatrix}.$$

A generalized modal matrix for $\mathbf{B}$ is found to be

$$\mathbf{M} = \begin{bmatrix} 1 & 0 & 4 \\ 0 & 3t & 0 \\ -1 & 0 & -1 \end{bmatrix}$$

Thus,

$$\mathbf{J} = \mathbf{M}^{-1}\mathbf{B}\mathbf{M} = \begin{bmatrix} -t & 0 & 0 \\ 0 & 2t & 1 \\ 0 & 0 & 2t \end{bmatrix}.$$

$$e^{\mathbf{A}t} = e^{\mathbf{B}} = \mathbf{M}e^{\mathbf{J}}\mathbf{M}^{-1}$$

$$= \begin{bmatrix} 1 & 0 & 4 \\ 0 & 3t & 0 \\ -1 & 0 & -1 \end{bmatrix} \begin{bmatrix} e^{-t} & 0 & 0 \\ 0 & e^{2t} & e^{2t} \\ 0 & 0 & e^{2t} \end{bmatrix} \frac{1}{3} \begin{bmatrix} -1 & 0 & -4 \\ 0 & 1/t & 0 \\ 1 & 0 & 1 \end{bmatrix}$$

$$= \frac{1}{3} \begin{bmatrix} -e^{-t} + 4e^{2t} & 0 & -4e^{-t} + 4e^{2t} \\ 3te^{2t} & 3e^{2t} & 3te^{2t} \\ e^{-t} - e^{2t} & 0 & 4e^{-t} - e^{2t} \end{bmatrix}. \tag{39}$$

Note that once again this derivation is not valid at $t = 0$ due to the $1/t$ term in $\mathbf{M}^{-1}$. However, if we consider the case $t = 0$ separately, we find that we are justified in using (39) for all t.

We now have two methods available to calculate $e^{\mathbf{A}t}$ and other functions of a matrix. The question naturally arises as to which method is best. In general, the method developed in Chapter 6 is by far the quicker and should be the one most often employed. Note that in order to use this method, we only need the eigenvalues, while the method given in this chapter requires the knowledge of both the eigenvalues and a canonical basis. On the other hand, if a canonical basis is available, or if such a set of vectors is easy to obtain, then the method given in this chapter yields $e^{\mathbf{A}t}$ very quickly and should be the method employed.

PROBLEMS 8.9

1. Find a canonical basis for the $\mathbf{B} = \mathbf{A}t$ given in Example 2 and hence verify that the matrix $\mathbf{M}$ is a valid generalized modal matrix for $\mathbf{B}$.

Find $e^{\mathbf{A}t}$ for the following matrices $\mathbf{A}$:

2. $\begin{bmatrix} 2 & 1 \\ 0 & 2 \end{bmatrix}$

3. $\begin{bmatrix} -1 & 1 & 0 \\ 0 & -1 & 1 \\ 0 & 0 & -1 \end{bmatrix}$

4. $\begin{bmatrix} 4 & 1 & 0 \\ 0 & 4 & 0 \\ 0 & 0 & 4 \end{bmatrix}$

5. $\begin{bmatrix} 2 & 1 & 0 \\ 0 & 2 & 0 \\ 0 & 0 & -1 \end{bmatrix}$

6. $\begin{bmatrix} 2 & 3 & 0 \\ -1 & -2 & 0 \\ 1 & 1 & 1 \end{bmatrix}$

7. $\begin{bmatrix} 3 & 1 & 0 \\ -1 & 1 & 0 \\ 1 & 2 & 2 \end{bmatrix}$

8. $\begin{bmatrix} 5 & -2 & 2 \\ 2 & 0 & 1 \\ -7 & 5 & -2 \end{bmatrix}.$

APPENDIX TO CHAPTER 8

We now prove the validity of Eq. (29) which states that if $\mathbf{A}$ is an arbitrary $n \times n$ matrix and $\mathbf{M}$ is a generalized modal matrix for $\mathbf{A}$, then $\mathbf{AM} = \mathbf{MJ}$ where $\mathbf{J}$ is a matrix in Jordan canonical form. In actuality, we only will prove this result for a special 4×4 matrix, but the reasoning used in this particular case can be extended easily to cover any arbitrary case.

We assume that the canonical basis used to form $\mathbf{M}$ consisted of one chain containing three vectors $\{\mathbf{x}_3, \mathbf{x}_2, \mathbf{x}_1\}$ and one chain containing one vector $\{\mathbf{y}_1\}$. The three-element chain corresponds to the eigenvalue λ_1 while $\mathbf{y}_1$ corresponds to the eigenvalue λ_2; λ_1 and λ_2 can be either equal or distinct.

Since $\mathbf{x}_1$ and $\mathbf{y}_1$ are both generalized eigenvectors of rank 1, they are themselves eigenvectors, hence $\mathbf{A}\mathbf{x}_1 = \lambda_1\mathbf{x}_1$ and $\mathbf{A}\mathbf{y}_1 = \lambda_2\mathbf{y}_1$. Furthermore, since $\mathbf{x}_2$ and $\mathbf{x}_1$ belong to the chain generated by $\mathbf{x}_3$, it follows that

$$\mathbf{x}_2 = (\mathbf{A} - \lambda_1\mathbf{I})\mathbf{x}_3$$
$$\mathbf{x}_1 = (\mathbf{A} - \lambda_1\mathbf{I})\mathbf{x}_2.$$

Hence,

$$\mathbf{x}_2 = \mathbf{A}\mathbf{x}_3 - \lambda_1\mathbf{x}_3$$
$$\mathbf{x}_1 = \mathbf{A}\mathbf{x}_2 - \lambda_1\mathbf{x}_2$$

or, by a rearrangement of terms,

$$\mathbf{A}\mathbf{x}_3 = \lambda_1\mathbf{x}_3 + \mathbf{x}_2$$
$$\mathbf{A}\mathbf{x}_2 = \lambda_1\mathbf{x}_2 + \mathbf{x}_1.$$

Thus,

$$\begin{aligned}\mathbf{AM} = \mathbf{A}[\mathbf{y}_1\ \mathbf{x}_1\ \mathbf{x}_2\ \mathbf{x}_3] &= [\mathbf{A}\mathbf{y}_1\ \mathbf{A}\mathbf{x}_1\ \mathbf{A}\mathbf{x}_2\ \mathbf{A}\mathbf{x}_3] \\ &= [\lambda_2\mathbf{y}_1\ \ \lambda_1\mathbf{x}_1\ \ \lambda_1\mathbf{x}_2 + \mathbf{x}_1\ \ \lambda_1\mathbf{x}_3 + \mathbf{x}_2] \\ &= [\mathbf{y}_1\ \mathbf{x}_1\ \mathbf{x}_2\ \mathbf{x}_3]\begin{bmatrix}\lambda_2 & 0 & 0 & 0 \\ 0 & \lambda_1 & 1 & 0 \\ 0 & 0 & \lambda_1 & 1 \\ 0 & 0 & 0 & \lambda_1\end{bmatrix}.\end{aligned}$$

Defining

$$\mathbf{J} = \begin{bmatrix}\lambda_2 & 0 & 0 & 0 \\ 0 & \lambda_1 & 1 & 0 \\ 0 & 0 & \lambda_1 & 1 \\ 0 & 0 & 0 & \lambda_1\end{bmatrix},$$

which is a matrix in Jordan canonical form, we obtain $\mathbf{AM} = \mathbf{MJ}$, the desired result.

We are now in a position to make some general observations which the reader should be able to prove. First, if $\{\mathbf{x}_m, \mathbf{x}_{m-1}, \ldots, \mathbf{x}_1\}$ is a chain generated by $\mathbf{x}_m$, then it follows from (15) that $\mathbf{x}_j = (\mathbf{A} - \lambda\mathbf{I})\mathbf{x}_{j+1}$ which, in turn, implies that $\mathbf{A}\mathbf{x}_{j+1} = \lambda\mathbf{x}_{j+1} + \mathbf{x}_j$.

Next, each complete chain of more than one vector in length that goes into composing $\mathbf{M}$ will give rise to an $\mathbf{S}_k$ submatrix in $\mathbf{J}$. Thus, for example, if a canonical basis for **A** contains a chain of three elements corresponding to the eigenvalue λ, the matrix **J** that is similar to **A** must contain a submatrix

$$\mathbf{S}_k = \begin{bmatrix} \lambda & 1 & 0 \\ 0 & \lambda & 1 \\ 0 & 0 & \lambda \end{bmatrix}.$$

The order of $\mathbf{S}_k$ is identical to the length of the chain.

The chains consisting of only one vector give rise collectively to the **D** submatrix in **J**. Thus, if there were no chains of one vector in a canonical basis for **A**, then **J** would contain no **D** submatrix, while if a canonical basis for **A** contained four one vector chains, then **J** would contain a **D** submatrix of order 4×4. In this latter case, the elements on the main diagonal of **D** would be the eigenvalues corresponding to the one element chains.

Finally, by rearranging the order in which whole chains are placed into **M**, we merely rearrange the order in which the corresponding $\mathbf{S}_k$ submatrices appear in **J**. For example, suppose that the characteristic equation of **A** is $(\lambda - 1)(\lambda - 2)(\lambda - 3)^2(\lambda - 4)^2$. Furthermore, assume that $\lambda = 3$ gives rise to the chain $\{\mathbf{z}_2, \mathbf{z}_1\}$, $\lambda = 4$ gives rise to the chain $\{\mathbf{w}_2, \mathbf{w}_1\}$, and the eigenvalues $\lambda = 1$ and $\lambda = 2$ correspond respectively to the eigenvectors $\mathbf{x}_1$ and $\mathbf{y}_1$. Then, if we choose $\mathbf{M} = [\mathbf{x}_1\ \mathbf{y}_1\ \mathbf{z}_1\ \mathbf{z}_2\ \mathbf{w}_1\ \mathbf{w}_2]$, it will follow that

$$\mathbf{M}^{-1}\mathbf{A}\mathbf{M} = \begin{bmatrix} 1 & 0 & 0 & 0 & 0 & 0 \\ 0 & 2 & 0 & 0 & 0 & 0 \\ 0 & 0 & 3 & 1 & 0 & 0 \\ 0 & 0 & 0 & 3 & 0 & 0 \\ 0 & 0 & 0 & 0 & 4 & 1 \\ 0 & 0 & 0 & 0 & 0 & 4 \end{bmatrix}$$

while if we pick $\mathbf{M} = [\mathbf{y}_1\ \mathbf{x}_1\ \mathbf{w}_1\ \mathbf{w}_2\ \mathbf{z}_1\ \mathbf{z}_2]$, it will follow that

$$\mathbf{M}^{-1}\mathbf{A}\mathbf{M} = \begin{bmatrix} 2 & 0 & 0 & 0 & 0 & 0 \\ 0 & 1 & 0 & 0 & 0 & 0 \\ 0 & 0 & 4 & 1 & 0 & 0 \\ 0 & 0 & 0 & 4 & 0 & 0 \\ 0 & 0 & 0 & 0 & 3 & 1 \\ 0 & 0 & 0 & 0 & 0 & 3 \end{bmatrix}.$$

Chapter 9 SPECIAL MATRICES

9.1 INTRODUCTION

By now it should be apparent that the eigenvalues and the eigenvectors (including the generalized eigenvectors) of a matrix serve to characterize the matrix almost completely. With a knowledge of these properties, we can calculate the determinant of the matrix and hence determine its invertibility, find its Jordan canonical form, and compute functions of the matrix. On the other hand, it also should be apparent that, practically speaking, it is impossible to obtain information of this sort and numerical methods must be employed for most matrices. To convince himself of this fact, the student is invited to write down a 4×4 matrix with no zero elements and then try to compute its eigenvalues.

Fortunately, however, there are certain classes of matrices, for example real symmetric matrices, whose structures are such that their eigenvalues and eigenvectors are particularly simple. We devote this entire chapter to the study of these matrices. It should be noted, however, that the matrices we are about to discuss appear so frequently in engineering and scientific problems that, even if they did not possess such appealing properties, they would still be worthy of our consideration.

Before beginning our study, we note that up until this point all matrices under consideration have been real; that is, they have not contained complex elements. (Note, however, that even though the matrices themselves have been real, the eigenvalues and eigenvectors often were complex. Example 2 of Section 5.3 is a case in point.) This restriction has been for convenience only and we now remedy the situation by henceforth allowing matrices to contain both real and complex elements. All the analysis of the previous eight chapters still remains valid; that is, eigenvalues, eigenvectors, inverses, functions of matrices, and Jordan canonical forms for complex matrices are found by exactly the same methods that were developed for real matrices.

EXAMPLE 1 Find $e^{\mathbf{A}t}$ for

$$\mathbf{A} = \begin{bmatrix} i & 1-i \\ 0 & i \end{bmatrix}.$$

Solution Define

$$\mathbf{B} = \mathbf{A}t = \begin{bmatrix} it & t-it \\ 0 & it \end{bmatrix}.$$

The eigenvalues of $\mathbf{B}$ are $\lambda_1 = \lambda_2 = it$. We could use the methods derived in Chapter 6 to compute $e^{\mathbf{A}t}$; we choose instead to use the methods of Chapter 8. A generalized modal matrix for $\mathbf{B}$ is found to be

$$\mathbf{M} = \begin{bmatrix} t-it & 0 \\ 0 & 1 \end{bmatrix}.$$

Then,

$$\mathbf{J} = \mathbf{M}^{-1}\mathbf{B}\mathbf{M} = \begin{bmatrix} it & 1 \\ 0 & it \end{bmatrix}$$

and

$$e^{\mathbf{A}t} = e^{\mathbf{B}} = \mathbf{M}e^{\mathbf{J}}\mathbf{M}^{-1} = \mathbf{M}\begin{bmatrix} e^{it} & e^{it} \\ 0 & e^{it} \end{bmatrix}\mathbf{M}^{-1} = e^{it}\begin{bmatrix} 1 & t-it \\ 0 & 1 \end{bmatrix}.$$

In working with complex numbers, we will frequently make use of the concept of conjugation.

Definition 1 The *complex conjugate* of a number $x = a + ib$ (where a and b are real) is the number $\bar{x} = a - ib$.

Thus, for example, the complex conjugates of $4 + 5i$, $-i$, $t - 2i$, 3, and e^{it}, t real, are $4 - 5i$, i, $t + 2i$, 3, and e^{-it}. Furthermore, it follows from the definition that the complex conjugate of a real number is itself while the complex conjugate of a pure imaginary number is its negative.

The complex conjugate of a matrix $\mathbf{A}$ is defined to be the matrix $\bar{\mathbf{A}}$ obtained by conjugating every element in $\mathbf{A}$. Thus, for the $\mathbf{A}$ given in Example 1, we have that

$$\bar{\mathbf{A}} = \begin{bmatrix} -i & 1+i \\ 0 & -i \end{bmatrix}.$$

In the following sections, we will have occasion to use certain properties of complex numbers and complex matrices. We list these properties here, and leave their verification as an exercise for the reader. Note that all properties pertaining to matrices are equally valid for vectors since vectors are a special case of matrices. Thus, for example, if

$$\mathbf{x} = \begin{bmatrix} -2i \\ 4 \\ 1+i \end{bmatrix}, \quad \text{then} \quad \bar{\mathbf{x}} = \begin{bmatrix} 2i \\ 4 \\ -1-i \end{bmatrix}.$$

(C1) $x\bar{x}$ is always real and positive, except when $x = 0$, whereupon $x\bar{x} = 0$.

(C2) x is real if and only if $x = \bar{x}$.

(C3) $\mathbf{A}$ is a real matrix if and only if $\mathbf{A} = \bar{\mathbf{A}}$.

(C4) $x + \bar{x}$ is real; $\mathbf{A} + \bar{\mathbf{A}}$ is a real matrix.

(C5) $\overline{xy} = \bar{x}\bar{y}$.

(C6) $\overline{\mathbf{Ax}} = \mathbf{A}\bar{\mathbf{x}}$ if $\mathbf{A}$ is real.

(C7) $\bar{\bar{x}} = x$; $\bar{\bar{\mathbf{A}}} = \mathbf{A}$.

(C8) $\overline{x + y} = \bar{x} + \bar{y}$

9.2 INNER PRODUCTS

The inner product is the basic tool that we will use in the ensuing sections. To any two vectors $\mathbf{x}$ and $\mathbf{y}$ of the same dimension, we associate a scalar, called the *inner product* of $\mathbf{x}$ and $\mathbf{y}$ and denoted by $\langle \mathbf{x}, \mathbf{y} \rangle$, by first conjugating $\mathbf{y}$, then multiplying together the corresponding elements of $\mathbf{x}$ and $\bar{\mathbf{y}}$, and finally summing the results.

EXAMPLE 1 Find $\langle \mathbf{x}, \mathbf{y} \rangle$ if

$$\mathbf{x} = \begin{bmatrix} 1 \\ 2i \\ 3 \end{bmatrix} \quad \text{and} \quad \mathbf{y} = \begin{bmatrix} -4i \\ 5 \\ 6+i \end{bmatrix}.$$

Solution

$$\bar{\mathbf{y}} = \begin{bmatrix} 4i \\ 5 \\ 6-i \end{bmatrix},$$

hence

$$\begin{aligned} \langle \mathbf{x}, \mathbf{y} \rangle &= 1(4i) + 2i(5) + 3(6-i) \\ &= 18 + 11i. \end{aligned}$$

EXAMPLE 2 Find $\langle \mathbf{x}, \mathbf{y} \rangle$ if

$$\mathbf{x} = \begin{bmatrix} 1+i \\ 2-3i \end{bmatrix} \quad \text{and} \quad \mathbf{y} = \begin{bmatrix} 5 \\ -4 \end{bmatrix}.$$

Solution Here $\mathbf{y}$ is a real vector, hence $\bar{\mathbf{y}} = \mathbf{y}$ and

$$\langle \mathbf{x}, \mathbf{y} \rangle = (1+i)(5) + (2-3i)(-4) = -3 + 17i.$$

From Example 2, we see that if $\mathbf{y}$ is real, then $\langle \mathbf{x}, \mathbf{y} \rangle$ is computed simply by multiplying the corresponding elements of $\mathbf{x}$ and $\mathbf{y}$ together and then summing. Thus, if the student is already familiar with the dot product of vectors, he will undoubtedly recognize that the definition for the inner product reduces to that of the dot product when $\mathbf{y}$ is real.

By relying on our knowledge of matrix multiplication, transposition, and determinants, we can give a formal definition for the inner product of two column vectors solely in terms of matrix notation. By slightly altering the definition, we can obtain a similar result for new vectors.

Definition 1 If $\mathbf{x}$ and $\mathbf{y}$ are column vectors of the same dimension, then $\langle \mathbf{x}, \mathbf{y} \rangle = \det(\mathbf{x}' \bar{\mathbf{y}})$.

Thus, if

$$\mathbf{x} = \begin{bmatrix} x_1 \\ x_2 \\ \vdots \\ x_n \end{bmatrix} \quad \text{and} \quad \mathbf{y} = \begin{bmatrix} y_1 \\ y_2 \\ \vdots \\ y_n \end{bmatrix},$$

then

$$\mathbf{x}' \bar{\mathbf{y}} = [x_1 \ x_2 \ \dots \ x_n] \begin{bmatrix} \bar{y}_1 \\ \bar{y}_2 \\ \vdots \\ \bar{y}_n \end{bmatrix} = [x_1\bar{y}_1 + x_2\bar{y}_2 + \cdots + x_n\bar{y}_n]$$

and

$$\langle \mathbf{x}, \mathbf{y} \rangle = \det(\mathbf{x}' \, \bar{\mathbf{y}}) = x_1 \bar{y}_1 + x_2 \bar{y}_2 + \cdots + x_n \bar{y}_n.$$

From the definition of the inner product and properties of complex numbers, the inner product can be shown to possess the following properties[27]:

(I1) $\langle \mathbf{x}, \mathbf{x} \rangle$ is real and positive if $\mathbf{x} \neq \mathbf{0}$; $\langle \mathbf{x}, \mathbf{x} \rangle = 0$ if and only if $\mathbf{x} = \mathbf{0}$.

(I2) $\langle \mathbf{x}, \mathbf{y} \rangle = \overline{\langle \mathbf{y}, \mathbf{x} \rangle}$

(I3) $\langle \lambda\mathbf{x}, \mathbf{y} \rangle = \lambda \langle \mathbf{x}, \mathbf{y} \rangle$ where λ is any scalar, real or complex.

(I4) $\langle \mathbf{x} + \mathbf{z}, \mathbf{y} \rangle = \langle \mathbf{x}, \mathbf{y} \rangle + \langle \mathbf{z}, \mathbf{y} \rangle$.

We will only prove (I2) here and leave the proofs of the other properties as exercises for the student (see Problems 3 and 4). Let

$$\mathbf{x} = \begin{bmatrix} x_1 \\ x_2 \\ \vdots \\ x_n \end{bmatrix} \quad \text{and} \quad \mathbf{y} = \begin{bmatrix} y_1 \\ y_2 \\ \vdots \\ y_n \end{bmatrix}.$$

Then

$$\langle \mathbf{y}, \mathbf{x} \rangle = y_1 \bar{x}_1 + y_2 \bar{x}_2 + \cdots + y_n \bar{x}_n;$$

hence,

$$\begin{aligned} \overline{\langle \mathbf{y}, \mathbf{x} \rangle} &= \overline{y_1 \bar{x}_1 + y_2 \bar{x}_2 + \cdots + y_n \bar{x}_n} \\ &= \bar{y}_1 \bar{\bar{x}}_1 + \bar{y}_2 \bar{\bar{x}}_2 + \cdots + \bar{y}_n \bar{\bar{x}}_n \qquad \left\{\begin{array}{l}\text{(Properties (C5) and} \\ \text{(C8) of Section 9.1)}\end{array}\right. \\ &= \bar{y}_1 x_1 + \bar{y}_2 x_2 + \cdots + \bar{y}_n x_n \\ &= x_1 \bar{y}_1 + x_2 \bar{y}_2 + \cdots + x_n \bar{y}_n = \langle \mathbf{x}, \mathbf{y} \rangle. \end{aligned}$$

Properties (I1)–(I4) can be used to generate other properties of the inner product. For example, (I1) and (I3) can be used to show that

(I5) $\langle \mathbf{0}, \mathbf{y} \rangle = 0$,

while (I2) and (I3) can be combined to prove that

(I6) $\langle \mathbf{x}, \lambda\mathbf{y} \rangle = \bar{\lambda} \langle \mathbf{x}, \mathbf{y} \rangle$.

EXAMPLE 3 Verify (I6) for $\mathbf{x} = \begin{bmatrix} i \\ 2 + 3i \end{bmatrix}$, $\mathbf{y} = \begin{bmatrix} 4 \\ 5 \end{bmatrix}$, and $\lambda = 1 - i$.

[27] In reality, these properties are used to define an inner product. The function given by Definition 1 is then shown to satisfy these properties.

Solution

$$\lambda \mathbf{y} = \begin{bmatrix} 4 - 4i \\ 5 - 5i \end{bmatrix},$$

hence

$$\langle \mathbf{x}, \lambda \mathbf{y} \rangle = i(4 + 4i) + (2 + 3i)(5 + 5i) = -9 + 29i.$$

But

$$\langle \mathbf{x}, \mathbf{y} \rangle = i(4) + (2 + 3i)5 = 10 + 19i,$$

hence

$$\bar{\lambda}\langle \mathbf{x}, \mathbf{y} \rangle = (1 + i)(10 + 19i) = -9 + 29i = \langle \mathbf{x}, \lambda \mathbf{y} \rangle.$$

We are now in a position to generalize the notion of "magnitude of a vector" introduced in Section 1.7.

Definition 2 The *magnitude* of a vector $\mathbf{x}$, designated by $\|\mathbf{x}\|$, is $\sqrt{\langle \mathbf{x}, \mathbf{x} \rangle}$.

We note from Property (I1) that the quantity $\langle \mathbf{x}, \mathbf{x} \rangle \geq 0$, hence the magnitude of a vector is always a nonnegative real number. Furthermore, if $\mathbf{x}$ is a real vector, then the above definition reduces to Definition 2 of Section 1.7.

EXAMPLE 4 Find $\|\mathbf{x}\|$ if

$$\mathbf{x} = \begin{bmatrix} 1 + 2i \\ -3i \end{bmatrix}.$$

Solution

$$\bar{\mathbf{x}} = \begin{bmatrix} 1 - 2i \\ 3i \end{bmatrix},$$

hence $\langle \mathbf{x}, \mathbf{x} \rangle = (1 + 2i)(1 - 2i) + (-3i)(3i) = 14$

and

$$\|\mathbf{x}\| = \sqrt{\langle \mathbf{x}, \mathbf{x} \rangle} = \sqrt{14}.$$

The concept of a normalized vector is identical to that defined in Section 1.7; that is, a nonzero vector is said to be *normalized* if it is divided by its magnitude. Thus, it follows (see Problem 7), that the vector $\mathbf{x}$ is normalized if and only if $\langle \mathbf{x}, \mathbf{x} \rangle = 1$.

PROBLEMS 9.2

1. Find $\langle \mathbf{x}, \mathbf{y} \rangle$ and $\langle \mathbf{y}, \mathbf{x} \rangle$ if

(a) $\mathbf{x} = \begin{bmatrix} i \\ -i \\ 1 \end{bmatrix}, \quad \mathbf{y} = \begin{bmatrix} 3 \\ 1 - i \\ -1 \end{bmatrix},$ (b) $\mathbf{x} = \begin{bmatrix} 1 \\ 2 \\ 1 \end{bmatrix}, \quad \mathbf{y} = \begin{bmatrix} 1 \\ -3 \\ 5 \end{bmatrix},$

(c)
$$\mathbf{x} = \begin{bmatrix} i \\ 1-i \\ 2 \\ 1+i \end{bmatrix}, \qquad \mathbf{y} = \begin{bmatrix} -i \\ 1+i \\ -3i \\ 2 \end{bmatrix}.$$

2. Find $\mathbf{x}$ if $\langle \mathbf{x}, \mathbf{a} \rangle \, \mathbf{b} = \mathbf{c}$, where

$$\mathbf{a} = \begin{bmatrix} 1 \\ 3 \\ -1 \end{bmatrix}, \qquad \mathbf{b} = \begin{bmatrix} 2 \\ 1 \\ 1 \end{bmatrix}, \qquad \text{and} \qquad \mathbf{c} = \begin{bmatrix} 3 \\ 0 \\ -1 \end{bmatrix}.$$

3. Use Property (C1) of Section 9.1 to prove (I1).
4. Prove properties (I3)–(I6).
5. Verify property (I6) for

$$\mathbf{x} = \mathbf{y} = \begin{bmatrix} 1 \\ 2 \end{bmatrix} \qquad \text{and} \qquad \lambda = i.$$

6. Normalize the following vectors:

(a) $\begin{bmatrix} 1 \\ i \end{bmatrix}$ (b) $\begin{bmatrix} 1 \\ 1-i \\ 2i \end{bmatrix}$ (c) $\begin{bmatrix} -i \\ 1+i \\ 1-i \end{bmatrix}$ (d) $\begin{bmatrix} 1 \\ 2 \\ i \\ 1-i \end{bmatrix}$.

7. Prove that a vector $\mathbf{x}$ is normalized if and only if $\langle \mathbf{x}, \mathbf{x} \rangle = 1$.

8. Prove that $\overline{\langle \mathbf{x}, \mathbf{y} \rangle} = \langle \bar{\mathbf{x}}, \bar{\mathbf{y}} \rangle$.
9. Let $\mathbf{x} = \sum_{i=1}^{n} c_i \mathbf{x}_i$ and let $\mathbf{y} = \sum_{j=1}^{m} d_j \mathbf{y}_j$ where $\mathbf{x}$, $\mathbf{y}$, $\mathbf{x}_i$, and $\mathbf{y}_j$ represent n-dimensional vectors and c_i and d_j represent scalars ($i = 1, 2, \ldots, n$; $j = 1, 2, \ldots, m$). Use Properties (I3), (I4), and (I6) to prove that

$$\langle \mathbf{x}, \mathbf{y} \rangle = \sum_{i=1}^{n} \sum_{j=1}^{m} c_i \, \bar{d}_j \langle \mathbf{x}_i, \mathbf{y}_j \rangle.$$

9.3 ORTHONORMAL VECTORS

Definition 1 Two vectors $\mathbf{x}$ and $\mathbf{y}$ are *orthogonal* (or perpendicular) if $\langle \mathbf{x}, \mathbf{y} \rangle = 0$.

Thus, given the vectors

$$\mathbf{x} = \begin{bmatrix} 1 \\ i \\ 1-i \end{bmatrix}, \qquad \mathbf{y} = \begin{bmatrix} -i \\ i \\ i \end{bmatrix}, \qquad \mathbf{z} = \begin{bmatrix} 1 \\ 1 \\ 0 \end{bmatrix},$$

we see that $\mathbf{x}$ is orthogonal to $\mathbf{y}$ and $\mathbf{y}$ is orthogonal to $\mathbf{z}$ since $\langle \mathbf{x}, \mathbf{y} \rangle = \langle \mathbf{y}, \mathbf{z} \rangle = 0$; but the vectors $\mathbf{x}$ and $\mathbf{z}$ are not orthogonal since $\langle \mathbf{x}, \mathbf{z} \rangle = 1 + i \neq 0$. In particular, as a direct consequence of Property (I5) of Section 9.2, we have that the zero vector is orthogonal to every vector.

A set of vectors is called an *orthogonal set* if each vector in the set is orthogonal to every other vector in the set. The set given above is not an orthogonal set since $\mathbf{z}$ is not orthogonal to $\mathbf{x}$ whereas the set given by $\{\mathbf{x}, \mathbf{y}, \mathbf{z}\}$

$$\mathbf{x} = \begin{bmatrix} 1 \\ 0 \\ 0 \end{bmatrix}, \qquad \mathbf{y} = \begin{bmatrix} 0 \\ i \\ 0 \end{bmatrix}, \qquad \mathbf{z} = \begin{bmatrix} 0 \\ 0 \\ 1 - i \end{bmatrix},$$

is an orthogonal set since each vector is orthogonal to every other vector.

Definition 2 A set of vectors is *orthonormal* if it is an orthogonal set having the property that every vector is a unit vector (a vector of magnitude 1).

The set of vectors

$$\left\{ \begin{bmatrix} 1 \\ 0 \\ 0 \end{bmatrix}, \quad \begin{bmatrix} 0 \\ 1 \\ 0 \end{bmatrix}, \quad \begin{bmatrix} 0 \\ 0 \\ 1 \end{bmatrix} \right\}$$

is an example of an orthonormal set.

Definition 2 can be simplified if we make use of the Kronecker delta, δ_{ij}, defined by

$$\delta_{ij} = \begin{cases} 1 & \text{if } i = j, \\ 0 & \text{if } i \neq j. \end{cases} \tag{1}$$

A set of vectors $\{\mathbf{x}_1, \mathbf{x}_2, \ldots, \mathbf{x}_n\}$ is an orthonormal set if and only if

$$\langle \mathbf{x}_i, \mathbf{x}_j \rangle = \delta_{ij} \qquad \text{for all } i \text{ and } j, \quad i, j = 1, 2, \ldots, n. \tag{2}$$

The importance of orthonormal sets is that they are almost equivalent to linearly independent sets. However, since orthonormal sets have the additional structure of an inner product associated with them, they are often more convenient to work with than the latter. We devote the remaining portion of this section to showing the equivalence of these two concepts. The utility of orthonormality will become self-evident in later sections.

Theorem 1 An orthonormal set of vectors is linearly independent.

Proof Let $\{\mathbf{x}_1, \mathbf{x}_2, \ldots, \mathbf{x}_n\}$ be an orthonormal set and consider the vector equation

$$c_1\mathbf{x}_1 + c_2\mathbf{x}_2 + \cdots + c_n\mathbf{x}_n = \mathbf{0} \tag{3}$$

where the c_j's ($j = 1, 2, \ldots, n$) are constants. The set of vectors will be linearly independent if the only constants that satisfy (3) are $c_1 = c_2 = \cdots = c_n = 0$. Take the inner product of both sides of (3) with $\mathbf{x}_1$. Thus,

$$\langle c_1\mathbf{x}_1 + c_2\mathbf{x}_2 + \cdots + c_n\mathbf{x}_n, \mathbf{x}_1\rangle = \langle \mathbf{0}, \mathbf{x}_1\rangle.$$

Using properties (I3), (I4), and (I5), we have

$$c_1\langle \mathbf{x}_1, \mathbf{x}_1\rangle + c_2\langle \mathbf{x}_2, \mathbf{x}_1\rangle + \cdots + c_n\langle \mathbf{x}_n, \mathbf{x}_1\rangle = 0.$$

Finally, noting that $\langle \mathbf{x}_i, \mathbf{x}_1\rangle = \delta_{i1}$, we obtain $c_1 = 0$. Now taking the inner product of both sides of (3) with $\mathbf{x}_2, \mathbf{x}_3, \ldots, \mathbf{x}_n$, successively, we obtain $c_2 = 0$, $c_3 = 0, \ldots, c_n = 0$. Combining these results, we find that $c_1 = c_2 = \cdots = c_n = 0$, which implies the theorem.

Theorem 2 For every linearly independent set of vectors $\{\mathbf{x}_1, \mathbf{x}_2, \ldots, \mathbf{x}_n\}$, there exists an orthonormal set of vectors $\{\mathbf{e}_1, \mathbf{e}_2, \ldots, \mathbf{e}_n\}$ such that each $\mathbf{e}_j$ ($j = 1, 2, \ldots, n$) is a linear combination of $\mathbf{x}_1, \mathbf{x}_2, \ldots, \mathbf{x}_j$.

Proof First define new vectors $\mathbf{y}_1, \mathbf{y}_2, \ldots, \mathbf{y}_n$ by

$$\mathbf{y}_1 = \mathbf{x}_1$$

$$\mathbf{y}_2 = \mathbf{x}_2 - \frac{\langle \mathbf{x}_2, \mathbf{y}_1\rangle}{\langle \mathbf{y}_1, \mathbf{y}_1\rangle}\mathbf{y}_1$$

$$\mathbf{y}_3 = \mathbf{x}_3 - \frac{\langle \mathbf{x}_3, \mathbf{y}_1\rangle}{\langle \mathbf{y}_1, \mathbf{y}_1\rangle}\mathbf{y}_1 - \frac{\langle \mathbf{x}_3, \mathbf{y}_2\rangle}{\langle \mathbf{y}_2, \mathbf{y}_2\rangle}\mathbf{y}_2$$

and, in general,

$$\mathbf{y}_j = \mathbf{x}_j - \sum_{k=1}^{j-1} \frac{\langle \mathbf{x}_j, \mathbf{y}_k\rangle}{\langle \mathbf{y}_k, \mathbf{y}_k\rangle}\mathbf{y}_k \qquad j = 2, 3, \ldots, n). \tag{4}$$

Each $\mathbf{y}_j$ is a linear combination of $\mathbf{x}_1, \mathbf{x}_2, \ldots, \mathbf{x}_j$ ($j = 1, 2, \ldots, n$). Since the $\mathbf{x}$'s are linearly independent, and the coefficient of the $\mathbf{x}_j$ term in (4) is unity, it follows that $\mathbf{y}_j$ is not the zero vector (see Problem 4). Furthermore, it can be shown that the $\mathbf{y}_j$ terms form an orthogonal set (see Problem 5), hence the only property that the $\mathbf{y}_j$ terms lack in order to be the required set is that their magnitudes may not be one. We remedy this situation by defining

$$\mathbf{e}_j = \frac{\mathbf{y}_j}{\|\mathbf{y}_j\|}.$$

The desired set is $\{\mathbf{e}_1, \mathbf{e}_2, \ldots, \mathbf{e}_n\}$.

The process used to construct the $\mathbf{e}_j$ terms is called the *Gram–Schmidt orthonormalization process.*

EXAMPLE 1 Use the Gram–Schmidt orthonormalization process to construct an orthonormal set of vectors from the linearly independent set $\{\mathbf{x}_1, \mathbf{x}_2\}$, where

$$\mathbf{x}_1 = \begin{bmatrix} 1 \\ i \end{bmatrix} \quad \text{and} \quad \mathbf{x}_2 = \begin{bmatrix} 1-i \\ 1 \end{bmatrix}.$$

Solution

$$\mathbf{y}_1 = \mathbf{x}_1 = \begin{bmatrix} 1 \\ i \end{bmatrix}.$$

Now $\langle \mathbf{x}_2, \mathbf{y}_1 \rangle = (1-i)(1) + 1(-i) = 1 - 2i$ and $\langle \mathbf{y}_1, \mathbf{y}_1 \rangle = 1(1) + i(-i) = 2$; hence,

$$\mathbf{y}_2 = \mathbf{x}_2 - \frac{\langle \mathbf{x}_2, \mathbf{y}_1 \rangle}{\langle \mathbf{y}_1, \mathbf{y}_1 \rangle}\mathbf{y}_1 = \mathbf{x}_2 - \frac{(1-2i)}{2}\mathbf{y}_1$$

$$= \begin{bmatrix} 1-i \\ 1 \end{bmatrix} - \begin{bmatrix} (1-2i)/2 \\ (2+i)/2 \end{bmatrix} = \begin{bmatrix} 1/2 \\ -i/2 \end{bmatrix}.$$

To obtain $\mathbf{e}_1$ and $\mathbf{e}_2$, we first note that $\langle \mathbf{y}_1, \mathbf{y}_1 \rangle = 2$ and

$$\langle \mathbf{y}_2, \mathbf{y}_2 \rangle = (1/2)(1/2) + (-i/2)(i/2) = 1/2;$$

hence, $\|\mathbf{y}_1\| = \sqrt{\langle \mathbf{y}_1, \mathbf{y}_1 \rangle} = \sqrt{2}$ and $\|\mathbf{y}_2\| = \sqrt{\langle \mathbf{y}_2, \mathbf{y}_2 \rangle} = \sqrt{1/2}$.

Thus,

$$\mathbf{e}_1 = \frac{\mathbf{y}_1}{\|\mathbf{y}_1\|} = \frac{1}{\sqrt{2}}\begin{bmatrix} 1 \\ i \end{bmatrix}$$

and

$$\mathbf{e}_2 = \frac{\mathbf{y}_2}{\|\mathbf{y}_2\|} = \sqrt{2}\begin{bmatrix} 1/2 \\ -i/2 \end{bmatrix}.$$

EXAMPLE 2 Use the Gram–Schmidt orthonormalization process to construct an orthonormal set of vectors from the linearly independent set $\{\mathbf{x}_1, \mathbf{x}_2, \mathbf{x}_3, \mathbf{x}_4\}$ where

$$\mathbf{x}_1 = \begin{bmatrix} 1 \\ i \\ 0 \\ 0 \end{bmatrix}, \quad \mathbf{x}_2 = \begin{bmatrix} i \\ 0 \\ 2 \\ 0 \end{bmatrix}, \quad \mathbf{x}_3 = \begin{bmatrix} 0 \\ 1 \\ 0 \\ 1 \end{bmatrix}, \quad \text{and} \quad \mathbf{x}_4 = \begin{bmatrix} 0 \\ 0 \\ 0 \\ i \end{bmatrix}.$$

Solution

$$\mathbf{y}_1 = \mathbf{x}_1 = \begin{bmatrix} 1 \\ i \\ 0 \\ 0 \end{bmatrix}.$$

$$\mathbf{y}_2 = \mathbf{x}_2 - \frac{\langle \mathbf{x}_2, \mathbf{y}_1 \rangle}{\langle \mathbf{y}_1, \mathbf{y}_1 \rangle} \mathbf{y}_1 = \mathbf{x}_2 - \frac{i}{2} \mathbf{y}_1 = \begin{bmatrix} i/2 \\ 1/2 \\ 2 \\ 0 \end{bmatrix}.$$

$$\mathbf{y}_3 = \mathbf{x}_3 - \frac{\langle \mathbf{x}_3, \mathbf{y}_1 \rangle}{\langle \mathbf{y}_1, \mathbf{y}_1 \rangle} \mathbf{y}_1 - \frac{\langle \mathbf{x}_3, \mathbf{y}_2 \rangle}{\langle \mathbf{y}_2, \mathbf{y}_2 \rangle} \mathbf{y}_2$$

$$= \mathbf{x}_3 - \frac{(-i)}{2} \mathbf{y}_1 - \frac{(1/2)}{9/2} \mathbf{y}_2 = \begin{bmatrix} (4/9)i \\ 4/9 \\ -2/9 \\ 1 \end{bmatrix}.$$

$$\mathbf{y}_4 = \mathbf{x}_4 - \frac{\langle \mathbf{x}_4, \mathbf{y}_1 \rangle}{\langle \mathbf{y}_1, \mathbf{y}_1 \rangle} \mathbf{y}_1 - \frac{\langle \mathbf{x}_4, \mathbf{y}_2 \rangle}{\langle \mathbf{y}_2, \mathbf{y}_2 \rangle} \mathbf{y}_2 - \frac{\langle \mathbf{x}_4, \mathbf{y}_3 \rangle}{\langle \mathbf{y}_3, \mathbf{y}_3 \rangle} \mathbf{y}_3$$

$$= \mathbf{x}_4 - \frac{0}{2} \mathbf{y}_1 - \frac{0}{(9/2)} \mathbf{y}_2 - \frac{i}{(13/9)} \mathbf{y}_3$$

$$= \begin{bmatrix} 4/13 \\ (-4/13)i \\ (2/13)i \\ (4/13)i \end{bmatrix}.$$

Thus,

$$\mathbf{e}_1 = \sqrt{\frac{1}{2}} \begin{bmatrix} 1 \\ i \\ 0 \\ 0 \end{bmatrix}, \qquad \mathbf{e}_2 = \sqrt{\frac{2}{9}} \begin{bmatrix} i/2 \\ 1/2 \\ 2 \\ 0 \end{bmatrix}, \qquad \mathbf{e}_3 = \sqrt{\frac{9}{13}} \begin{bmatrix} (4/9)i \\ 4/9 \\ -2/9 \\ 1 \end{bmatrix},$$

$$\mathbf{e}_4 = \sqrt{\frac{13}{4}} \begin{bmatrix} 4/13 \\ -(4/13)i \\ (2/13)i \\ (4/13)i \end{bmatrix}.$$

PROBLEMS 9.3

1. Find x so that

$$\begin{bmatrix} 3 \\ x \\ i \end{bmatrix} \quad \text{is orthogonal to} \quad \begin{bmatrix} 1-i \\ 2 \\ -i \end{bmatrix}.$$

2. Find x and y so that

$$\begin{bmatrix} x \\ y \\ 3 \end{bmatrix}$$

is orthogonal to both

$$\begin{bmatrix} 0 \\ i \\ 1-i \end{bmatrix} \quad \text{and} \quad \begin{bmatrix} i \\ 1 \\ 4 \end{bmatrix}.$$

3. Use the Gram–Schmidt orthonormalization process to construct an orthonormal set from

$$\text{(a)} \quad \mathbf{x}_1 = \begin{bmatrix} 1 \\ 2 \\ 1 \end{bmatrix}, \quad \mathbf{x}_2 = \begin{bmatrix} 1 \\ 0 \\ 1 \end{bmatrix}, \quad \mathbf{x}_3 = \begin{bmatrix} 1 \\ 0 \\ 2 \end{bmatrix};$$

$$\text{(b)} \quad \mathbf{x}_1 = \begin{bmatrix} i \\ 0 \\ 1 \\ 0 \end{bmatrix}, \quad \mathbf{x}_2 = \begin{bmatrix} 0 \\ i \\ 0 \\ -1 \end{bmatrix}, \quad \mathbf{x}_3 = \begin{bmatrix} 0 \\ 1 \\ 0 \\ 1 \end{bmatrix}, \quad \mathbf{x}_4 = \begin{bmatrix} 0 \\ 0 \\ 1-i \\ 0 \end{bmatrix}.$$

4. Prove that no **y**-vector in the Gram-Schmidt orthonormalization process is zero.

5. Prove that the **y**-vectors in the Gram-Schmidt orthonormalization process form an orthogonal set. (Hint: first show that $\langle \mathbf{y}_2, \mathbf{y}_1 \rangle = 0$, hence $\mathbf{y}_2$ must be orthogonal to $\mathbf{y}_1$. Then use induction.)

6. Find a three-dimensional unit vector that is orthogonal to both

$$\begin{bmatrix} 1 \\ 2 \\ 1 \end{bmatrix} \quad \text{and} \quad \begin{bmatrix} i \\ 1 \\ 1-i \end{bmatrix}.$$

7. Let $\{\mathbf{x}_1, \mathbf{x}_2, \ldots, \mathbf{x}_n\}$ be an orthonormal set of vectors. Prove that

$$\left\langle \sum_{i=1}^{n} c_i \mathbf{x}_i, \ \sum_{j=1}^{n} c_j \mathbf{x}_j \right\rangle = \sum_{i=1}^{n} c_i \bar{c}_i .$$

(Hint: use Problem 9 of Section 9.2)

8. Show that an orthogonal set of vectors need not be linearly independent. Compare this result with Theorem 1.

9.4 SELF-ADJOINT MATRICES

We begin our study of special matrices with the "adjoint matrix." *Warning*: this will not be the same matrix as defined in Section 3.1. The use of the same word here is unfortunate, but since it is in such common usage throughout engineering and mathematics, it would be unwise to adopt another. In fact, if the student is ever confronted with the word "adjoint", he should assume, unless it has been clearly stipulated to the contrary, that it refers to the matrix defined below.

Definition 1 Let $\mathbf{A}$ be an $m \times n$ matrix. The *adjoint of* $\mathbf{A}$, designated by $\mathbf{A}^*$, is an $n \times m$ matrix satisfying the property that

$$\langle \mathbf{x}, \mathbf{A}\mathbf{y} \rangle = \langle \mathbf{A}^*\mathbf{x}, \mathbf{y} \rangle \tag{5}$$

for all m-dimensional vectors $\mathbf{x}$ and n-dimensional vectors $\mathbf{y}$.

In other words, the inner product of the vector $\mathbf{x}$ with the vector $\mathbf{Ay}$ is equal to the inner product of the vector $\mathbf{A}^*\mathbf{x}$ with the vector $\mathbf{y}$ for every m-dimensional vector $\mathbf{x}$ and n-dimensional vector $\mathbf{y}$. Note that although $\mathbf{A}$ need not be square, its order must be such that the dimensions of $\mathbf{Ax}$ and $\mathbf{A}^*\mathbf{y}$ are identical to $\mathbf{y}$ and $\mathbf{x}$, so that the inner products given in (5) are well defined.

EXAMPLE 1 Verify that the adjoint of

$$\mathbf{A} = \begin{bmatrix} 1 & 2 & 3 \\ 4 & 5 & 6 \end{bmatrix} \quad \text{is} \quad \mathbf{A}^* = \begin{bmatrix} 1 & 4 \\ 2 & 5 \\ 3 & 6 \end{bmatrix}.$$

Solution Since the order of $\mathbf{A}$ is 2×3, $\mathbf{x}$ must be 2-dimensional and $\mathbf{y}$ must be 3-dimensional. Designate the arbitrary vectors $\mathbf{x}$ and $\mathbf{y}$ by

$$\begin{bmatrix} x_1 \\ x_2 \end{bmatrix} \quad \text{and} \quad \begin{bmatrix} y_1 \\ y_2 \\ y_3 \end{bmatrix}.$$

Then,

$$\mathbf{Ay} = \begin{bmatrix} (y_1 + 2y_2 + 3y_3) \\ (4y_1 + 5y_2 + 6y_3) \end{bmatrix} \quad \text{and} \quad \mathbf{A}^*\mathbf{x} = \begin{bmatrix} x_1 + 4x_2 \\ 2x_1 + 5x_2 \\ 3x_1 + 6x_2 \end{bmatrix}.$$

Hence,

$$\begin{aligned} \langle \mathbf{x}, \mathbf{Ay} \rangle &= x_1\overline{(y_1 + 2y_2 + 3y_3)} + x_2\overline{(4y_1 + 5y_2 + 6y_3)} \\ &= x_1\bar{y}_1 + 2x_1\bar{y}_2 + 3x_1\bar{y}_3 + 4x_2\bar{y}_1 + 5x_2\bar{y}_2 + 6x_2\bar{y}_3 \end{aligned}$$

and

$$\begin{aligned} \langle \mathbf{A}^*\mathbf{x}, \mathbf{y} \rangle &= (x_1 + 4x_2)\bar{y}_1 + (2x_1 + 5x_2)\bar{y}_2 + (3x_1 + 6x_2)\bar{y}_3 \\ &= x_1\bar{y}_1 + 4x_2\bar{y}_1 + 2x_1\bar{y}_2 + 5x_2\bar{y}_2 + 3x_1\bar{y}_3 + 6x_2\bar{y}_3. \end{aligned}$$

After rearranging terms, we find that $\langle \mathbf{x}, \mathbf{Ay} \rangle = \langle \mathbf{A}^*\mathbf{x}, \mathbf{y} \rangle$ where $\mathbf{x}$ and $\mathbf{y}$ are arbitrary. Thus, (5) is satisfied and $\mathbf{A}^*$, as given, is the adjoint of $\mathbf{A}$.

EXAMPLE 2 Determine whether

$$\mathbf{B} = \begin{bmatrix} 1 & i \\ i & 1 \end{bmatrix} \quad \text{is the adjoint of} \quad \mathbf{A} = \begin{bmatrix} 1 & i \\ i & 1 \end{bmatrix}.$$

Solution Since the order of **A** is 2 × 2, both **x** and **y** must be 2-dimensional vectors. Designate **x** and **y** by

$$\begin{bmatrix} x_1 \\ x_2 \end{bmatrix} \quad \text{and} \quad \begin{bmatrix} y_1 \\ y_2 \end{bmatrix}.$$

Then,

$$\begin{aligned} \langle \mathbf{x}, \mathbf{Ay} \rangle &= x_1\overline{(y_1 + iy_2)} + x_2\overline{(iy_1 + y_2)} \\ &= x_1\bar{y}_1 - ix_1\bar{y}_2 - ix_2\bar{y}_1 + x_2\bar{y}_2 \end{aligned}$$

and

$$\begin{aligned} \langle \mathbf{Bx}, \mathbf{y} \rangle &= (x_1 + ix_2)\bar{y}_1 + (ix_1 + x_2)\bar{y}_2 \\ &= x_1\bar{y}_1 + ix_2\bar{y}_1 + ix_1\bar{y}_2 + x_2\bar{y}_2 . \end{aligned}$$

Thus, we see that $\langle \mathbf{x}, \mathbf{Ay} \rangle$ is not always equal to $\langle \mathbf{Bx}, \mathbf{y} \rangle$ (for example, choose

$$\mathbf{x} = \mathbf{y} = \begin{bmatrix} 1 \\ 1 \end{bmatrix});$$

hence, **B** is not the adjoint of **A**.

Theorem 1 The adjoint matrix has the following properties:

(A1) $(\mathbf{A}^*)^* = \mathbf{A}$

(A2) $(\mathbf{A} + \mathbf{B})^* = \mathbf{A}^* + \mathbf{B}^*$

(A3) $(\mathbf{AB})^* = \mathbf{B}^*\mathbf{A}^*$

(A4) $(\lambda\mathbf{A})^* = \bar{\lambda}\mathbf{A}^*$

Proof We shall prove only (A3) and leave the others as an exercise for the student (see Problem 3). Let **x** and **y** be arbitrary. Then

$$\langle (\mathbf{AB})^*\mathbf{x}, \mathbf{y} \rangle = \langle \mathbf{x}, \mathbf{ABy} \rangle = \langle \mathbf{A}^*\mathbf{x}, \mathbf{By} \rangle = \langle \mathbf{B}^*\mathbf{A}^*\mathbf{x}, \mathbf{y} \rangle,$$

hence

$$\langle (\mathbf{AB})^*\mathbf{x}, \mathbf{y} \rangle - \langle \mathbf{B}^*\mathbf{A}^*\mathbf{x}, \mathbf{y} \rangle = 0,$$

or

$$\langle [(\mathbf{AB})^* - \mathbf{B}^*\mathbf{A}^*]\mathbf{x}, \mathbf{y} \rangle = 0 \qquad \{(\text{Property (I4) of Section 9.2}).$$

But since this is true for every **y**, we note that in particular it must be true for $\mathbf{y} = [(\mathbf{AB})^* - \mathbf{B}^*\mathbf{A}^*]\mathbf{x}$. Thus,

$$\langle [(\mathbf{AB})^* - \mathbf{B}^*\mathbf{A}^*]\mathbf{x}, [(\mathbf{AB})^* - \mathbf{B}^*\mathbf{A}^*]\mathbf{x} \rangle = 0.$$

Now since both arguments of the inner product are equal, it follows from Property (I1) of Section 9.2 that $[(\mathbf{AB})^* - \mathbf{B}^*\mathbf{A}^*]\mathbf{x} = \mathbf{0}$. However, since $\mathbf{x}$ is also arbitrary, it follows from Lemma 1 of Section 7.7 that $(\mathbf{AB})^* - \mathbf{B}^*\mathbf{A}^* = \mathbf{0}$, from which (A3) immediately follows.

It was not accidental that the adjoint of the matrix given in Example 1 was the transpose of $\mathbf{A}$. In fact, for real matrices, that is matrices having no complex elements, this will always be the case.

Theorem 2 The adjoint of a *real matrix* $\mathbf{A}$ exists, is unique, and equals the transpose of $\mathbf{A}$; that is, $\mathbf{A}^* = \mathbf{A}'$.[28]

We defer discussion of $\mathbf{A}^*$ for the case where $\mathbf{A}$ has complex elements until Section 9.7. Suffice to say here that for complex matrices the adjoint exists, is unique, but does not equal the transpose (see Example 2).

Definition 2 A matrix $\mathbf{A}$ is *self-adjoint* if it equals its own adjoint, that is, $\mathbf{A} = \mathbf{A}^*$.

An immediate consequence of Definition 2 is that a self-adjoint matrix must be square (for an $m \times n$ matrix to equal an $n \times m$ (see Definition 1), n must equal m). It follows from (5) that an $n \times n$ matrix is self-adjoint if and only if

$$\langle \mathbf{x}, \mathbf{Ay} \rangle = \langle \mathbf{Ax}, \mathbf{y} \rangle \tag{6}$$

for all n-dimensional vectors $\mathbf{x}$ and $\mathbf{y}$.

Combining Definition 2 with Theorem 2, we have

Theorem 3 A *real matrix* is self-adjoint if and only if it is symmetric, that is, $\mathbf{A} = \mathbf{A}'$.

Again we defer discussion of self-adjoint complex matrices until Section 9.7.

EXAMPLE 3 Verify that

$$\mathbf{A} = \begin{bmatrix} 1 & 2 \\ 2 & 3 \end{bmatrix}$$

is self-adjoint.

Solution Proceeding exactly as we did in Example 1, we find that

$$\langle \mathbf{x}, \mathbf{Ay} \rangle = \langle \mathbf{Ax}, \mathbf{y} \rangle = x_1\bar{y}_1 + 2x_1\bar{y}_2 + 2x_2\bar{y}_1 + 3x_2\bar{y}_2;$$

hence, (6) is satisfied which implies that $\mathbf{A}$ is self-adjoint.

[28] For a proof of Theorem 2, see E. Nering, "Linear Algebra and Matrix Theory," p. 155. Wiley, New York, 1963.

EXAMPLE 4 Verify that

$$\mathbf{A} = \begin{bmatrix} 1 & i \\ -i & 2 \end{bmatrix}$$

is self-adjoint.

Solution By direct calculation, we find that

$$\langle \mathbf{Ax}, \mathbf{y} \rangle = \langle \mathbf{x}, \mathbf{Ay} \rangle = x_1\bar{y}_1 - i(x_1\bar{y}_2) + i(x_2\bar{y}_1) + 2x_2\bar{y}_2;$$

hence (6) is satisfied so **A** is self-adjoint.

Note in Example 4 that although **A** is self-adjoint, it is not symmetric. Why does this example not violate Theorem 3?

PROBLEMS 9.4

1. Which of the following matrices are self-adjoint?

(a) $\begin{bmatrix} 1 & 2 \\ 3 & 4 \end{bmatrix}$ (b) $\begin{bmatrix} 1 & 1-i \\ 1+i & 2 \end{bmatrix}$ (c) $\begin{bmatrix} i & 1 \\ 1 & 2 \end{bmatrix}$

(d) $\begin{bmatrix} 1 & 3 \\ 3 & -1 \end{bmatrix}$ (e) $\begin{bmatrix} 1 & i & 1 \\ -i & 0 & 1 \\ 1 & 1 & i \end{bmatrix}$ (f) $\begin{bmatrix} 1 & 2 & 3 \\ 2 & 5 & 6 \\ 3 & 6 & 7 \end{bmatrix}$

2. Prove that $\langle \mathbf{Ax}, \mathbf{y} \rangle = \langle \mathbf{x}, \mathbf{A}^*\mathbf{y} \rangle$ (Hint: first use Property (I2) of Section 9.2 and then the definition of the adjoint.)

3. Prove the remainder of Theorem 1.

4. Prove Theorem 2 for the case of 2×2 matrices.

9.5 REAL SYMMETRIC MATRICES

In this section and the next we restrict our attention to real matrices. We return to the general case of complex matrices in Section 9.7. Note, however, that although we require the matrices themselves to be real *a priori*, we place no such restrictions on either the eigenvalues or eigenvectors.

We begin our discussion with those real matrices that are self-adjoint, or equivalently (see Theorem 3 of Section 9.4), those real matrices that are symmetric.

Theorem 1 The eigenvalues of a real symmetric matrix are real.

Proof Let **A** be a real symmetric matrix, λ an eigenvalue of **A**, and **x** an eigenvector of **A** corresponding to λ. Thus, $\mathbf{Ax} = \lambda\mathbf{x}$. Combining this result with (6) and the properties of the inner product, we find that

$$\lambda\langle \mathbf{x}, \mathbf{x}\rangle = \langle \lambda\mathbf{x}, \mathbf{x}\rangle = \langle \mathbf{A}\mathbf{x}, \mathbf{x}\rangle = \langle \mathbf{x}, \mathbf{A}\mathbf{x}\rangle$$
$$= \langle \mathbf{x}, \lambda\mathbf{x}\rangle = \bar{\lambda}\langle \mathbf{x}, \mathbf{x}\rangle,$$

hence

$$\lambda\langle \mathbf{x}, \mathbf{x}\rangle - \bar{\lambda}\langle \mathbf{x}, \mathbf{x}\rangle = 0,$$

or

$$(\lambda - \bar{\lambda})\langle \mathbf{x}, \mathbf{x}\rangle = 0.$$

Since $\mathbf{x}$ is an eigenvector, $\mathbf{x} \neq \mathbf{0}$, hence, $\langle \mathbf{x}, \mathbf{x}\rangle \neq 0$ (Property (I1) of Section 9.2). Thus, it follows that $\lambda - \bar{\lambda} = 0$ or $\lambda = \bar{\lambda}$, which implies that λ is real (Property (C2) of Section 9.1).

Theorem 2 The eigenvectors of a real symmetric matrix can always be chosen to be real.

Proof Let $\mathbf{A}$, $\mathbf{x}$, and λ be the same as in the previous proof. We first assume that $\mathbf{x}$ is not pure imaginary (that is, $\mathbf{x} = \mathbf{a} + i\mathbf{b}$, $\mathbf{a}$ and $\mathbf{b}$ real, $\mathbf{a} \neq \mathbf{0}$) and define a new vector $\mathbf{y} = \mathbf{x} + \bar{\mathbf{x}}$. Then $\mathbf{y}$ is real (Property (C4) of Section 9.1), nonzero, and an eigenvector of $\mathbf{A}$ corresponding to the eigenvalue λ. To substantiate this last claim, we note that $\mathbf{A}\mathbf{x} = \lambda\mathbf{x}$, hence by conjugating both sides, we obtain $\overline{\mathbf{A}\mathbf{x}} = \overline{\lambda\mathbf{x}}$. However, since λ is a real number (Theorem 1) and $\mathbf{A}$ is a real matrix, it follows that $\overline{\lambda\mathbf{x}} = \bar{\lambda}\bar{\mathbf{x}} = \lambda\bar{\mathbf{x}}$ and $\overline{\mathbf{A}\mathbf{x}} = \mathbf{A}\bar{\mathbf{x}}$, hence $\mathbf{A}\bar{\mathbf{x}} = \lambda\bar{\mathbf{x}}$. Thus, $\mathbf{A}\mathbf{y} = \mathbf{A}(\mathbf{x} + \bar{\mathbf{x}}) = \mathbf{A}\mathbf{x} + \mathbf{A}\bar{\mathbf{x}} = \lambda\mathbf{x} + \lambda\bar{\mathbf{x}} = \lambda(\mathbf{x} + \bar{\mathbf{x}}) = \lambda\mathbf{y}$ which implies that $\mathbf{y}$ is an eigenvector of $\mathbf{A}$ corresponding to the eigenvalue λ. If $\mathbf{x}$ is pure imaginary, then $\mathbf{y} = i\mathbf{x}$ is the desired eigenvector (see Problem 4).

As an example of Theorem 2, consider the real symmetric matrix

$$\mathbf{A} = \begin{bmatrix} 3 & 4 \\ 4 & -3 \end{bmatrix} \qquad \text{and the vector} \qquad \mathbf{x} = \begin{bmatrix} 2+4i \\ 1+2i \end{bmatrix}$$

which is an eigenvector of $\mathbf{A}$ corresponding to the eigenvalue $\lambda = 5$. By following the procedure given in Theorem 2, we obtain

$$\mathbf{y} = \mathbf{x} + \bar{\mathbf{x}} = \begin{bmatrix} 2+4i \\ 1+2i \end{bmatrix} + \begin{bmatrix} 2-4i \\ 1-2i \end{bmatrix} = \begin{bmatrix} 4 \\ 2 \end{bmatrix}$$

as a real eigenvector of $\mathbf{A}$ corresponding to the eigenvalue $\lambda = 5$.

For the remainder of this book, we will always assume that the procedures outlined in the proof of Theorem 2 have been performed and that the eigenvectors under consideration which correspond to a real symmetric matrix are themselves real. Furthermore, we note that if $\mathbf{A}$ is real symmetric, then $\mathbf{A} - \lambda\mathbf{I}$, where λ is an eigenvalue of $\mathbf{A}$, is also real symmetric, hence it follows that $(\mathbf{A} - \lambda\mathbf{I})$ is self-adjoint. This, in turn, implies that

$$\langle \mathbf{x}, (\mathbf{A} - \lambda\mathbf{I})\mathbf{y}\rangle = \langle (\mathbf{A} - \lambda\mathbf{I})\mathbf{x}, \mathbf{y}\rangle \tag{7}$$

for all vectors $\mathbf{x}$ and $\mathbf{y}$ of appropriate dimension.

Theorem 3 A real symmetric matrix is diagonalizable.

Proof Let **A** be a real symmetric matrix. Then, from our work in Chapter 8, we know that **A** is diagonalizable if it possesses n linearly independent eigenvectors, or equivalently, if no generalized eigenvector of **A** has rank greater than 1.

Assume that **x** is a generalized eigenvector of rank 2 corresponding to the eigenvalue λ. (By identical reasoning as that used to prove Theorem 2, we can assume that **x** is real (see Problem 5).) Then, it must be the case (Definition 1 of Section 8.4) that

$$(\mathbf{A} - \lambda\mathbf{I})^2\mathbf{x} = \mathbf{0}$$

and (8)

$$(\mathbf{A} - \lambda\mathbf{I})\mathbf{x} \neq \mathbf{0}.$$

But,

$$\begin{aligned} 0 = \langle \mathbf{x}, \mathbf{0} \rangle &= \langle \mathbf{x}, (\mathbf{A} - \lambda\mathbf{I})^2\mathbf{x} \rangle \\ &= \langle \mathbf{x}, (\mathbf{A} - \lambda\mathbf{I})(\mathbf{A} - \lambda\mathbf{I})\mathbf{x} \rangle \\ &= \langle (\mathbf{A} - \lambda\mathbf{I})\mathbf{x}, (\mathbf{A} - \lambda\mathbf{I})\mathbf{x} \rangle \qquad \{\text{from (7)}\}. \end{aligned}$$

Using Property (I1) of Section 9.2, we conclude that $(\mathbf{A} - \lambda\mathbf{I})\mathbf{x} = 0$, which contradicts (8). Something is wrong! Since all our steps are valid, it must be our assumption which is incorrect, hence, **A** cannot possess a generalized eigenvector of rank 2.

This conclusion, in turn, implies that **A** has no generalized eigenvector $\mathbf{x}_m$ of rank m, $m > 1$. For if it did, we then could form a chain from $\mathbf{x}_m$ and obtain a generalized eigenvector of rank 2. But we have just shown that **A** cannot possess such a vector; hence, we are led to the conclusion that **A** cannot possess a generalized eigenvector of rank m, $m > 1$. This in turn implies that all generalized eigenvectors of **A** are of rank 1.

Theorem 4 Eigenvectors of a real symmetric matrix corresponding to distinct eigenvalues are orthogonal.

Proof Let **A** be a real symmetric matrix with eigenvectors **x** and **y** which correspond respectively to the different eigenvalues λ and μ. Thus, $\mathbf{Ax} = \lambda\mathbf{x}$ and $\mathbf{Ay} = \mu\mathbf{y}$. Then, recalling that **x**, **y**, λ, and μ are all real (see Theorems 1 and 2) and **A** is self-adjoint, we have

$$\begin{aligned} (\lambda - \mu)\langle \mathbf{x}, \mathbf{y} \rangle &= \lambda\langle \mathbf{x}, \mathbf{y} \rangle - \mu\langle \mathbf{x}, \mathbf{y} \rangle \\ &= \langle \lambda\mathbf{x}, \mathbf{y} \rangle - \langle \mathbf{x}, \bar{\mu}\mathbf{y} \rangle \\ &= \langle \mathbf{Ax}, \mathbf{y} \rangle - \langle \mathbf{x}, \mu\mathbf{y} \rangle \\ &= \langle \mathbf{x}, \mathbf{Ay} \rangle - \langle \mathbf{x}, \mathbf{Ay} \rangle = 0. \end{aligned}$$

But $\lambda \neq \mu$; hence it follows that $\langle \mathbf{x}, \mathbf{y} \rangle = 0$, which implies that **x** and **y** are orthogonal.

EXAMPLE 1 Verify Theorem 4 for

$$\mathbf{A} = \begin{bmatrix} 3 & 4 \\ 4 & -3 \end{bmatrix}.$$

Solution The eigenvalues of $\mathbf{A}$ are $\lambda_1 = 5$ and $\lambda_2 = -5$. An eigenvector corresponding to $\lambda_1 = 5$ is found to be

$$\mathbf{x} = \begin{bmatrix} 2 \\ 1 \end{bmatrix}$$

while an eigenvector corresponding to $\lambda_2 = -5$ is found to be

$$\mathbf{y} = \begin{bmatrix} 1 \\ -2 \end{bmatrix}.$$

Then, $\langle \mathbf{x}, \mathbf{y} \rangle = 2(1) + (1)(-2) = 0$, which implies that $\mathbf{x}$ and $\mathbf{y}$ are orthogonal and verifies Theorem 4.

From Theorem 3, we know that a real symmetric matrix is diagonalizable. It follows, therefore, that every real symmetric matrix possesses a set of n linearly independent eigenvectors. The question now arises whether such a set of eigenvectors can be chosen to be orthonormal. Note that Theorem 4 does not answer this question since it gives no information about eigenvectors corresponding to the same eigenvalue.

Definition 1 A set of vectors $\{\mathbf{x}_1, \mathbf{x}_2, \ldots, \mathbf{x}_p\}$ is a *complete orthonormal set of eigenvectors* for the $n \times n$ matrix $\mathbf{A}$ (real or complex) if (1) the set is an orthonormal set, (2) each vector in the set is an eigenvector of $\mathbf{A}$, and (3) the set contains exactly n vectors, that is, $p = n$.

Note that Theorem 1 of Section 9.3 guarantees that a complete orthonormal set is also a linearly independent set.

Theorem 5 Every real symmetric $n \times n$ matrix possesses a complete orthonormal set of eigenvectors.

Proof Let $\{\mathbf{x}_1, \mathbf{x}_2, \ldots, \mathbf{x}_r\}$ be a set of r linearly independent eigenvectors corresponding to the same eigenvalue λ_i. Use the Gram–Schmidt orthonormalization process on this set to construct the orthonormal set

$$\{\mathbf{e}_1, \mathbf{e}_2, \ldots, \mathbf{e}_r\}.$$

Since each $\mathbf{e}_j$ $(j = 1, 2, \ldots, r)$ is simply a linear combination of the vectors $\mathbf{x}_1, \mathbf{x}_2, \ldots, \mathbf{x}_r$, it must also be an eigenvector of $\mathbf{A}$ corresponding to λ_i (see Problem 6). Theorem 4 guarantees that each $\mathbf{e}_j$ will be orthogonal to every eigenvector that corresponds to an eigenvalue distinct from λ_i; hence, taking into consideration Theorem 3, which guarantees the existence of n-linearly independent eigenvectors, the result immediately follows.

In essence, Theorem 5 states that the columns of a modal matrix of a real symmetric matrix are not only linearly independent eigenvectors, but also can be chosen in such a manner that they form an orthonormal set.

EXAMPLE 2 Find an orthonormal set of eigenvectors for

$$\mathbf{A} = \begin{bmatrix} 9 & -2 & -2 & -4 & 0 \\ -2 & 11 & 0 & 2 & 0 \\ -2 & 0 & 7 & -2 & 0 \\ -4 & 2 & -2 & 9 & 0 \\ 0 & 0 & 0 & 0 & 3 \end{bmatrix}.$$

Solution **A** is real and symmetric. Its eigenvalues are $\lambda_1 = \lambda_2 = 9$, $\lambda_3 = \lambda_4 = 3$, $\lambda_5 = 15$. Linearly independent eigenvectors corresponding to $\lambda = 9$ are found to be

$$\mathbf{x}_1 = \begin{bmatrix} 1 \\ 1 \\ -1 \\ 0 \\ 0 \end{bmatrix} \quad \text{and} \quad \mathbf{x}_2 = \begin{bmatrix} 1 \\ 2 \\ 0 \\ -1 \\ 0 \end{bmatrix}.$$

Using the Gram–Schmidt orthonormalization process on these vectors, we obtain

$$\mathbf{e}_1 = \begin{bmatrix} 1/\sqrt{3} \\ 1/\sqrt{3} \\ -1/\sqrt{3} \\ 0 \\ 0 \end{bmatrix} \quad \text{and} \quad \mathbf{e}_2 = \begin{bmatrix} 0 \\ -1/\sqrt{3} \\ -1/\sqrt{3} \\ 1/\sqrt{3} \\ 0 \end{bmatrix}.$$

Linearly independent eigenvectors corresponding $\lambda = 3$ are found to be

$$\mathbf{x}_3 = \begin{bmatrix} 0 \\ 0 \\ 0 \\ 0 \\ 1 \end{bmatrix} \quad \text{and} \quad \mathbf{x}_4 = \begin{bmatrix} 1 \\ 0 \\ 1 \\ 1 \\ 0 \end{bmatrix}.$$

Using the Gram–Schmidt orthonormalization process on these vectors, we obtain

$$\mathbf{e}_3 = \begin{bmatrix} 0 \\ 0 \\ 0 \\ 0 \\ 1 \end{bmatrix} \quad \text{and} \quad \mathbf{e}_4 = \begin{bmatrix} 1/\sqrt{3} \\ 0 \\ 1/\sqrt{3} \\ 1/\sqrt{3} \\ 0 \end{bmatrix}.$$

An eigenvector corresponding to $\lambda = 15$, is

$$\mathbf{x}_5 = \begin{bmatrix} -1 \\ 1 \\ 0 \\ 1 \\ 0 \end{bmatrix}.$$

Normalizing this vector, we obtain

$$\mathbf{e}_5 = \begin{bmatrix} -1/\sqrt{3} \\ 1/\sqrt{3} \\ 0 \\ 1/\sqrt{3} \\ 0 \end{bmatrix}.$$

The set $\{\mathbf{e}_1, \mathbf{e}_2, \mathbf{e}_3, \mathbf{e}_4, \mathbf{e}_5\}$ is the required set.

PROBLEMS 9.5

1. Find an orthonormal set of eigenvectors and verify Theorems 2 and 4 for the following matrices (the eigenvalues are given below each matrix):

(a) $\begin{bmatrix} 2 & -2 & 1 \\ -2 & -1 & 2 \\ 1 & 2 & 2 \end{bmatrix}$ $(3,\ 3,\ -3)$

(b) $\begin{bmatrix} 1 & 2 & 0 & 2 \\ 2 & 0 & -1 & -1 \\ 0 & -1 & 2 & 1 \\ 2 & -1 & 1 & 0 \end{bmatrix}$ $(3,\ 3,\ -3,\ 0)$

(c) $\begin{bmatrix} -1 & 0 & 2 & 2 \\ 0 & -1 & 2 & -2 \\ 2 & 2 & 1 & 0 \\ 2 & -2 & 0 & 1 \end{bmatrix}$ $(3,\ 3,\ -3,\ -3)$

(d) $\begin{bmatrix} 3 & 0 & 0 & 0 \\ 0 & 1 & -2 & 2 \\ 0 & -2 & 1 & 2 \\ 0 & 2 & 2 & 1 \end{bmatrix}.$ $(3,\ 3,\ 3,\ -3)$

2. Show that

$$\begin{bmatrix} 3 & i \\ i & 1 \end{bmatrix}$$

has an eigenvector of rank 2. Why does this result not violate Theorem 3?

3. Prove that the eigenvalues of a self-adjoint matrix are real.

4. Let $\mathbf{x}$ be an eigenvector of a real symmetric matrix $\mathbf{A}$ such that every component of $\mathbf{x}$ is pure imaginary. Prove that $\mathbf{y} = i\mathbf{x}$ is a real eigenvector of $\mathbf{A}$ corresponding to the same eigenvalue to which $\mathbf{x}$ corresponds.

5. Prove that if a generalized eigenvector of rank 2 exists for a real symmetric matrix, then it can be chosen to be real. (Note that Theorem 3 implies that such a vector does not exist.)

6. Let $\mathbf{x}_1, \mathbf{x}_2, \ldots, \mathbf{x}_r$ be eigenvectors of a matrix $\mathbf{A}$ corresponding to the same eigenvalue. Prove that $\mathbf{y} = c_1\mathbf{x}_1 + c_2\mathbf{x}_2 + \cdots + c_r\mathbf{x}_r$ (the c's are scalars) is also an eigenvector of $\mathbf{A}$ corresponding to λ providing that the c's are such that $\mathbf{y}$ is not the zero vector.
7. Let $\mathbf{A}$ be an $n \times n$ real symmetric matrix. Prove that every n-dimensional vector $\mathbf{x}$ can be written as $\mathbf{x} = c_1\mathbf{x}_1 + c_2\mathbf{x}_2 + \cdots + c_n\mathbf{x}_n$ where $\{\mathbf{x}_1, \mathbf{x}_2, \ldots, \mathbf{x}_n\}$ is a complete orthonormal set of eigenvectors for $\mathbf{A}$ and c_j $(j = 1, 2, \ldots, n)$ is a scalar. (Hint: use Theorem 5 of this Section and Theorem 2 of Section 4.8.)
8. Let $\mathbf{A}$, $\mathbf{x}$, and $\mathbf{x}_j$ $(j = 1, 2, \ldots, n)$ be the same as in Problem 7, and suppose that $\mathbf{x}_j$ corresponds to the eigenvalue λ_j.
   (a) Show that

$$\mathbf{x} = \sum_{i=1}^{n} c_i \mathbf{x}_i = \sum_{j=1}^{n} c_j \mathbf{x}_j.$$

   (b) Show that

$$\langle \mathbf{Ax}, \mathbf{x} \rangle = \sum_{i=1}^{n} \sum_{j=1}^{n} c_i \bar{c}_j \langle \mathbf{Ax}_i, \mathbf{x}_j \rangle$$

$$= \sum_{i=1}^{n} c_i \bar{c}_i \lambda_i$$

   (Hint: use Problem 9 of Section 9.2.)

9.6 ORTHOGONAL MATRICES

Definition 1 A real matrix $\mathbf{P}$ is *orthogonal* if $\mathbf{P}^{-1} = \mathbf{P}'$ (that is, if the inverse of $\mathbf{P}$ equals the transpose of $\mathbf{P}$).

EXAMPLE 1 Verify that

$$\mathbf{P} = \begin{bmatrix} 1/\sqrt{3} & 1/\sqrt{3} & 1/\sqrt{3} \\ 1/\sqrt{2} & 0 & -1/\sqrt{2} \\ 1/\sqrt{6} & -2/\sqrt{6} & 1/\sqrt{6} \end{bmatrix}$$

is orthogonal.

Solution By direct calculation, we find that $\mathbf{P'P} = \mathbf{PP'} = \mathbf{I}$, hence, it follows from the definition of the inverse that $\mathbf{P}' = \mathbf{P}^{-1}$. Thus, $\mathbf{P}$ is orthogonal.

Theorem 1 The determinant of an orthogonal matrix is ± 1.

Proof Let $\mathbf{P}$ be orthogonal, thus $\mathbf{P}^{-1} = \mathbf{P}'$ or $\mathbf{PP}' = \mathbf{I}$. Taking determinants of both sides of this equation and using Property 8 of Section 2.3, we obtain det(**P**) det(**P**′) = det(**I**). But det(**P**′) = det(**P**) (Property 7 of Section 2.3) and det(**I**) = 1, hence, it follows that $[\det(\mathbf{P})]^2 = 1$, which in turn implies the desired result.

Theorem 2 A real matrix is orthogonal if and only if its columns (and rows), considered as vectors, form an orthonormal set.

Proof We must first prove that a matrix is orthogonal if its columns form an orthonormal set and then prove the converse, that the columns of a matrix form an orthonormal set if the matrix is orthogonal. We will prove the first part here and leave the converse as an exercise for the student.

Assume that the columns of a matrix **P** form an orthonormal set. Then **P** may be represented by $\mathbf{P} = [\mathbf{x}_1 \ \mathbf{x}_2 \ \dots \ \mathbf{x}_n]$ where $\mathbf{x}_j$ is the jth column of **P** having the property that $\langle \mathbf{x}_i, \mathbf{x}_j \rangle = \delta_{ij}$ (see (2)). It now follows from the definition of the inner product (note that $\mathbf{x}_j$ is real, hence $\langle \mathbf{x}_i, \mathbf{x}_j \rangle = \langle \mathbf{x}_j, \mathbf{x}_i \rangle$) that **P**′**P** can be written as

$$\mathbf{P}'\mathbf{P} = \mathbf{P}\mathbf{P}' = \begin{bmatrix} \langle \mathbf{x}_1, \mathbf{x}_1 \rangle & \langle \mathbf{x}_1, \mathbf{x}_2 \rangle & \cdots & \langle \mathbf{x}_1, \mathbf{x}_n \rangle \\ \langle \mathbf{x}_2, \mathbf{x}_1 \rangle & \langle \mathbf{x}_2, \mathbf{x}_2 \rangle & \cdots & \langle \mathbf{x}_2, \mathbf{x}_n \rangle \\ \vdots & \vdots & & \vdots \\ \langle \mathbf{x}_n, \mathbf{x}_1 \rangle & \langle \mathbf{x}_n, \mathbf{x}_2 \rangle & \cdots & \langle \mathbf{x}_n, \mathbf{x}_n \rangle \end{bmatrix}.$$

Since $\langle \mathbf{x}_i, \mathbf{x}_j \rangle = \delta_{ij}$, we obtain $\mathbf{P}'\mathbf{P} = \mathbf{P}\mathbf{P}' = \mathbf{I}$ from which it follows that $\mathbf{P}' = \mathbf{P}^{-1}$. Thus, **P** is orthogonal.

If we now apply Theorem 2 to the results of the previous section (in particular to Theorems 3 and 5) then we obtain the following important conclusion.

Theorem 3 For every $n \times n$ real symmetric matrix **A** there exists an $n \times n$ real orthogonal matrix **P** such that $\mathbf{P}^{-1}\mathbf{A}\mathbf{P} = \mathbf{D}$, or, equivalently, such that $\mathbf{P}'\mathbf{A}\mathbf{P} = \mathbf{D}$, where **D** is a diagonal matrix.

EXAMPLE 2 Verify Theorem 3 for the matrix given in Example 1 of the previous section.

Solution In that example, we found a complete orthonormal set of a eigenvectors given by $\mathbf{e}_1, \mathbf{e}_2, \mathbf{e}_3, \mathbf{e}_4, \mathbf{e}_5$. Define

$$\mathbf{P} = [\mathbf{e}_1 \ \mathbf{e}_2 \ \mathbf{e}_3 \ \mathbf{e}_4 \ \mathbf{e}_5] = \begin{bmatrix} 1/\sqrt{3} & 0 & 0 & 1/\sqrt{3} & -1/\sqrt{3} \\ 1/\sqrt{3} & -1/\sqrt{3} & 0 & 0 & 1/\sqrt{3} \\ -1/\sqrt{3} & -1/\sqrt{3} & 0 & 1/\sqrt{3} & 0 \\ 0 & 1/\sqrt{3} & 0 & 1/\sqrt{3} & 1/\sqrt{3} \\ 0 & 0 & 1 & 0 & 0 \end{bmatrix}.$$

Since the columns of $\mathbf{P}$ form an orthonormal set, it follows from Theorem 2 that $\mathbf{P}$ is orthogonal. By direct calculation, we find that

$$\mathbf{P}'\mathbf{A}\mathbf{P} = \begin{bmatrix} 9 & 0 & 0 & 0 & 0 \\ 0 & 9 & 0 & 0 & 0 \\ 0 & 0 & 3 & 0 & 0 \\ 0 & 0 & 0 & 3 & 0 \\ 0 & 0 & 0 & 0 & 15 \end{bmatrix}$$

which verifies Theorem 3.

We conclude this section with one final note: orthogonal matrices leave inner products invariant. That is,

$$\langle \mathbf{Px}, \mathbf{Py} \rangle = \langle \mathbf{x}, \mathbf{P}^*\mathbf{Py} \rangle = \langle \mathbf{x}, \mathbf{P}'\mathbf{Py} \rangle = \langle \mathbf{x}, \mathbf{P}^{-1}\mathbf{Py} \rangle = \langle \mathbf{x}, \mathbf{y} \rangle$$

or

$$\langle \mathbf{Px}, \mathbf{Py} \rangle = \langle \mathbf{x}, \mathbf{y} \rangle \tag{9}$$

Geometrically, (9) implies that orthogonal transformations correspond to a rotation of the coordinate axes around the origin.

EXAMPLE 3 Verify (9) for the matrix

$$\mathbf{P} = \begin{bmatrix} 1/\sqrt{2} & 1/\sqrt{2} \\ -1/\sqrt{2} & 1/\sqrt{2} \end{bmatrix}.$$

Solution Since the columns of $\mathbf{P}$ form an orthonormal set of vectors, $\mathbf{P}$ is orthogonal. Designate $\mathbf{x}$ and $\mathbf{y}$ by

$$\begin{bmatrix} x_1 \\ x_2 \end{bmatrix} \quad \text{and} \quad \begin{bmatrix} y_1 \\ y_2 \end{bmatrix}.$$

Note that $\mathbf{x}$ and $\mathbf{y}$ can both be complex. Then, $\langle \mathbf{x}, \mathbf{y} \rangle = x_1\bar{y}_1 + x_2\bar{y}_2$,

$$\mathbf{Px} = \begin{bmatrix} (x_1/\sqrt{2} + x_2/\sqrt{2}) \\ (-x_1/\sqrt{2} + x_2/\sqrt{2}) \end{bmatrix}, \quad \text{and} \quad \mathbf{Py} = \begin{bmatrix} (y_1/\sqrt{2} + y_2/\sqrt{2}) \\ (-y_1/\sqrt{2} + y_2/\sqrt{2}) \end{bmatrix}.$$

Thus,

$$\begin{aligned} \langle \mathbf{Px}, \mathbf{Py} \rangle &= (x_1/\sqrt{2} + x_2/\sqrt{2})\overline{(y_1/\sqrt{2} + y_2/\sqrt{2})} \\ &\quad + (-x_1/\sqrt{2} + x_2/\sqrt{2})\overline{(-y_1/\sqrt{2} + y_2/\sqrt{2})} \\ &= x_1\bar{y}_1 + x_2\bar{y}_2 = \langle \mathbf{x}, \mathbf{y} \rangle. \end{aligned}$$

PROBLEMS 9.6

1. Determine which of the following matrices are orthogonal:

(a) $\begin{bmatrix} 1/\sqrt{2} & -1/\sqrt{2} \\ 1/\sqrt{2} & 1/\sqrt{2} \end{bmatrix}$ (b) $\begin{bmatrix} 3/\sqrt{8} & i/\sqrt{8} \\ i/\sqrt{8} & -3/\sqrt{8} \end{bmatrix}$

(c) $\begin{bmatrix} \cos\theta & -\sin\theta \\ \sin\theta & \cos\theta \end{bmatrix}$ (d) $\begin{bmatrix} 1/2 & -1/2 & -1/2 & 1/2 \\ 1/2 & -1/2 & 1/2 & -1/2 \\ 1/\sqrt{2} & 1/\sqrt{2} & 0 & 0 \\ 0 & 0 & 1/\sqrt{2} & 1/\sqrt{2} \end{bmatrix}.$

2. Verify Theorem 3 for

$$\mathbf{A} = \begin{bmatrix} 2 & -2 & 1 \\ -2 & -1 & 2 \\ 1 & 2 & 2 \end{bmatrix}.$$

3. Let **A** be a real matrix and **P** be a real orthogonal matrix. Prove that $\mathbf{P}^{-1}\mathbf{AP}$ is symmetric if and only if **A** is symmetric.
4. Prove that a real matrix **P** is orthogonal if and only if its inverse equals its adjoint.
5. Show by example that it is possible for a real matrix to be both symmetric and orthogonal.

9.7 HERMITIAN MATRICES

We now return to the general case of complex matrices. From our previous work, it would seem advisable to first determine the adjoint of a complex matrix (from Example 2 of Section 9.4, we conclude that the adjoint of a complex matrix is not its transpose as was the case with real matrices), and then ascertain whether or not self-adjoint complex matrices have properties similar to their real counterparts. We begin our discussion by defining the complex conjugate transpose matrix.

Definition 1 If **A** is an $n \times p$ matrix then the *complex conjugate transpose* of **A** is a $p \times n$ matrix, denoted by $\overline{\mathbf{A}}'$, obtained by first taking the complex conjugate of each element of **A** and then transposing the result.

Thus, if

$$\mathbf{A} = \begin{bmatrix} 2+i & 3 & -i \\ 1 & 1-i & 2i \end{bmatrix}, \qquad \text{then} \qquad \overline{\mathbf{A}}' = \begin{bmatrix} 2-i & 1 \\ 3 & 1+i \\ i & -2i \end{bmatrix}.$$

Definition 2 A matrix is *Hermitian* if it is equal to its own complex conjugate transpose.

An example of a Hermitian matrix is

$$\mathbf{A} = \begin{bmatrix} 1 & 2-i & 4i \\ 2+i & 3 & -1-i \\ -4i & -1+i & 4 \end{bmatrix}.$$

It follows immediately from Definition 2 that Hermitian matrices must be square and that the main diagonal elements must be real (see Problem 2). Furthermore, we note that neither Definition 1 nor Definition 2 requires the matrix under consideration to be complex; that is, the concepts of the complex conjugate transpose and the Hermitian matrix are equally applicable to real matrices. We leave it as an exercise for the student, however, to show that for real matrices these concepts reduce to those of the transpose and real symmetric matrix respectively. Thus, we may think of the Hermitian matrix as a generalization of the real symmetric matrix and the complex conjugate transpose as the generalization of the transpose. This analogy is strengthened further by the following theorem.

Theorem 1 The adjoint $\mathbf{A}^*$ of a matrix $\mathbf{A}$ (real or complex) exists, is unique and equals the complex conjugate transpose of $\mathbf{A}$. That is, $\mathbf{A}^* = \bar{\mathbf{A}}'$.[29]

Combining this theorem with Definition 2, we have a proof of

Theorem 2 A matrix (real or complex) is self-adjoint if and only if it is Hermitian.

EXAMPLE 1 Verify that

$$\mathbf{A} = \begin{bmatrix} 1 & -2-i \\ -2+i & 3 \end{bmatrix}$$

is self-adjoint.

Solution Designate the arbitrary 2-dimensional vectors $\mathbf{x}$ and $\mathbf{y}$ by

$$\begin{bmatrix} x_1 \\ x_2 \end{bmatrix} \quad \text{and} \quad \begin{bmatrix} y_1 \\ y_2 \end{bmatrix}.$$

Then by direct calculation we find that

$$\langle \mathbf{Ax}, \mathbf{y} \rangle = \langle \mathbf{x}, \mathbf{Ay} \rangle = x_1\bar{y}_1 + (-2-i)x_2\bar{y}_1 + (-2+i)x_1\bar{y}_2 + 3x_2\bar{y}_2.$$

[29] For a proof of Theorem 1, see F. R. Gantmacher, "The Theory of Matrices," p. 265. Chelsea, New York, 1960.

Thus (6) is satisfied, which implies **A** is self-adjoint.

Note that if **A** is real, Theorems 1 and 2 reduce to Theorems 2 and 3 of Section 9.4. Furthermore, by directing our attention to the theorems in Section 9.5, we note that the proofs of all those theorems, with the exception of Theorem 2, did not depend on the fact that the matrix involved was real symmetric, but rather on the fact that the matrix was self-adjoint. Hence, those theorems remain equally valid for self-adjoint complex matrices or, equivalently, Hermitian matrices. We incorporate those results, as they pertain to Hermitian matrices, into one master theorem.

Theorem 3 Let **A** be a Hermitian matrix. The eigenvalues of **A** are real, **A** is diagonalizable, eigenvectors corresponding to distinct eigenvalues are orthogonal, and **A** possesses a complete orthonormal set of eigenvectors.

EXAMPLE 2 Find a complete orthonormal set of eigenvectors for

$$\mathbf{A} = \begin{bmatrix} 1 & -2i & 0 & 2i \\ 2i & -1 & -2 & 0 \\ 0 & -2 & 1 & -2 \\ -2i & 0 & -2 & -1 \end{bmatrix}.$$

Solution The eigenvalues of this Hermitian matrix are $\lambda_1 = \lambda_2 = 3$, $\lambda_3 = \lambda_4 = -3$. Two linearly independent eigenvectors corresponding to $\lambda = 3$ are found to be

$$\mathbf{x}_1 = \begin{bmatrix} i \\ -1 \\ 1 \\ 0 \end{bmatrix} \quad \text{and} \quad \mathbf{x}_2 = \begin{bmatrix} 2i \\ -1 \\ 0 \\ 1 \end{bmatrix}.$$

Using the Gram–Schmidt orthonormalization process on these vectors we obtain

$$\mathbf{e}_1 = \begin{bmatrix} i/\sqrt{3} \\ -1/\sqrt{3} \\ 1/\sqrt{3} \\ 0 \end{bmatrix} \quad \text{and} \quad \mathbf{e}_2 = \begin{bmatrix} i/\sqrt{3} \\ 0 \\ -1/\sqrt{3} \\ 1/\sqrt{3} \end{bmatrix}.$$

Two linearly independent eigenvectors corresponding to $\lambda = -3$ are found to be

$$\mathbf{x}_3 = \begin{bmatrix} -i \\ -1 \\ 0 \\ 1 \end{bmatrix} \quad \text{and} \quad \mathbf{x}_4 = \begin{bmatrix} i \\ 2 \\ 1 \\ 0 \end{bmatrix}.$$

Using the Gram–Schmidt orthonormalization process on these two vectors, we obtain

$$\mathbf{e}_3 = \begin{bmatrix} -i/\sqrt{3} \\ -1/\sqrt{3} \\ 0 \\ 1/\sqrt{3} \end{bmatrix} \quad \text{and} \quad \mathbf{e}_4 = \begin{bmatrix} 0 \\ 1/\sqrt{3} \\ 1/\sqrt{3} \\ 1/\sqrt{3} \end{bmatrix}.$$

The set $\{\mathbf{e}_1, \mathbf{e}_2, \mathbf{e}_3, \mathbf{e}_4\}$ is the required set.

Theorem 4 A matrix **A** is Hermitian if and only if $\langle \mathbf{Ax}, \mathbf{x} \rangle$ is real for all (real and complex) vectors **x**.

Proof If **A** is Hermitian, hence self-adjoint, it follows (see Property (I2) of Section 9.2) that $\langle \mathbf{Ax}, \mathbf{x} \rangle = \langle \mathbf{x}, \mathbf{A}^*\mathbf{x} \rangle = \langle \mathbf{x}, \mathbf{Ax} \rangle = \overline{\langle \mathbf{Ax}, \mathbf{x} \rangle}$. Hence, from Property (C2) of Section 9.1, we conclude that $\langle \mathbf{Ax}, \mathbf{x} \rangle$ is real. (Recall that the inner product of two vectors is itself a number.) We leave the converse, **A** is Hermitian if $\langle \mathbf{Ax}, \mathbf{x} \rangle$ is real, as an exercise for the student (See Problem 5).

In conclusion, we note that since a real symmetric matrix is just a special case of a Hermitian matrix, Theorem 4 implies the following:

Corollary 1 A real matrix is symmetric if and only if $\langle \mathbf{Ax}, \mathbf{x} \rangle$ is real for all (real and complex) vectors **x**.

EXAMPLE 3 Verify Theorem 4 for the Hermitian matrix

$$\mathbf{A} = \begin{bmatrix} 2 & 1-i \\ 1+i & -1 \end{bmatrix}.$$

Solution Designate **x** by

$$\begin{bmatrix} x_1 \\ x_2 \end{bmatrix}.$$

Then,

$$\mathbf{Ax} = \begin{bmatrix} 2x_1 + (1-i)x_2 \\ (1+i)x_1 - x_2 \end{bmatrix},$$

and

$$\begin{aligned} \langle \mathbf{Ax}, \mathbf{x} \rangle &= [2x_1 + (1+ -i)x_2]\bar{x}_1 + [(1+i)x_1 - x_2]\bar{x}_2 \\ &= 2x_1\bar{x}_1 + (1-i)\bar{x}_1x_2 + (1+i)x_1\bar{x}_2 - x_2\bar{x}_2 \\ &= 2x_1\bar{x}_1 + [(1-i)\bar{x}_1x_2 + \overline{(1-i)\bar{x}_1x_2}] - x_2\bar{x}_2. \end{aligned}$$

Since the quantity inside the brackets is of the form $a + \bar{a}$ and all other terms are of the form $a\bar{a}$, we have, from Properties (C1) and (C4) of Section 9.1, that $\langle \mathbf{Ax}, \mathbf{x} \rangle$ is the sum of real numbers, hence, is itself real.

PROBLEMS 9.7

1. Determine which of the following matrices are Hermitian and find a complete orthonormal set of eigenvectors for those matrices:

(a) $\begin{bmatrix} 1 & i \\ -i & 1 \end{bmatrix}$ (b) $\begin{bmatrix} 1 & i \\ i & 1 \end{bmatrix}$ (c) $\begin{bmatrix} 1 & 2i & -i \\ -2i & i & 1 \\ i & 1 & 2 \end{bmatrix}$

(d) $\begin{bmatrix} 4 & 0 & 3i \\ 0 & 5 & 0 \\ -3i & 0 & -4 \end{bmatrix}$ (e) $\begin{bmatrix} 1 & i & 0 & -i \\ -i & 2 & 1 & 0 \\ 0 & 1 & 1 & 1 \\ i & 0 & 1 & 2 \end{bmatrix}$

(f) $\begin{bmatrix} 1 & i & 0 & -i \\ -i & 1 & 0 & -i \\ 0 & 0 & 0 & 0 \\ i & -1 & 0 & 1 \end{bmatrix}.$

2. Prove that the main diagonal elements of a Hermitian matrix must be real.

3. Prove that $\mathbf{A}^* = \overline{\mathbf{A}}'$ for a general 2×2 complex matrix.

4. Prove the equality

$$\langle \mathbf{Ax}, \mathbf{y} \rangle = \tfrac{1}{4}[\langle \mathbf{A}(\mathbf{x}+\mathbf{y}), (\mathbf{x}+\mathbf{y}) \rangle - \langle \mathbf{A}(\mathbf{x}-\mathbf{y}), (\mathbf{x}-\mathbf{y}) \rangle]$$
$$+ \frac{i}{4}[\langle \mathbf{A}(\mathbf{x}+i\mathbf{y}), (\mathbf{x}+i\mathbf{y}) \rangle - \langle \mathbf{A}(\mathbf{x}-i\mathbf{y}), (\mathbf{x}-i\mathbf{y}) \rangle].$$

5. Use Problem 4 to prove that $\mathbf{A}$ is Hermitian if $\langle \mathbf{Ax}, \mathbf{x} \rangle$ is real for all vectors $\mathbf{x}$.

9.8 UNITARY MATRICES

Definition 1 A matrix $\mathbf{U}$ is *unitary* if $\mathbf{U}^{-1} = \overline{\mathbf{U}}'$ (that is, if the inverse of $\mathbf{U}$ equals the complex conjugate transpose of $\mathbf{U}$).

EXAMPLE 1 Verify that

$$\mathbf{U} = \begin{bmatrix} i/2 & -i/\sqrt{3} & 5/\sqrt{60} \\ 1/2 & -i/\sqrt{3} & (-4+3i)/\sqrt{60} \\ (1-i)/2 & 1/\sqrt{3} & (3-i)/\sqrt{60} \end{bmatrix}$$

is unitary.

Solution By direct calculation, we find that $\overline{\mathbf{U}}'\mathbf{U} = \mathbf{U}\overline{\mathbf{U}}' = \mathbf{I}$, hence, it follows from the definition of the inverse that $\overline{\mathbf{U}}' = \mathbf{U}^{-1}$. Thus, $\mathbf{U}$ is unitary.

We note that if $\mathbf{U}$ is a real matrix, then $\overline{\mathbf{U}}' = \mathbf{U}'$, and the concept of a unitary matrix reduces to that of an orthogonal matrix. Thus, we can think of a unitary matrix as a generalization of an orthogonal matrix. Furthermore, many of the properties of orthogonal matrices discussed in Section 9.6 remain valid in this more general case. As such, we just state the results here, and refer the student to their appropriate counterparts in Section 9.6 for the proofs.

Theorem 1 A matrix is unitary if and only if its columns (and rows), considered as vectors, form an orthonormal set.

Theorem 2 For every $n \times n$ Hermitian matrix $\mathbf{A}$, there exists an $n \times n$ unitary matrix $\mathbf{U}$ such that $\mathbf{U}^{-1}\mathbf{AU} = \mathbf{D}$, or equivalently, such that $\overline{\mathbf{U}}'\mathbf{AU} = \mathbf{D}$, where $\mathbf{D}$ is a real diagonal matrix.

EXAMPLE 2 Verify Theorem 2 for the matrix given in Example 2 of the previous section.

Solution In that example, we found a complete orthonormal set of eigenvectors to be $\{\mathbf{e}_1, \mathbf{e}_2, \mathbf{e}_3, \mathbf{e}_4\}$. Define $\mathbf{U} = [\mathbf{e}_1\ \mathbf{e}_2\ \mathbf{e}_3\ \mathbf{e}_4]$. Since the columns of $\mathbf{U}$ form an orthonormal set, it follows from Theorem 1 that $\mathbf{U}$ is unitary. By direct calculation, we find that

$$\overline{\mathbf{U}}'\mathbf{AU} = \begin{bmatrix} 3 & 0 & 0 & 0 \\ 0 & 3 & 0 & 0 \\ 0 & 0 & -3 & 0 \\ 0 & 0 & 0 & -3 \end{bmatrix}$$

which verifies Theorem 2.

Once again, as was the case with orthogonal matrices, we find that unitary matrices leave the inner product invariant. That is,

$$\langle \mathbf{Ux}, \mathbf{Uy} \rangle = \langle \mathbf{x}, \mathbf{y} \rangle. \tag{10}$$

EXAMPLE 3 Verify (10) for the matrix

$$\mathbf{U} = \begin{bmatrix} -1/\sqrt{2} & i/\sqrt{2} \\ -i/\sqrt{2} & 1/\sqrt{2} \end{bmatrix}.$$

Solution Since the columns of $\mathbf{U}$ form an orthonormal set of vectors, $\mathbf{U}$ is unitary. Designate $\mathbf{x}$ and $\mathbf{y}$ by

$$\begin{bmatrix} x_1 \\ x_2 \end{bmatrix} \quad \text{and} \quad \begin{bmatrix} y_1 \\ y_2 \end{bmatrix}.$$

Then,

$$\langle \mathbf{x}, \mathbf{y} \rangle = x_1\bar{y}_1 + x_2\bar{y}_2,$$

$$\mathbf{Ux} = \begin{bmatrix} ((-1/\sqrt{2})x_1 + (i/\sqrt{2})x_2) \\ ((-i/\sqrt{2})x_1 + (1/\sqrt{2})x_2) \end{bmatrix} \quad \text{and} \quad \mathbf{Uy} = \begin{bmatrix} ((-1/\sqrt{2})y_1 + (i/\sqrt{2})y_2) \\ ((-i/\sqrt{2})y_1 + (1/\sqrt{2})y_2) \end{bmatrix}.$$

Thus,

$$\begin{aligned} \langle \mathbf{Ux}, \mathbf{Uy} \rangle &= ((-1/\sqrt{2})x_1 + (i/\sqrt{2})x_2)\overline{((-1/\sqrt{2})y_1 + (i/\sqrt{2})y_2)} \\ &\quad + ((-i/\sqrt{2})x_1 + (1/\sqrt{2})x_2)\overline{((-i/\sqrt{2})y_1 + (1/\sqrt{2})y_2)} \\ &= ((-1/\sqrt{2})x_1 + (i/\sqrt{2})x_2)((-1/\sqrt{2})\bar{y}_1 - (i/\sqrt{2})\bar{y}_2) \\ &\quad + ((-i/\sqrt{2})x_1 + (1/\sqrt{2})x_2)((i/\sqrt{2})\bar{y}_1 + (1/\sqrt{2})\bar{y}_2) \\ &= x_1\bar{y}_1 + x_2\bar{y}_2 = \langle \mathbf{x}, \mathbf{y} \rangle. \end{aligned}$$

PROBLEMS 9.8

1. Determine which of the following matrices are unitary:

(a) $\begin{bmatrix} 1/\sqrt{2} & -i/\sqrt{2} \\ i/\sqrt{2} & -1/\sqrt{2} \end{bmatrix}$ (b) $\begin{bmatrix} 1/\sqrt{2} & -i/\sqrt{2} \\ i/\sqrt{2} & 1/\sqrt{2} \end{bmatrix}$

(c) $\begin{bmatrix} 1/\sqrt{2} & -1/\sqrt{2} \\ 1/\sqrt{2} & 1/\sqrt{2} \end{bmatrix}$ (d) $1/\sqrt{3}\begin{bmatrix} 1 & i & -i & 0 \\ 0 & 1 & 1 & 1 \\ i & 0 & -1 & 1 \\ -i & -1 & 0 & 1 \end{bmatrix}.$

2. Verify Theorem 2 for

$$\mathbf{A} = \begin{bmatrix} 1 & i & 0 & -i \\ -i & 1 & 0 & -1 \\ 0 & 0 & 0 & 0 \\ i & -1 & 0 & 1 \end{bmatrix}.$$

3. Prove that a matrix **U** is unitary if and only if its inverse equals its adjoint.

4. Let **U** be a unitary matrix. Prove that $\mathbf{U}^{-1}\mathbf{AU}$ is Hermitian if and only if **A** is Hermitian.

5. Prove Eq. (10).

6. Show by example that it is possible for a matrix to be both Hermitian and unitary.

7. What is the relationship between the eigenvalues of $\mathbf{U}^{-1}\mathbf{AU}$ and those of **A**?

9.9 SUMMARY

We now summarize some of the more important results of the previous five sections. In what follows, the student will note that we generally differentiate between real symmetric matrices and Hermitian matrices, and between orthogonal matrices and unitary matrices. We do this for convenience only, because such a classification is actually superfluous. Recall that real symmetric matrices are special cases of Hermitian matrices (that is, those Hermitian matrices that are real), while orthogonal matrices are special cases of unitary matrices (for the same reason).

Given a matrix $\mathbf{A}$, its adjoint $\mathbf{A}^*$ is defined by the relation $\langle \mathbf{x}, \mathbf{Ay} \rangle = \langle \mathbf{A}^*\mathbf{x}, \mathbf{y} \rangle$ for all vectors $\mathbf{x}$ and $\mathbf{y}$. The adjoint of a real matrix is its transpose while the adjoint of a complex matrix is its complex conjugate transpose. A matrix is called self-adjoint if it equals its own adjoint. Thus, a real matrix is self-adjoint if and only if it is symmetric while a complex matrix is self-adjoint if and only if it is Hermitian.

Self-adjoint matrices (real and complex) have real eigenvalues and possess a complete orthonormal set of eigenvectors. Furthermore, self-adjoint matrices are diagonalizable. For a real matrix, the modal matrix can be chosen to be an orthogonal matrix while for a complex matrix, the modal matrix can be chosen to be a unitary matrix.

Both orthogonal and unitary matrices are defined by the property that their inverses must equal their adjoints. This property requires that the inverse of a real orthogonal matrix be equal to its transpose while the inverse of a unitary matrix be equal to its complex conjugate transpose. Both orthogonal and unitary matrices have the property that they leave the inner product invariant.

9.10 POSITIVE DEFINITE MATRICES

We know from Theorem 4 of Section 9.7 that if $\mathbf{A}$ is Hermitian then $\langle \mathbf{Ax}, \mathbf{x} \rangle$ must be real. If in particular the quantity $\langle \mathbf{Ax}, \mathbf{x} \rangle$ is nonnegative for all $\mathbf{x}$, then the Hermitian matrix $\mathbf{A}$ is called *nonnegative definite*. If the quantity $\langle \mathbf{Ax}, \mathbf{x} \rangle$ is always positive for all nonzero vectors $\mathbf{x}$ (note that if $\mathbf{x} = \mathbf{0}$, then $\langle \mathbf{Ax}, \mathbf{x} \rangle = 0$), then the Hermitian matrix is called positive definite.

Definition 1 An $n \times n$ Hermitian matrix is *positive definite*, designated by $\mathbf{A} > 0$, if $\langle \mathbf{Ax}, \mathbf{x} \rangle$ is positive for all nonzero n-dimensional (real or complex) vectors $\mathbf{x}$.

EXAMPLE 1 Verify that

$$\mathbf{A} = \begin{bmatrix} 2 & i \\ -i & 2 \end{bmatrix}$$

is positive definite.

Solution Designate $\mathbf{x}$ by

$$\begin{bmatrix} x_1 \\ x_2 \end{bmatrix}.$$

Then,

$$\begin{aligned}\langle \mathbf{Ax}, \mathbf{x} \rangle &= 2x_1\bar{x}_1 + i\bar{x}_1x_2 - ix_1\bar{x}_2 + 2x_2\bar{x}_2 \\ &= x_1\bar{x}_1 + (x_1 + ix_2)(\bar{x}_1 - i\bar{x}_2) + x_2\bar{x}_2 \\ &= x_1\bar{x}_1 + (x_1 + ix_2)\overline{(x_1 + ix_2)} + x_2\bar{x}_2.\end{aligned}$$

But each of these three terms is in the form of a complex number times its conjugate and therefore is nonnegative (see Property (C1) of Section 9.1). Thus, the sum of these terms is positive (unless $x_1 = x_2 = 0$) which implies that $\mathbf{A}$ is positive definite.

Theorem 1 A Hermitian matrix is positive definite if and only if its eigenvalues are positive.

Proof We will prove only that the eigenvalues of a positive definite matrix are positive and leave the converse, if the eigenvalues of a Hermitian matrix are positive then the matrix is positive definite, as an exercise for the student (see Problem 4). Let $\mathbf{A}$ be a positive definite Hermitian matrix and $\mathbf{x}$ an eigenvector of $\mathbf{A}$ corresponding to the eigenvalue λ. Then

$$\langle \mathbf{Ax}, \mathbf{x} \rangle = \langle \lambda\mathbf{x}, \mathbf{x} \rangle = \lambda\langle \mathbf{x}, \mathbf{x} \rangle$$

or

$$\lambda = \frac{\langle \mathbf{Ax}, \mathbf{x} \rangle}{\langle \mathbf{x}, \mathbf{x} \rangle}. \tag{11}$$

Since $\mathbf{x}$ is an eigenvector, it cannot be zero, hence $\langle \mathbf{x}, \mathbf{x} \rangle$ is positive. Furthermore, since $\mathbf{A}$ is positive definite and $\mathbf{x} \neq \mathbf{0}$, it follows that $\langle \mathbf{Ax}, \mathbf{x} \rangle$ is also positive. Combining these results with (11), we find that λ must be positive.

By similar reasoning, we can also prove that a Hermitian matrix is nonnegative definite if and only if its eigenvalues are nonnegative.

EXAMPLE 2 Is

$$\mathbf{A} = \begin{bmatrix} 4 & 1 \\ 1 & 1 \end{bmatrix}$$

positive definite?

Solution Note that **A** is Hermitian (since **A** is real it is also symmetric). The eigenvalues of **A** are

$$\lambda_1 = \frac{5 + \sqrt{13}}{2} \quad \text{and} \quad \lambda_2 = \frac{5 - \sqrt{13}}{2}$$

which are both positive. Thus, by Theorem 1, **A** is positive definite.

EXAMPLE 3 Is the matrix

$$\mathbf{A} = \begin{bmatrix} 6 & 2i \\ 6i & -1 \end{bmatrix}$$

positive definite?

Solution Although the eigenvalues of **A** are both positive $\lambda_1 = 2$ and $\lambda_2 = 3$, the matrix is not Hermitian and hence cannot be positive definite. In particular, if we choose

$$\mathbf{x} = \begin{bmatrix} 1 \\ 1 \end{bmatrix},$$

we find that $\langle \mathbf{Ax}, \mathbf{x} \rangle = 5 + 8i$, which is not even real.

PROBLEMS 9.10

1. Determine which of the following matrices are positive definite. For those matrices that are not positive definite, produce a vector which will verify that conclusion.

(a) $\begin{bmatrix} 5 & 3i \\ 3i & -5 \end{bmatrix}$ (b) $\begin{bmatrix} 3 & 2i \\ -2i & 2 \end{bmatrix}$ (c) $\begin{bmatrix} 1 & 0 \\ 2i & 0 \end{bmatrix}$

(d) $\begin{bmatrix} 5 & 2 \\ 2 & 1 \end{bmatrix}$ (e) $\begin{bmatrix} 1 & 2i & i \\ 0 & 2 & -i \\ 0 & i & 3 \end{bmatrix}$ (f) $\begin{bmatrix} 1 & 0 & i \\ 0 & 2 & 0 \\ i & 0 & 5 \end{bmatrix}$

2. Prove that the elements on the main diagonal of a positive definite matrix must be positive.

3. A Hermitian matrix is *negative definite* if and only if $\langle \mathbf{Ax}, \mathbf{x} \rangle$ is negative for all nonzero vectors $\mathbf{x}$. Prove that $\mathbf{A}$ is negative definite if and only if $-\mathbf{A}$ is positive definite.

4. Using Problem 8 of Section 9.5, complete the proof of Theorem 1.

5. Prove that a Hermitian matrix is negative definite if and only if its eigenvalues are negative.

6. Let $\mathbf{A}$ be a positive definite matrix. Define $\langle \mathbf{x}, \mathbf{y} \rangle_1 = \langle \mathbf{Ax}, \mathbf{y} \rangle$, where $\langle \mathbf{Ax}, \mathbf{y} \rangle$ designates our old inner product. Show that $\langle \mathbf{x}, \mathbf{y} \rangle_1$ satisfies Properties (I1)–(I4) of Section 9.2, thereby also defining an inner product. Thus, we see that positive definite matrices can be used to generate different inner products.

ANSWERS AND HINTS TO SELECTED PROBLEMS

Chapter 1

SECTION 1.2

1. $\begin{bmatrix} 13 & -1 \\ -12 & 18 \end{bmatrix}$

2. $\begin{bmatrix} 3 & 1 & 1 \\ -12 & -3 & 0 \\ -9 & -1 & -35/3 \end{bmatrix}$

3. $\begin{bmatrix} -\theta^3 + 6\theta^2 + \theta & 6\theta - 6 \\ 21 & -\theta^4 - 2\theta^2 - \theta + 6/\theta \end{bmatrix}$

4. The addition is not defined since the matrices are not of the same order.

SECTION 1.4

1. $\mathbf{AB} = \begin{bmatrix} 8 & 0 \\ 5 & 3 \end{bmatrix}, \quad \mathbf{BA} = \begin{bmatrix} 8 & 10 & 0 \\ 2 & -2 & -3 \\ 2 & 10 & 5 \end{bmatrix}$

2. $\mathbf{AB} = \begin{bmatrix} 13 & 15 & 6 \\ -12 & -13 & -18 \\ 5 & 15 & 7 \end{bmatrix}, \quad \mathbf{BA} = \begin{bmatrix} 0 & 33 & 20 \\ -5 & 4 & -1 \\ -4 & -6 & 3 \end{bmatrix}$

3. $\mathbf{AB} = \begin{bmatrix} 6 & 3 & 5 \\ 3 & 1 & 2 \end{bmatrix}$; **BA** is not defined.

8. (a) 3×3 (b) 3×3 (c) 2×2 (d) not defined.

9. $\begin{bmatrix} 1 & 1 & 1 \\ 2 & 1 & 3 \\ 1 & 1 & 0 \end{bmatrix} \begin{bmatrix} x \\ y \\ z \end{bmatrix} = \begin{bmatrix} 2 \\ 4 \\ 0 \end{bmatrix}$

10. $\begin{bmatrix} 5 & 3 & 2 & 4 \\ 1 & 1 & 0 & 1 \\ 3 & 2 & 2 & 0 \\ 1 & 1 & 2 & 3 \end{bmatrix} \begin{bmatrix} x \\ y \\ z \\ w \end{bmatrix} = \begin{bmatrix} 5 \\ 0 \\ -3 \\ 4 \end{bmatrix}$

SECTION 1.5

3. Consider $(\mathbf{AA}')'$ and use Properties (5) and (1).

6. (a) The (1, 1) element of **B** multiplies each element in the first column of **A**, the (2, 2) element of **B** multiplies each element in the second column of **A**, etc.

(b) The (1, 1) element of **B** multiplies each element in the first row of **A**, the (2,2) element of **B** multiplies each element in the second row of **A**, etc.

7. $$\mathbf{A}^3 = \begin{bmatrix} 40 & 9 & 3 \\ 27 & 4 & 3 \\ 45 & 9 & 4 \end{bmatrix}$$

8. $$\mathbf{A}^3 = \begin{bmatrix} 1 & 0 & 0 \\ 0 & 8 & 0 \\ 0 & 0 & 27 \end{bmatrix}$$

SECTION 1.6

1. (a), (b), and (d) are submatrices.

2. $$\left[\begin{array}{ccc|c} 4 & 5 & -1 & 9 \\ \hline 15 & 10 & 4 & 22 \\ 1 & 1 & 5 & 9 \end{array}\right]$$

3. Partition **A** and **B** into four 2×2 submatrices each. Then,

$$\mathbf{AB} = \left[\begin{array}{cc|cc} 11 & 9 & 0 & 0 \\ 4 & 6 & 0 & 0 \\ \hline 0 & 0 & 2 & 1 \\ 0 & 0 & 4 & -1 \end{array}\right]$$

SECTION 1.7

1. $p = 1$

2. $$\begin{bmatrix} -4/3 \\ -1 \\ -8/3 \\ 1/3 \end{bmatrix}$$

3. (a) $\sqrt{15}$

(b) $\sqrt{39}$

4. (a) $[3/\sqrt{10} \quad 0 \quad 1/\sqrt{10}]$

(b) $[6/\sqrt{95} \quad -7/\sqrt{95} \quad 1/\sqrt{95} \quad 0 \quad -3/\sqrt{95}]$

Chapter 2

SECTION 2.1

1. 1 **2.** 5

3. $|\mathbf{A}| = -5$, $|\mathbf{B}| = 10$, $|\mathbf{AB}| = -50$. Thus, $|\mathbf{AB}| = |\mathbf{A}|\,|\mathbf{B}|$.

SECTION 2.2

1. 22 **2.** -7 **3.** -11 **4.** 0

SECTION 2.3

2. For an upper triangular matrix, expand by the first column at each step.

3. Use the third column to simplify both the first and second columns.

6. Factor the numbers -1, 2, 2, and 3 from the third row, second row, first column and second column respectively.

7. Factor a five from the third row. Then use this new third row to simplify the second row and the new second row to simplify the first row.

SECTION 2.4

1. -311 **2.** 0 **3.** 152 **4.** 2187

SECTION 2.5

1. $x = 1, y = -4, z = 5.$ **2.** $x = y = z = 0.$

3. $x = 1, y = 2, z = 5, w = -3.$

Chapter 3

SECTION 3.1

1. $\begin{bmatrix} 4 & -1 \\ -3 & 1 \end{bmatrix}$ **2.** $\begin{bmatrix} 4 & -6 \\ -6 & 12 \end{bmatrix}$ **3.** $\begin{bmatrix} 9 & -5 & -2 \\ 5 & -3 & -1 \\ -36 & 21 & 8 \end{bmatrix}$

4. $1/12\begin{bmatrix} 3 & -1 & -8 \\ 0 & 4 & 2 \\ 0 & 0 & 6 \end{bmatrix}$

5. $\begin{bmatrix} 1 & 0 & 0 & 0 \\ 2 & -1 & 0 & 0 \\ -8 & 3 & 1/2 & 0 \\ -25 & 10 & 2 & -1 \end{bmatrix}$

6. $1/17\begin{bmatrix} 1 & 7 & -2 \\ 7 & -2 & 3 \\ -2 & 3 & 4 \end{bmatrix}$

7. $\begin{bmatrix} 1/\lambda_1 & 0 & 0 & 0 \\ 0 & 1/\lambda_2 & 0 & 0 \\ 0 & 0 & 1/\lambda_3 & 0 \\ 0 & 0 & 0 & 1/\lambda_4 \end{bmatrix}$

SECTION 3.2

1. $\mathbf{x} = 2, \quad y = -1$

2. $l = 1 \quad p = 3$

3. $x = y = z = 1$

4. $l = 1 \quad m = -2, \quad n = 0$

SECTION 3.3

8. $1/125\begin{bmatrix} -11 & -2 \\ 2 & -11 \end{bmatrix}$

9. First show that $(\mathbf{BA}^{-1})' = \mathbf{A}^{-1}\mathbf{B}'$ and that $(\mathbf{A}^{-1}\mathbf{B}')^{-1} = (\mathbf{B}')^{-1}\mathbf{A}$.

SECTION 3.4

1. $1/9\begin{bmatrix} 1 & 1 \\ 2 & -7 \end{bmatrix}$

2. $1/3\begin{bmatrix} -2 & 2 & -1 \\ -4 & 1 & 1 \\ 5 & -2 & 1 \end{bmatrix}$

3. $\begin{bmatrix} -1 & 1 & 2 \\ 0 & -1 & 1 \\ 1 & 0 & -2 \end{bmatrix}$

4. $1/3\begin{bmatrix} 1 & 1 & 1 & -5 \\ -1 & 2 & 5 & -4 \\ 1 & -2 & -8 & 7 \\ 1 & -2 & -5 & 7 \end{bmatrix}$

Chapter 4

SECTION 4.1

1. (a) No (b) Yes **2.** (a) Yes (b) No (c) Yes

SECTION 4.3

1. $x = 1, y = 1, z = 2$

2. $x = -6z, y = 7z, z$ arbitrary

3. $x = y = 1$

4. $r = t + 13/7$, $s = 2t + 15/7$, t arbitrary

5. $l = (1/5)(-n + 1)$, $m = (1/5)(3n - 5p - 3)$, n, p arbitrary

SECTION 4.4

1. Dependent; $3(1\ 3) + 2(2\ -1) - 7(1\ 1) = (0\ 0)$

2. Independent

3. Dependent;

$$2\begin{bmatrix}2\\1\\1\\3\end{bmatrix} + 1\begin{bmatrix}4\\-1\\2\\-1\end{bmatrix} - 1\begin{bmatrix}8\\1\\4\\5\end{bmatrix} = \begin{bmatrix}0\\0\\0\\0\end{bmatrix}$$

4. Dependent; $2(2\ 1\ 1) - 1(3\ -1\ 4) - 1(1\ 3\ -2) = (0\ 0\ 0)$.

5. Yes;

$$\begin{bmatrix}2\\1\\2\end{bmatrix} = -2\begin{bmatrix}1\\1\\0\end{bmatrix} + 1\begin{bmatrix}1\\0\\-1\end{bmatrix} + 3\begin{bmatrix}1\\1\\1\end{bmatrix}$$

6. No.

SECTION 4.5

1. 2 **2.** 2 **3.** 1 **4.** 2

5. 3 **6.** 0 **7.** Yes **8.** No

SECTION 4.6

1. Consistent with no arbitrary unknowns; $x = 2/3$, $y = 1/3$

2. Inconsistent

3. Consistent with one arbitrary unknown; $x = (1/2)(3 - 2z)$, $y = -1/2$

4. Consistent with two arbitrary unknowns; $x = (1/7)(11 - 5z - 2w)$, $y = (1/7)(1 - 3z + 3w)$

5. Consistent with no arbitrary unknowns; $x = y = 1$, $z = -1$

6. $\mathbf{A}$ is a submatrix of $\mathbf{A}^b$.

SECTION 4.7

5. $x = 5/2$, $y = -1/2 + z$; z arbitrary

6. $x = (1/4)(9 - 5z + 3w)$, $y = (1/4)(7 - 3z + w)$; z, w both arbitrary

SECTION 4.8

1. Dependent **2.** Independent **3.** All x

4. $x = -1$ **5.** No x

Chapter 5

SECTION 5.1

1. (a), (b), and (d) **2.** $-2, -1, 2$ respectively

SECTION 5.2

1. 1, 1, 0 **2.** $\pm 4i$ **3.** $\pm i$

4. $3t$, $9t$ **5.** 0, 2, 2 **6.** 2, 4, $1 \pm \sqrt{5}\, i$

SECTION 5.3

1. $\lambda_1 = 1, \quad \lambda_2 = -1; \qquad \mathbf{x}_1 = \begin{bmatrix} 1 \\ 1 \end{bmatrix}, \mathbf{x}_2 = \begin{bmatrix} 1 \\ 3 \end{bmatrix}$

2. $\lambda_1 = \sqrt{7}, \quad \lambda_2 = -\sqrt{7}; \qquad \mathbf{x}_1 = \begin{bmatrix} 1 \\ -2 + \sqrt{7} \end{bmatrix}, \quad \mathbf{x}_2 = \begin{bmatrix} 1 \\ -2 - \sqrt{7} \end{bmatrix}$

3. $\lambda_1 = 3t, \quad \lambda_2 = t, \qquad \mathbf{x}_1 = \begin{bmatrix} 1 \\ 1 \end{bmatrix}, \quad \mathbf{x}_2 = \begin{bmatrix} 1 \\ -1 \end{bmatrix}$

4. $\lambda_1 = 0, \quad \lambda_2 = 1, \quad \lambda_3 = 4; \qquad \mathbf{x}_1 = \begin{bmatrix} 0 \\ -2 \\ 1 \end{bmatrix}, \quad \mathbf{x}_2 = \begin{bmatrix} 1 \\ 0 \\ -1 \end{bmatrix}, \quad \mathbf{x}_3 = \begin{bmatrix} 2 \\ 0 \\ 1 \end{bmatrix}$

5. $\lambda_1 = 3, \quad \lambda_2 = 4, \quad \lambda_3 = 5; \qquad \mathbf{x}_1 = \begin{bmatrix} 1 \\ 0 \\ 0 \end{bmatrix}, \quad \mathbf{x}_2 = \begin{bmatrix} -1 \\ -1 \\ 1 \end{bmatrix}, \quad \mathbf{x}_3 = \begin{bmatrix} 1 \\ 0 \\ 2 \end{bmatrix}$

6. $\lambda_1 = 1, \lambda_2 = 2, \lambda_3 = 3, \lambda_4 = 4$

$$\mathbf{x}_1 = \begin{bmatrix} 10 \\ -6 \\ 11 \\ 4 \end{bmatrix}, \quad \mathbf{x}_2 = \begin{bmatrix} 1 \\ 0 \\ 0 \\ 0 \end{bmatrix}, \quad \mathbf{x}_3 = \begin{bmatrix} 2 \\ 0 \\ 1 \\ 0 \end{bmatrix}, \quad \mathbf{x}_4 = \begin{bmatrix} 2 \\ 0 \\ 1 \\ -1 \end{bmatrix}$$

SECTION 5.4

2. Consider

$$\mathbf{A} = \begin{bmatrix} 0 & 2 \\ 0 & 1 \end{bmatrix}$$

SECTION 5.5

1. $\begin{bmatrix} 1 \\ 0 \\ -1 \end{bmatrix}, \begin{bmatrix} 1 \\ -1 \\ 0 \end{bmatrix}, \begin{bmatrix} 1 \\ 0 \\ 1 \end{bmatrix}$

2. $\begin{bmatrix} 1 \\ 0 \\ -1 \end{bmatrix}, \begin{bmatrix} 1 \\ 0 \\ 1 \end{bmatrix}$

3. $\begin{bmatrix} 1 \\ 0 \\ 0 \\ 0 \end{bmatrix}, \begin{bmatrix} 0 \\ -1 \\ 1 \\ 0 \end{bmatrix}, \begin{bmatrix} -1 \\ -1 \\ 0 \\ 1 \end{bmatrix}$

Chapter 6

SECTION 6.1

1. (a) $\begin{bmatrix} 0 & -4 & 8 \\ 0 & 4 & -8 \\ 0 & 0 & 0 \end{bmatrix}, \quad \begin{bmatrix} 0 & 8 & -16 \\ 0 & -8 & 16 \\ 0 & 0 & 0 \end{bmatrix};$

(b) $\begin{bmatrix} 57 & 78 \\ 117 & 174 \end{bmatrix}, \quad \begin{bmatrix} 234 & 348 \\ 522 & 756 \end{bmatrix}$

2. $$p_k(\mathbf{A}) = \begin{bmatrix} p_k(\lambda_1) & 0 & 0 \\ 0 & p_k(\lambda_2) & 0 \\ 0 & 0 & p_k(\lambda_3) \end{bmatrix}.$$

5. $\cos \mathbf{A} = \sum_{k=0}^{\infty} \frac{(-1)^k}{(2k)!} x^{2k}$, $\begin{bmatrix} \cos 1 & 0 \\ 0 & \cos 2 \end{bmatrix}$

SECTION 6.2

1. $\begin{bmatrix} -2 & 1 \\ 3/2 & -1/2 \end{bmatrix}$

2. Since $a_0 = 0$, the inverse does not exist.

3. Since $a_0 = 0$, the inverse does not exist.

4. $\begin{bmatrix} -1/3 & -1/3 & 2/3 \\ -1/3 & 1/6 & 1/6 \\ 1/2 & 1/4 & -1/4 \end{bmatrix}$

5. $\begin{bmatrix} 1 & 0 & 0 & 0 \\ 0 & -1 & 0 & 0 \\ 0 & 0 & -1 & 0 \\ 0 & 0 & 0 & 1 \end{bmatrix}$

SECTION 6.3

1. $\begin{bmatrix} -2 & 3 \\ -1 & 2 \end{bmatrix}$

2. $\begin{bmatrix} 9 & -9 \\ 3 & -3 \end{bmatrix}$

3. $\begin{bmatrix} 0 & 1 \\ 0 & -1 \end{bmatrix}$

4. $\begin{bmatrix} 6 & -9 \\ 3 & -6 \end{bmatrix}$

5. $\begin{bmatrix} -(4^{78}) + 2(3^{78}) & -(4^{78}) + 3^{78} \\ 2(4^{78}) - 2(3^{78}) & 2(4^{78}) - (3^{78}) \end{bmatrix}$

6. $\begin{bmatrix} 2 & -4 & -3 \\ 0 & 0 & 0 \\ 1 & -5 & -2 \end{bmatrix}$

7. $\begin{bmatrix} 1 & 0 & (-4 + 4(2^{222}))/3 \\ 0 & 1 & (-2 + 2(2^{222}))/3 \\ 0 & 0 & 2^{222} \end{bmatrix}$

SECTION 6.4

2. $\begin{bmatrix} 4 & 1 & -3 \\ 0 & -1 & 0 \\ 5 & 1 & -4 \end{bmatrix}$

3. $\begin{bmatrix} 0 & 0 & 0 \\ 0 & 0 & 0 \\ 0 & 0 & 0 \end{bmatrix}$

4.
$$
\begin{aligned}
(5)^{10} - 3(5)^5 &= \alpha_5(5)^5 + \alpha_4(5)^4 + \alpha_3(5)^3 + \alpha_2(5)^2 + \alpha_1(5) + \alpha_0 \\
10(5)^9 - 15(5)^4 &= 5\alpha_5(5)^4 + 4\alpha_4(5)^3 + 3\alpha_3(5)^2 + 2\alpha_2(5) + \alpha_1 \\
90(5)^8 - 60(5)^3 &= 20\alpha_5(5)^3 + 12\alpha_4(5)^2 + 6\alpha_3(5) + 2\alpha_2 \\
720(5)^7 - 180(5)^2 &= 60\alpha_5(5)^2 + 24\alpha_4(5) + 6\alpha_3 \\
(2)^{10} - 3(2)^5 &= \alpha_5(2)^5 + \alpha_4(2)^4 + \alpha_3(2)^3 + \alpha_2(2)^2 + \alpha_1(2) + \alpha_0 \\
10(2)^9 - 15(2)^4 &= 5\alpha_5(2)^4 + 4\alpha_4(2)^3 + 3\alpha_3(2)^2 + 2\alpha_2(2) + \alpha_1
\end{aligned}
$$

SECTION 6.5

1. $1/7\begin{bmatrix} 3e^5 + 4e^{-2} & 3e^5 - 3e^{-2} \\ 4e^5 - 4e^{-2} & 4e^5 + 3e^{-2} \end{bmatrix}$ **2.** $e^3\begin{bmatrix} 2 & -1 \\ 1 & 0 \end{bmatrix}$

3. $e^2\begin{bmatrix} 0 & 1 & 3 \\ -1 & 2 & 5 \\ 0 & 0 & 1 \end{bmatrix}$

4. $1/16\begin{bmatrix} 12e^2 + 4e^{-2} & 4e^2 - 4e^{-2} & 38e^2 + 2e^{-2} \\ 12e^2 - 12e^{-2} & 4e^2 + 12e^{-2} & 46e^2 - 6e^{-2} \\ 0 & 0 & 16e^2 \end{bmatrix}$

5. $(1/5)\begin{bmatrix} -1 & 6 \\ 4 & 1 \end{bmatrix}$

6. (a) $\begin{bmatrix} \log(3/2) & \log(3/2) - \log(1/2) \\ 0 & \log(1/2) \end{bmatrix}$

(b) and (c) are not defined since they possess eigenvalues having absolute value greater than 1.

(d) $\begin{bmatrix} 0 & 0 \\ 0 & 0 \end{bmatrix}$

SECTION 6.6

1. $1/7\begin{bmatrix} 3e^{8t} + 4e^t & 4e^{8t} - 4e^t \\ 3e^{8t} - 3e^t & 4e^{8t} + 3e^t \end{bmatrix}$

2. $\begin{bmatrix} (2/\sqrt{3})\sinh\sqrt{3}t + \cosh\sqrt{3}t & (1/\sqrt{3})\sinh\sqrt{3}t \\ (-1/\sqrt{3})\sinh\sqrt{3}t & (-2/\sqrt{3})\sinh\sqrt{3}t + \cosh\sqrt{3}t \end{bmatrix}$

Note:

$$\sinh \sqrt{3}t = \frac{e^{\sqrt{3}t} - e^{-\sqrt{3}t}}{2} \quad \text{and} \quad \cosh \sqrt{3}t = \frac{e^{\sqrt{3}t} + e^{-\sqrt{3}t}}{2}$$

3. $e^{3t}\begin{bmatrix} 1+t & t \\ -t & 1-t \end{bmatrix}$

4. $\begin{bmatrix} 1 & t & t^2/2 \\ 0 & 1 & t \\ 0 & 0 & 1 \end{bmatrix}$

5. $$1/12\begin{bmatrix} 12e^t & 0 & 0 \\ -9e^t + 14e^{3t} - 5e^{-3t} & 8e^{3t} + 4e^{-3t} & 4e^{3t} - 4e^{-3t} \\ -24e^t + 14e^{3t} + 10e^{-3t} & 8e^{3t} - 8e^{-3t} & 4e^{3t} + 8e^{-3t} \end{bmatrix}$$

SECTION 6.7

1. $\begin{bmatrix} (1/2)\sin 2t + \cos 2t & (-1/2)\sin 2t \\ (5/2)\sin 2t & (-1/2)\sin 2t + \cos 2t \end{bmatrix}$

2. $\begin{bmatrix} \sqrt{2}\sin\sqrt{2}t + \cos\sqrt{2}t & -\sqrt{2}\sin\sqrt{2}t \\ (3/\sqrt{2})\sin\sqrt{2}t & -\sqrt{2}\sin\sqrt{2}t + \cos\sqrt{2}t \end{bmatrix}$

3. $e^{4t}\begin{bmatrix} -\sin t + \cos t & \sin t \\ 2\sin t & \sin t + \cos t \end{bmatrix}$

SECTION 6.8

4. $e^{\mathbf{A}} = \begin{bmatrix} e & e-1 \\ 0 & 1 \end{bmatrix}$, $\quad e^{\mathbf{B}} = \begin{bmatrix} 1 & e-1 \\ 0 & e \end{bmatrix}$, $\quad e^{\mathbf{A}}e^{\mathbf{B}} = \begin{bmatrix} e & 2e^2 - 2e \\ 0 & e \end{bmatrix}$,

$e^{\mathbf{B}}e^{\mathbf{A}} = \begin{bmatrix} e & 2e-2 \\ 0 & e \end{bmatrix}$, $\quad e^{\mathbf{A}+\mathbf{B}} = \begin{bmatrix} e & 2e \\ 0 & e \end{bmatrix}$

5. $\mathbf{A} = \begin{bmatrix} 1 & 0 \\ 0 & 2 \end{bmatrix}$, $\quad \mathbf{B} = \begin{bmatrix} 3 & 0 \\ 0 & 4 \end{bmatrix}$. Also see Problem 6.

SECTION 6.9

1. (a) $\begin{bmatrix} -\sin t & 2t \\ 2 & e^{(t-1)} \end{bmatrix}$

(b) $\begin{bmatrix} 6t^2 e^{t^3} & 2t-1 & 0 \\ 2t+3 & 2\cos 2t & 1 \\ -18\cos^2(3t^2)\sin(3t^2) & 0 & 1/t \end{bmatrix}$

4. $\begin{bmatrix} \sin t & (1/3)t^3 - t \\ t^2 & e^{(t-1)} \end{bmatrix}$

Chapter 7

SECTION 7.1

1. $\mathbf{x}(t) = \begin{bmatrix} y(t) \\ z(t) \end{bmatrix}, \quad \mathbf{A}(t) = \begin{bmatrix} 3 & 2 \\ 4 & 1 \end{bmatrix}, \quad \mathbf{f}(t) = \begin{bmatrix} 0 \\ 0 \end{bmatrix}; \quad \mathbf{c} = \begin{bmatrix} 1 \\ 1 \end{bmatrix}, \quad t_0 = 0$

2. $$\mathbf{x}(t) = \begin{bmatrix} r(t) \\ s(t) \\ u(t) \end{bmatrix}, \quad \mathbf{A}(t) = \begin{bmatrix} t^2 & -3 & -\sin t \\ 1 & -1 & 0 \\ 2 & e^t & t^2 - 1 \end{bmatrix}, \quad \mathbf{f}(t) = \begin{bmatrix} \sin t \\ t^2 - 1 \\ \cos t \end{bmatrix},$$

$$\mathbf{c} = \begin{bmatrix} 4 \\ -2 \\ 5 \end{bmatrix}, \quad t_0 = 1$$

3. Only (c)

SECTION 7.2

1. $$\mathbf{x}(t) = \begin{bmatrix} x_1(t) \\ x_2(t) \end{bmatrix}, \quad \mathbf{A}(t) = \begin{bmatrix} 0 & 1 \\ -2 & 3 \end{bmatrix}, \quad \mathbf{f}(t) = \begin{bmatrix} 0 \\ e^{-t} \end{bmatrix},$$

$$\mathbf{c} = \begin{bmatrix} 2 \\ 2 \end{bmatrix}, \quad t_0 = 1$$

2. $$\mathbf{x}(t) = \begin{bmatrix} x_1(t) \\ x_2(t) \\ x_3(t) \end{bmatrix}, \quad \mathbf{A}(t) = \begin{bmatrix} 0 & 1 & 0 \\ 0 & 0 & 1 \\ 1/4 & 0 & -t/4 \end{bmatrix}, \quad \mathbf{f}(t) = \begin{bmatrix} 0 \\ 0 \\ 0 \end{bmatrix},$$

$$\mathbf{c} = \begin{bmatrix} 2 \\ 1 \\ -205 \end{bmatrix}, \quad t_0 = -1$$

3. $$\mathbf{x}(t) = \begin{bmatrix} x_1(t) \\ x_2(t) \\ x_3(t) \\ x_4(t) \end{bmatrix}, \quad \mathbf{A}(t) = \begin{bmatrix} 0 & 1 & 0 & 0 \\ 0 & 0 & 1 & 0 \\ 0 & 0 & 0 & 1 \\ 0 & e^{-t} & -te^{-t} & 0 \end{bmatrix}, \quad \mathbf{f}(t) = \begin{bmatrix} 0 \\ 0 \\ 0 \\ 1 \end{bmatrix},$$

$$\mathbf{c} = \begin{bmatrix} 1 \\ 2 \\ \pi \\ e^3 \end{bmatrix}, \quad t_0 = 0$$

4.

$$\mathbf{x}(t)=\begin{bmatrix}x_1(t)\\x_2(t)\\x_3(t)\\x_4(t)\\x_5(t)\\x_6(t)\end{bmatrix},\qquad \mathbf{A}(t)=\begin{bmatrix}0&1&0&0&0&0\\0&0&1&0&0&0\\0&0&0&1&0&0\\0&0&0&0&1&0\\0&0&0&0&0&1\\0&0&0&0&-4&0\end{bmatrix},\qquad \mathbf{f}(t)=\begin{bmatrix}0\\0\\0\\0\\0\\t^2-t\end{bmatrix},$$

$$\mathbf{c}=\begin{bmatrix}2\\1\\0\\2\\1\\0\end{bmatrix},\qquad t_0=\pi$$

SECTION 7.3

1.

$$\mathbf{x}(t)=\begin{bmatrix}x_1(t)\\x_2(t)\\y_1(t)\\y_2(t)\\y_3(t)\\y_4(t)\end{bmatrix},\qquad \mathbf{A}(t)=\begin{bmatrix}0&1&0&0&0&0\\0&2&0&0&0&1\\0&0&0&1&0&0\\0&0&0&0&1&0\\0&0&0&0&0&1\\t&0&-t&0&1&0\end{bmatrix},\qquad \mathbf{f}(t)=\begin{bmatrix}0\\-t\\0\\0\\0\\-e^t\end{bmatrix},$$

$$\mathbf{c}=\begin{bmatrix}2\\0\\0\\3\\9\\4\end{bmatrix},\qquad t_0=-1$$

2.

$$\mathbf{x}(t)=\begin{bmatrix}x_1(t)\\x_2(t)\\x_3(t)\\y_1(t)\\y_2(t)\end{bmatrix},\qquad \mathbf{A}(t)=\begin{bmatrix}0&1&0&0&0\\0&0&1&0&0\\1&0&0&-1&1\\0&0&0&0&1\\-1&0&1&0&2\end{bmatrix},\qquad \mathbf{f}(t)=\begin{bmatrix}0\\0\\0\\0\\0\end{bmatrix},$$

$$\mathbf{c}=\begin{bmatrix}21\\4\\-5\\5\\7\end{bmatrix},\qquad t_0=0$$

3.
$$\mathbf{x}(t)=\begin{bmatrix} x_1(t)\\ y_1(t)\\ y_2(t)\\ z_1(t)\end{bmatrix},\qquad \mathbf{A}(t)=\begin{bmatrix} 0&1&0&0\\ 0&0&1&0\\ 0&0&0&1\\ 1&1&0&0\end{bmatrix},\qquad \mathbf{f}(t)=\begin{bmatrix} -2\\ 0\\ -2\\ 0\end{bmatrix},$$

$$\mathbf{c}=\begin{bmatrix} 1\\ 2\\ 17\\ 0\end{bmatrix},\qquad t_0=\pi$$

4.
$$\mathbf{x}(t)=\begin{bmatrix} x_1(t)\\ x_2(t)\\ y_1(t)\\ y_2(t)\\ z_1(t)\\ z_2(t)\end{bmatrix},\qquad \mathbf{A}(t)=\begin{bmatrix} 0&1&0&0&0&0\\ 0&0&1&0&1&0\\ 0&0&0&1&0&0\\ 1&0&1&0&0&0\\ 0&0&0&0&0&1\\ 1&0&0&0&-1&0\end{bmatrix},\qquad \mathbf{f}(t)=\begin{bmatrix} 0\\ 2\\ 0\\ -1\\ 0\\ 1\end{bmatrix},$$

$$\mathbf{c}=\begin{bmatrix} 4\\ -4\\ 5\\ -5\\ 9\\ -9\end{bmatrix},\qquad t_0=20$$

SECTION 7.4

3. (a)
$$e^{-3t}\begin{bmatrix} 1&-t&t^2/2\\ 0&1&-t\\ 0&0&1\end{bmatrix},$$
(b)
$$e^{3(t-2)}\begin{bmatrix} 1&(t-2)&(t-2)^2/2\\ 0&1&(t-2)\\ 0&0&1\end{bmatrix},$$

(c)
$$e^{3(t-s)}\begin{bmatrix} 1&(t-s)&(t-s)^2/2\\ 0&1&(t-s)\\ 0&0&1\end{bmatrix},$$
(d)
$$e^{-3(t-2)}\begin{bmatrix} 1&-(t-2)&(t-2)^2/\\ 0&1&-(t-2)\\ 0&0&1\end{bmatrix}$$

SECTION 7.5

1. $x(t)=5e^{(t-2)}-3e^{-(t-2)}, \qquad y(t)=5e^{(t-2)}-e^{-(t-2)}$

2. $x(t)=2e^{(t-1)}-1, \qquad y(t)=2e^{(t-1)}-1$

3. $x(t)=k_3e^t+3k_4e^{-t}, \qquad y(\bar{t})=k_3e^t+k_4e^{-t}$

4. $x(t) = k_3 e^t + 3k_4 e^{-t} - 1, \qquad y(t) = k_3 e^t + k_4 e^{-t} - 1$

5. $x(t) = \cos 2t - (1/6) \sin 2t + (1/3) \sin t$

6. $x(t) = t^4/24 + (5/4)t^2 - (2/3)t + 3/8$

7. $x(t) = (4/9)e^{2t} + (5/9)e^{-t} - (1/3)te^{-t}$

8. $x(t) = -8 \cos t - 6 \sin t + 8 + 6t$

$y(t) = \quad 4 \cos t - 2 \sin t - 3$

SECTION 7.7

2. First show that

$$\Phi'(t_1, t_0)\left[\int_{t_0}^{t_1} \Phi(t_1, s)\Phi'(t_1, s)\, ds\right]^{-1} \Phi(t_1, t_0)$$
$$= \left[\Phi(t_0, t_1) \int_{t_0}^{t_1} \Phi(t_1, s)\Phi'(t_1, s)\, ds \Phi'(t_0, t_1)\right]^{-1}$$
$$= \left[\int_{t_0}^{t_1} \Phi(t_0, t_1)\Phi(t_1, s)[\Phi(t_0, t_1)\Phi(t_1, s)]'\, ds\right]^{-1}$$

Chapter 8

SECTION 8.1

2. If **PA** = **BP**, then $\mathbf{P} = \begin{bmatrix} a & b \\ 0 & 0 \end{bmatrix}$ which is singular

3. (a) Yes (b) No (c) No (d) No

SECTION 8.2

1. Yes; $\mathbf{M} = \begin{bmatrix} 3 & 1 \\ 1 & 1 \end{bmatrix}, \quad \mathbf{D} = \begin{bmatrix} 1 & 0 \\ 0 & -1 \end{bmatrix}$

2. Yes; $\mathbf{M} = \begin{bmatrix} 2+i & 2-i \\ 1 & 1 \end{bmatrix}, \quad \mathbf{D} = \begin{bmatrix} i & 0 \\ 0 & -i \end{bmatrix}$

3. No

4. Yes; $\mathbf{M} = \begin{bmatrix} -10 & 0 & 0 \\ 1 & 1 & -3 \\ 8 & 2 & 1 \end{bmatrix}$, $\mathbf{D} = \begin{bmatrix} 1 & 0 & 0 \\ 0 & 3 & 0 \\ 0 & 0 & -4 \end{bmatrix}$

5. Yes; $\mathbf{M} = \begin{bmatrix} 1 & 0 & 1 \\ -2 & -2 & 0 \\ 0 & 1 & 1 \end{bmatrix}$, $\mathbf{D} = \begin{bmatrix} 3 & 0 & 0 \\ 0 & 3 & 0 \\ 0 & 0 & 7 \end{bmatrix}$ **6.** No

SECTION 8.3

1. $\begin{bmatrix} 2-(2^{27}) & -1+2^{27} \\ 2-(2^{28}) & -1+2^{28} \end{bmatrix}$ **2.** $\begin{bmatrix} 89-(2^{17}) & -88+2^{17} \\ 176-(2^{18}) & -175+2^{18} \end{bmatrix}$

3. $\begin{bmatrix} 2e-e^2 & -e+e^2 \\ 2e-2e^2 & -e+2e^2 \end{bmatrix}$ **4.** $e^2\begin{bmatrix} \cos\sqrt{5} & (1/\sqrt{5})\sin\sqrt{5} \\ -\sqrt{5}\sin\sqrt{5} & \cos\sqrt{5} \end{bmatrix}$

5. $(1/70)\begin{bmatrix} 70e & 0 & 0 \\ -7e+25e^3-18e^{-4} & 10e^3+60e^{-4} & 30e^3-30e^{-4} \\ -56e+50e^3+6e^{-4} & 20e^3-20e^{-4} & 60e^3+10e^{-4} \end{bmatrix}$

6. $(-1/4)\begin{bmatrix} -2e^3-2e^7 & e^3-e^7 & 2e^3-2e^7 \\ 0 & -4e^3 & 0 \\ 2e^3-2e^7 & e^3-e^7 & -2e^3-2e^7 \end{bmatrix}$

9. $(-1/4)\begin{bmatrix} -2\sin 3-2\sin 7 & \sin 3-\sin 7 & 2\sin 3-2\sin 7 \\ 0 & -4\sin 3 & 0 \\ 2\sin 3-2\sin 7 & \sin 3-\sin 7 & -2\sin 3-2\sin 7 \end{bmatrix}$

SECTION 8.4

1. (a) Yes (b) No (c) Yes (d) Yes (e) No (f) No

2. $\begin{bmatrix} 0 \\ 1 \end{bmatrix}$ **3.** $\begin{bmatrix} 0 \\ 1 \\ 0 \end{bmatrix}$ **4.** $\begin{bmatrix} 0 \\ 0 \\ 1 \end{bmatrix}$ **5.** $\begin{bmatrix} 0 \\ 0 \\ 1 \end{bmatrix}$ **6.** $\begin{bmatrix} 1 \\ 0 \\ -1 \end{bmatrix}$

7. For $\lambda = 3$, $\mathbf{x}_3 = \begin{bmatrix} 1 \\ 0 \\ 0 \\ 0 \\ -1 \end{bmatrix}$, and for $\lambda = 4$, $\mathbf{x}_2 = \begin{bmatrix} 0 \\ 1 \\ 0 \\ 0 \\ 0 \end{bmatrix}$

SECTION 8.5

1.
$$\mathbf{x}_4 = \begin{bmatrix} 0 \\ 0 \\ 0 \\ 1 \end{bmatrix}, \quad \mathbf{x}_3 = \begin{bmatrix} -1 \\ 4 \\ 1 \\ 0 \end{bmatrix}, \quad \mathbf{x}_2 = \begin{bmatrix} 7 \\ -1 \\ 0 \\ 0 \end{bmatrix}, \quad \mathbf{x}_1 = \begin{bmatrix} -1 \\ 0 \\ 0 \\ 0 \end{bmatrix}$$

2. For $\lambda = 3$,
$$\mathbf{x}_3 = \begin{bmatrix} 1 \\ 0 \\ 0 \\ 0 \\ -1 \end{bmatrix}, \quad \mathbf{x}_2 = \begin{bmatrix} 0 \\ 0 \\ 0 \\ -2 \\ 0 \end{bmatrix}, \quad \mathbf{x}_1 = \begin{bmatrix} 0 \\ 0 \\ -2 \\ 0 \\ 0 \end{bmatrix}$$

For $\lambda = 4$,
$$\mathbf{x}_2 = \begin{bmatrix} 0 \\ 1 \\ 0 \\ 0 \\ 0 \end{bmatrix}, \quad \mathbf{x}_1 = \begin{bmatrix} 1 \\ 0 \\ 0 \\ 0 \\ 0 \end{bmatrix}$$

SECTION 8.6

1.
$$\mathbf{x}_2 = \begin{bmatrix} 0 \\ 1 \end{bmatrix}, \quad \mathbf{x}_1 = \begin{bmatrix} 1 \\ -1 \end{bmatrix}$$

2.
$$\mathbf{x}_1 = \begin{bmatrix} -1 \\ 1 \\ 1 \end{bmatrix} \text{ corresponds to } \lambda = 1 \text{ and } \mathbf{y}_2 = \begin{bmatrix} 0 \\ 0 \\ 1 \end{bmatrix},$$

$$\mathbf{y}_1 = \begin{bmatrix} 3 \\ 0 \\ -3 \end{bmatrix} \text{ correspond to } \lambda = 4$$

3.
$$\mathbf{x}_3 = \begin{bmatrix} 0 \\ 0 \\ 1 \end{bmatrix}, \quad \mathbf{x}_2 = \begin{bmatrix} -1 \\ 2 \\ 0 \end{bmatrix}, \quad \mathbf{x}_1 = \begin{bmatrix} 2 \\ 0 \\ 0 \end{bmatrix}$$

4.
$$\mathbf{x}_1 = \begin{bmatrix} 1 \\ -2 \\ 0 \end{bmatrix}, \quad \mathbf{y}_1 = \begin{bmatrix} 0 \\ -2 \\ 1 \end{bmatrix} \text{ correspond to } \lambda = 3 \text{ and}$$

$$\mathbf{z}_1 = \begin{bmatrix} 1 \\ 0 \\ 1 \end{bmatrix} \text{ corresponds to } \lambda = 7$$

5.

$$\mathbf{x}_3=\begin{bmatrix}0\\0\\0\\1\end{bmatrix},\qquad \mathbf{x}_2=\begin{bmatrix}-1\\1\\0\\0\end{bmatrix},\qquad \mathbf{x}_1=\begin{bmatrix}1\\0\\0\\0\end{bmatrix},\qquad \mathbf{y}_1=\begin{bmatrix}0\\-1\\1\\-1\end{bmatrix}$$

6.

$$\mathbf{x}_2=\begin{bmatrix}0\\1\\0\\0\end{bmatrix},\qquad \mathbf{x}_1=\begin{bmatrix}1\\0\\0\\0\end{bmatrix}\text{ correspond to }\lambda=3\text{ and}$$

$$\mathbf{y}_2=\begin{bmatrix}3\\1\\0\\-1\end{bmatrix},\qquad \mathbf{y}_1=\begin{bmatrix}-1\\-1\\-1\\0\end{bmatrix}\text{ correspond to }\lambda=4$$

7.

$$\mathbf{x}_4=\begin{bmatrix}0\\0\\0\\2\\-2\\1\end{bmatrix},\qquad \mathbf{x}_3=\begin{bmatrix}-1\\1\\2\\0\\0\\0\end{bmatrix},\qquad \mathbf{x}_2=\begin{bmatrix}3\\4\\0\\0\\0\\0\end{bmatrix},\qquad \mathbf{x}_1=\begin{bmatrix}4\\0\\0\\0\\0\\0\end{bmatrix}$$

correspond to $\lambda=4$, and

$$\mathbf{y}_2=\begin{bmatrix}-5\\-2\\0\\1\\1\\0\end{bmatrix},\qquad \mathbf{y}_1=\begin{bmatrix}3\\2\\1\\1\\0\\0\end{bmatrix}$$

correspond to $\lambda=5$.

SECTION 8.7

1. $\begin{bmatrix}2&1\\0&2\end{bmatrix}$

2. $\begin{bmatrix}1&0&0\\0&4&1\\0&0&4\end{bmatrix}$

3. $\begin{bmatrix}5&1&0\\0&5&1\\0&0&5\end{bmatrix}$

4. $\begin{bmatrix}2&0&0&0\\0&2&1&0\\0&0&2&1\\0&0&0&2\end{bmatrix}$

5. $\begin{bmatrix}4&1&0&0&0&0\\0&4&1&0&0&0\\0&0&4&1&0&0\\0&0&0&4&0&0\\0&0&0&0&5&1\\0&0&0&0&0&5\end{bmatrix}$

SECTION 8.8

1. $\begin{bmatrix} 16 & 32 & 24 \\ 0 & 16 & 32 \\ 0 & 0 & 16 \end{bmatrix}$

2. $\begin{bmatrix} 1 & 0 & 0 & 0 & 0 & 0 \\ 0 & 1 & 0 & 0 & 0 & 0 \\ 0 & 0 & 1 & -10 & 45 & -120 \\ 0 & 0 & 0 & 1 & -10 & 45 \\ 0 & 0 & 0 & 0 & 1 & -10 \\ 0 & 0 & 0 & 0 & 0 & 1 \end{bmatrix}$

3. $e^4\begin{bmatrix} 1 & 0 & 0 \\ 0 & 1 & 1 \\ 0 & 0 & 1 \end{bmatrix}$

4. $e^2\begin{bmatrix} 1 & 1 & 1 \\ 0 & 1 & 2 \\ 0 & 0 & 1 \end{bmatrix}$

5. $\begin{bmatrix} -1 & 0 & \pi^2/12 \\ 0 & -1 & 0 \\ 0 & 0 & -1 \end{bmatrix}$

6. $e^{\pi}\begin{bmatrix} 1 & \pi/3 & (\pi^2/12)-\pi \\ 0 & 1 & \pi/2 \\ 0 & 0 & 1 \end{bmatrix}$

7. $\begin{bmatrix} e & 0 & 0 \\ -e+2e^2 & 2e^2 & -e^2 \\ e^2 & e^2 & 0 \end{bmatrix}$

8. $\begin{bmatrix} e^2 & e^2 & 0 & 0 \\ 0 & e^2 & 0 & 0 \\ 0 & 0 & (e^{3/2}/\sqrt{27})\left(\sin\frac{\sqrt{27}}{2}+\sqrt{27}\cos\frac{\sqrt{27}}{2}\right) & e^{3/2}/\sqrt{27}\left(14\sin\frac{\sqrt{27}}{2}\right) \\ 0 & 0 & e^{3/2}/\sqrt{27}\left(-2\sin\frac{\sqrt{27}}{2}\right) & e^{3/2}/\sqrt{27}\left(-\sin\frac{\sqrt{27}}{2}+\sqrt{27}\cos\frac{\sqrt{27}}{2}\right) \end{bmatrix}$

SECTION 8.9

2. $e^{2t}\begin{bmatrix} 1 & t \\ 0 & 1 \end{bmatrix}$

3. $e^{-t}\begin{bmatrix} 1 & t & t^2/2 \\ 0 & 1 & t \\ 0 & 0 & 1 \end{bmatrix}$

4. $e^{4t}\begin{bmatrix} 1 & t & 0 \\ 0 & 1 & 0 \\ 0 & 0 & 1 \end{bmatrix}$

5. $\begin{bmatrix} e^{2t} & te^{2t} & 0 \\ 0 & e^{2t} & 0 \\ 0 & 0 & e^{-t} \end{bmatrix}$

6. $1/2\begin{bmatrix} -e^{-t}+3e^t & -3e^{-t}+3e^t & 0 \\ e^{-t}-e^t & 3e^{-t}-e^t & 0 \\ 2te^t & 2te^t & 2e^t \end{bmatrix}$

7. $$e^{2t}\begin{bmatrix} 1+t & t & 0 \\ -t & 1-t & 0 \\ t-(1/2)t^2 & 2t-(1/2)t^2 & 1 \end{bmatrix}$$

8. $$e^{t}\begin{bmatrix} 1+4t-t^2 & -2t+2t^2 & 2t \\ 2t-t^2/2 & 1-t+t^2 & t \\ -7t+(3/2)t^2 & 5t-3t^2 & 1-3t \end{bmatrix}$$

Chapter 9

SECTION 9.2

1. (a) $2i, -2i$ (b) $0, 0$ (c) $1+6i, 1-6i$

2. No vector $\mathbf{x}$ exists

6. (a) $\begin{bmatrix} 1/\sqrt{2} \\ i/\sqrt{2} \end{bmatrix}$ (b) $\begin{bmatrix} 1/\sqrt{7} \\ (1-i)/\sqrt{7} \\ 2i/\sqrt{7} \end{bmatrix}$ (c) $\begin{bmatrix} -i/\sqrt{5} \\ (1+i)/\sqrt{5} \\ (1-i)/\sqrt{5} \end{bmatrix}$

(d) $\begin{bmatrix} 1/\sqrt{8} \\ 2/\sqrt{8} \\ i/\sqrt{8} \\ (1-i)/\sqrt{8} \end{bmatrix}$

SECTION 9.3

1. $x = -(1+\frac{3}{2}i)$.

2. $x = -3-15i, \quad y = 3-3i$

3. (a) $$\mathbf{e}_1 = \begin{bmatrix} 1/\sqrt{6} \\ 2/\sqrt{6} \\ 1/\sqrt{6} \end{bmatrix}, \quad \mathbf{e}_2 = \begin{bmatrix} 1/\sqrt{3} \\ -1/\sqrt{3} \\ 1/\sqrt{3} \end{bmatrix}, \quad \mathbf{e}_3 = \begin{bmatrix} -1/\sqrt{2} \\ 0 \\ 1/\sqrt{2} \end{bmatrix}$$

(b) $$\mathbf{e}_1 = \begin{bmatrix} i/\sqrt{2} \\ 0 \\ 1/\sqrt{2} \\ 0 \end{bmatrix}, \quad \mathbf{e}_2 = \begin{bmatrix} 0 \\ i/\sqrt{2} \\ 0 \\ -1/\sqrt{2} \end{bmatrix}, \quad \mathbf{e}_3 = \begin{bmatrix} 0 \\ 1/2+i/2 \\ 0 \\ 1/2-i/2 \end{bmatrix},$$

$$\mathbf{e}_4 = \begin{bmatrix} -1/2-i/2 \\ 0 \\ 1/2-i/2 \\ 0 \end{bmatrix}.$$

6. $\begin{bmatrix} \pm 1/\sqrt{3} \\ \mp 1/\sqrt{3} \\ \pm 1/\sqrt{3} \end{bmatrix}$

8. Consider the vectors

$$\begin{bmatrix} 0 \\ 1 \end{bmatrix}, \quad \begin{bmatrix} 1 \\ 0 \end{bmatrix}, \quad \text{and} \quad \begin{bmatrix} 0 \\ 0 \end{bmatrix}$$

SECTION 9.4

1. (b), (d), and (f) are self-adjoint.

SECTION 9.5

(a)
$$\mathbf{e}_1 = \begin{bmatrix} 1/\sqrt{2} \\ 0 \\ 1/\sqrt{2} \end{bmatrix}, \quad \mathbf{e}_2 = \begin{bmatrix} -1/\sqrt{3} \\ 1/\sqrt{3} \\ 1/\sqrt{3} \end{bmatrix}, \quad \mathbf{e}_3 = \begin{bmatrix} 1/\sqrt{6} \\ 2/\sqrt{6} \\ -1/\sqrt{6} \end{bmatrix}.$$

(b)
$$\mathbf{e}_1 = \begin{bmatrix} 1/\sqrt{3} \\ 1/\sqrt{3} \\ -1/\sqrt{3} \\ 0 \end{bmatrix}, \quad \mathbf{e}_2 = \begin{bmatrix} 1/\sqrt{3} \\ 0 \\ 1/\sqrt{3} \\ 1/\sqrt{3} \end{bmatrix}, \quad \mathbf{e}_3 = \begin{bmatrix} -1/\sqrt{3} \\ 1/\sqrt{3} \\ 0 \\ 1/\sqrt{3} \end{bmatrix},$$

$$\mathbf{e}_4 = \begin{bmatrix} 0 \\ 1/\sqrt{3} \\ 1/\sqrt{3} \\ -1/\sqrt{3} \end{bmatrix}$$

(c)
$$\mathbf{e}_1 = \begin{bmatrix} 1/\sqrt{3} \\ 0 \\ 1/\sqrt{3} \\ 1/\sqrt{3} \end{bmatrix}, \quad \mathbf{e}_2 = \begin{bmatrix} 0 \\ 1/\sqrt{3} \\ 1/\sqrt{3} \\ -1/\sqrt{3} \end{bmatrix}, \quad \mathbf{e}_3 = \begin{bmatrix} 1/\sqrt{3} \\ 1/\sqrt{3} \\ -1/\sqrt{3} \\ 0 \end{bmatrix},$$

$$\mathbf{e}_4 = \begin{bmatrix} -1/\sqrt{3} \\ 1/\sqrt{3} \\ 0 \\ 1/\sqrt{3} \end{bmatrix}.$$

(d)

$$\mathbf{e}_1 = \begin{bmatrix} 1 \\ 0 \\ 0 \\ 0 \end{bmatrix}, \quad \mathbf{e}_2 = \begin{bmatrix} 0 \\ 1/\sqrt{2} \\ 0 \\ 1/\sqrt{2} \end{bmatrix}, \quad \mathbf{e}_3 = \begin{bmatrix} 0 \\ 1/\sqrt{6} \\ -2/\sqrt{6} \\ -1/\sqrt{6} \end{bmatrix}, \quad \mathbf{e}_4 = \begin{bmatrix} 0 \\ 1/\sqrt{3} \\ 1/\sqrt{3} \\ -1/\sqrt{3} \end{bmatrix}.$$

2. $\begin{bmatrix} 0 \\ 1 \end{bmatrix}$; the matrix is not real.

3. Follow the proof of Theorem 1.

SECTION 9.6

1. (a), (c), and (d) are orthogonal, (b) is not.

2.
$$\mathbf{P} = \begin{bmatrix} 1/\sqrt{2} & -1/\sqrt{3} & 1/\sqrt{6} \\ 0 & 1/\sqrt{3} & 2/\sqrt{6} \\ 1/\sqrt{2} & 1/\sqrt{3} & -1/\sqrt{6} \end{bmatrix}.$$

3. Show that $(\mathbf{P}^{-1}\mathbf{A}\mathbf{P})' = \mathbf{P}^{-1}\mathbf{A}'\mathbf{P}$.

5.
$$\begin{bmatrix} 1/\sqrt{3} & 1/\sqrt{3} & -1/\sqrt{3} & 0 \\ 1/\sqrt{3} & 0 & 1/\sqrt{3} & 1/\sqrt{3} \\ -1/\sqrt{3} & 1/\sqrt{3} & 0 & 1/\sqrt{3} \\ 0 & 1/\sqrt{3} & 1/\sqrt{3} & -1/\sqrt{3} \end{bmatrix}$$

SECTION 9.7

1. (a) Hermitian;

$$\mathbf{e}_1 = \begin{bmatrix} i/\sqrt{2} \\ -1/\sqrt{2} \end{bmatrix}, \quad \mathbf{e}_2 = \begin{bmatrix} -1/2 \\ i/2 \end{bmatrix}$$

(b) Not Hermitian

(c) Not Hermitian

(d) Hermitian;

$$\mathbf{e}_1 = \begin{bmatrix} 0 \\ 1 \\ 0 \end{bmatrix}, \quad \mathbf{e}_2 = \begin{bmatrix} -3/\sqrt{10} \\ 0 \\ i/\sqrt{10} \end{bmatrix}, \quad \mathbf{e}_3 = \begin{bmatrix} i/\sqrt{10} \\ 0 \\ -3/\sqrt{10} \end{bmatrix}.$$

(e) Hermitian;

$$\mathbf{e}_1 = \begin{bmatrix} 1/\sqrt{3} \\ 0 \\ i/\sqrt{3} \\ -i/\sqrt{3} \end{bmatrix}, \quad \mathbf{e}_2 = \begin{bmatrix} -i/\sqrt{3} \\ 1/\sqrt{3} \\ -1/\sqrt{3} \\ 0 \end{bmatrix}, \quad \mathbf{e}_3 = \begin{bmatrix} 0 \\ 1/\sqrt{3} \\ 1/\sqrt{3} \\ 1/\sqrt{3} \end{bmatrix},$$

$$\mathbf{e}_4 = \begin{bmatrix} i/\sqrt{3} \\ 1/\sqrt{3} \\ 0 \\ -1/\sqrt{3} \end{bmatrix}.$$

(f) Not Hermitian

4. Use the properties of the inner product.

5. Note that each of the four terms on the right-hand side of the equality in Problem 4 is of the form $\langle \mathbf{Az}, \mathbf{z} \rangle$, hence, is real. Thus, $\langle \mathbf{Az}, \mathbf{z} \rangle = \overline{\langle \mathbf{Az}, \mathbf{z} \rangle} = \langle \mathbf{z}, \mathbf{Az} \rangle = \langle \mathbf{A}^*\mathbf{z}, \mathbf{z} \rangle$. Use this result to show that $\langle \mathbf{Ax}, \mathbf{y} \rangle = \langle \mathbf{A}^*\mathbf{x}, \mathbf{y} \rangle$ and then conclude that $\mathbf{A} = \mathbf{A}^*$.

SECTION 9.8

1. (a), (c), and (d) are unitary, (b) is not.

2.
$$\mathbf{U} = 1/\sqrt{3}\begin{bmatrix} 1 & 0 & -i & i \\ 0 & 1 & 1 & 1 \\ i & 1 & -1 & 0 \\ -i & 1 & 0 & -1 \end{bmatrix}$$

4. Show that $(\mathbf{U}^{-1}\mathbf{AU})^* = \mathbf{U}^{-1}\mathbf{A}^*\mathbf{U}$.

6. $\begin{bmatrix} 1/\sqrt{2} & -i/\sqrt{2} \\ i/\sqrt{2} & -1/\sqrt{2} \end{bmatrix}$

7. Since the matrices are similar, their eigenvalues are identical (Theorem 1 of Section 8.1).

SECTION 9.10

1. (a), (c), (e), and (f) are not positive definite. Vectors are

$$\begin{bmatrix}1\\1\end{bmatrix}, \begin{bmatrix}1\\1\end{bmatrix}, \begin{bmatrix}1\\0\\1\end{bmatrix}, \begin{bmatrix}1\\0\\1\end{bmatrix}.$$

2. Consider the vectors $\mathbf{e}_j$ given in the Appendix to Chapter 7.

REFERENCES

On Matrices

FINKBEINER, D. T., "Introduction to Matrices and Linear Transformations." Freeman, San Francisco, 1960.

FRANKLIN, J. N., "Matrix Theory." Prentice-Hall, Englewood Cliffs, New Jersey, 1968.

FRIEDMAN, B., "Principles and Techniques of Applied Mathematics." Wiley, New York, 1956.

GANTMACHER, F. R., "The Theory of Matrices" Vol. I. Chelsea, New York, 1960.

GERE, J. M., AND WEAVER, JR., W., "Matrix Algebra for Engineers." Van Nostrand, Princeton, New Jersey, 1965.

LANCASTER, P., "Theory of Matrices." Academic Press, New York, 1969.

PIPES, L. A., "Matrix Methods for Engineers." Prentice-Hall, Englewood Cliffs, New Jersey, 1963.

SCHNEIDER, H., AND BARKER, G. P., "Matrices and Linear Algebra." Holt, New York, 1968.

STEIN, F. M., "Introduction to Matrices and Determinants." Wadsworth, Belmont, California, 1967.

On Linear Algebra

HALMOS, P. R., "Finite-Dimensional Vector Spaces." Van Nostrand, Princeton, New Jersey, 1958.

HOFFMAN, K., AND KUNZE, R., "Linear Algebra." Prentice-Hall, Englewood Cliffs, New Jersey, 1961.

NEARING, E. D., "Linear Algebra and Matrix Theory." Wiley, New York, 1963.

ZELINSKY, D., "A First Course in Linear Algebra." Academic Press, New York, 1968.

On Differential Equations

BOYCE, W. E., and DIPRIMA, R. C., "Elementary Differential Equations and Boundary Value Problems." Wiley, New York, 1969.

CODDINGTON, E. A., AND LEVINSON, N., "Theory of Ordinary Differential Equations." McGraw-Hill, New York, 1955.

RITGER, P. D., AND ROSE, N. J., "Differential Equations with Applications." McGraw-Hill, New York, 1968.

On Numerical Analysis

HOUSEHOLDER, A. S., "The Theory of Matrices in Numerical Analysis." Ginn (Blaisdell), Boston, New York, 1964.

RALSTON, A., "A First Course in Numerical Analysis." McGraw-Hill, New York, 1965.

INDEX

M4
N5
O6
P7
Q8
R9
S0
T1